Current Topics in Microbiology and Immunology

235

Editors

R.W. Compans, Atlanta/Georgia
M. Cooper, Birmingham/Alabama
J.M. Hogle, Boston/Massachusetts · Y. Ito, Kyoto
H. Koprowski, Philadelphia/Pennsylvania · F. Melchers, Basel
M. Oldstone, La Jolla/California · S. Olsnes, Oslo
M. Potter, Bethesda/Maryland · H. Saedler, Cologne
P.K. Vogt, La Jolla/California · H. Wagner, Munich

Springer
Berlin
Heidelberg
New York
Barcelona
Budapest
Hong Kong
London
Milan
Paris
Singapore
Tokyo

Marburg and Ebola Viruses

Edited by Hans-Dieter Klenk

With 34 Figures and 20 Tables

 Springer

Professor Dr. med. HANS-DIETER KLENK
Klinikum der Philipps-Universität Marburg
Institut für Virologie
Robert-Koch-Str. 17
D-35037 Marburg
Germany

Cover Illustration: Electron micrograph showing Marburg virus particles budding from the plasma membrane of a human endothelial cell (From Schnittler et al., Journal of Clinical Investigation 91: 1301–1309, 1993)

Cover Design: design & production GmbH, Heidelberg

ISSN 0070-217X
ISBN-13: 978-3-642-64193-0   e-ISBN-13: 978-3-642-59949-1
DOI: 10.1007/978-3-642-59949-1

This work is subject to copyright. All rights are reserved, whether the whole or part of the material is concerned, specifically the rights of translation, reprinting, reuse of illustrations, recitation, broadcasting, reproduction on microfilm or in any other way, and storage in data banks. Duplication of this publication or parts thereof is permitted only under the provisions of the German Copyright Law of September 9, 1965, in its current version, and permission for use must always be obtained from Springer-Verlag. Violations are liable for prosecution under the German Copyright Law.

© Springer-Verlag Berlin Heidelberg 1999
Softcover reprint of the hardcover 1st edition 1999
Library of Congress Catalog Card Number 15-12910

The use of general descriptive names, registered names, trademarks, etc. in this publication does not imply, even in the absence of a specific statement, that such names are exempt from the relevant protective laws and regulations and therefore free for general use.

Product liability: The publishers cannot guarantee the accuracy of any information about dosage and application contained in this book. In every individual case the user must check such information by consulting other relevant literature.

Typesetting: Scientific Publishing Services (P) Ltd, Madras

SPIN: 10520735   27/3020 – 5 4 3 2 1 0 – Printed on acid-free paper

Preface

Almost exactly 30 years have passed since the discovery of filoviruses. In August 1967, three laboratory workers at the Behringwerke, in Marburg, Germany, became ill with a hemorrhagic disease after processing organs from African green monkeys (*Cercopithecus aethiops*). In the course of the epidemic, 17 more persons engaged in this kind of work were admitted to the hospital, and two members of the medical staff became infected while attending these patients in the hospital. In November, the last patient was admitted. Six cases, including two patients with nosocomial infections, occurred at the same time in Frankfurt, Germany. Furthermore, in Belgrade, Yugoslavia, a veterinarian performing an autopsy on dead monkeys became infected along with his wife, who nursed him during the first few days of the illness. All together, there were 31 cases, including six nogocomial cases; there were seven deaths. To Gustav-Adolf Martini, of the Department of Internal Medicine at the University of Marburg, it was immediately clear that he was observing an unusual infectious disease. By the end of October, Werner Slenczka and Rudolf Siegert, of the Institute of Hygiene, had identified the causative agent as a new virus, which soon received the name of the town where it was discovered. The identification of Marburg virus 2 months after its first appearance is, even by present day standards, a remarkable accomplishment of medical virology.

Since then virologically confirmed cases of Marburg disease have been observed only sporadically in Zimbabwe, South Africa, and Kenya. However, another filovirus emerged in 1976, when two epidemics occurred simultaneously in Zaire and Sudan. The agent was isolated from patients in both countries and named after the Ebola river in northwestern Zaire. Over 500 cases were reported, with mortality rates of 88% in Zaire and 53% in Sudan. Sporadic outbreaks of Ebola disease occurred in both countries in 1977 and 1979. In 1994, the first Ebola case was observed in West Africa, in Cote d'Ivoire, when an ecologist was infected by examining a dead chimpanzee. Ebola virus re-emerged in Kikwit, Zaire, in 1995, with 315 cases and 245 deaths.

From 1994 to 1997 there were three outbreaks of Ebola virus disease in Gabon. The Reston strain of Ebola virus, which does not appear to be pathogenic for humans, was first isolated in 1989 from cynomolgus monkeys imported into the United States from the Philippines.

The following is thus clear from the recorded history of filovirus outbreaks: all have been self-limited so far, and the total number of human infections hitherto recorded does not exceed one thousand cases. However, because of the dramatic course of the disease, the excessive case-fatality rates, and the lack of immunoprophylactic and chemotherapeutic measures, re-emergence of these viruses regularly causes great public concern.

This volume contains articles from leading scientists in the filovirus field. The first three chapters, by H. Feldmann and M. Kiley, S. Becker and E. Mühlberger, and V. Volchkov, respectively, provide a comprehensive overview of the structure and replication of filoviruses, with special emphasis on protein processing. They show that filoviruses are typical nonsegmented negative-stranded RNA viruses, that differ, however, in some interesting molecular aspects from other members of this group. An account of the original Marburg virus outbreak, which includes a series of observations not previously published, is given by W. Slenczka. B. LeGuenno, P. Formenty, and C. Boesch report on recent Ebola virus outbreaks in Côte d'Ivoire and Liberia. This chapter addresses, in particular, the still unsolved problem of the virus reservoir and the value of ecology as a discipline that may help to identify the natural host. The next two chapters summarize our knowledge of the impact of filovirus infections on humans, information largely obtained during the Ebola virus outbreak in Kikwit in 1995. C.J. Peters and A.S. Khan concentrate on the clinical aspects of filovirus disease, whereas S.R. Zaki and C.S. Goldsmith focus on human pathology. Both articles also give a brief overview of the currently available diagnostic tools. An understanding of human filovirus disease depends to a large extent on studies employing experimental infection of animals. These studies are discussed in two comprehensive chapters by S. Fisher-Hoch, J.B. McCormick and by E.I. Ryabchikova, L.V. Kolesnikova, and S.V. Netesov. The animal studies have recently been complemented by tissue and cell culture studies indicating that the mononuclear phagocyte system and the endothelium play central roles in the pathogenesis and development of the disease-decisive symptoms of filoviral hemorrhagic fever. These data are reviewed in a contribution by H.-J. Schnittler and H. Feldmann. The final chapter, by M.G. Ignatyev, summarizes what is currently known about the immune response to filovirus infections and provides a critical review of evidence, suggesting that these infections may have immunosuppressive effects.

September 1998
Hans-Dieter Klenk
Marburg/Lahn

List of Contents

H. FELDMANN and M.P. KILEY
Classification, Structure, and Replication of Filoviruses... 1

S. BECKER and E. MÜHLBERGER
Co- and Posttranslational Modifications and Functions
of Marburg Virus Proteins 23

V.E. VOLCHKOV
Processing of the Ebola Virus Glycoprotein 35

W.G. SLENCZKA
The Marburg Virus Outbreak of 1967
and Subsequent Episodes 49

B. LE GUENNO, P. FORMENTY, and C. BOESCH
Ebola Virus Outbreaks in the Ivory Coast
and Liberia, 1994–1995........................ 77

C.J. PETERS and A.S. KHAN
Filovirus Diseases............................ 85

S.R. ZAKI and C.S. GOLDSMITH
Pathologic Features of Filovirus Infections in Humans .. 97

S. P. FISHER-HOCH and J. B. MCCORMICK
Experimental Filovirus Infections. 117

E.I. RYABCHIKOVA, L.V. KOLESNIKOVA, and S.V. NETESOV
Animal Pathology of Filoviral Infections 145

H.J. SCHNITTLER and H. FELDMANN
Molecular Pathogenesis of Filovirus Infections:
Role of Macrophages and Endothelial Cells 175

G.M. IGNATYEV
Immune Response to Filovirus Infections............ 205

Subject Index................................ 219

List of Contributors

(Their addresses can be found at the beginning of their respective chapters.)

BECKER, S.	23	LE GUENNO, B.L.	77
BOESCH, C.	77	MCCORMICK, J.B.	117
FELDMANN, H.	1, 175	MÜHLBERGER, E.	23
FISHER-HOCH, S.P.	117	NETESOV, S.V.	145
FORMENTY, P.	77	PETERS, C.J.	85
GOLDSMITH, C.S.	97	RYABCHIKOVA, E.I.	145
IGNATYEV, G.M.	205	SCHNITTLER, H. J.	175
KHAN, A.S.	85	SLENCZKA, W.G.	49
KILEY, M.P.	1	VOLCHKOV, V.E.	35
KOLESNIKOVA, L.V.	145	ZAKI, S.R.	97

Classification, Structure, and Replication of Filoviruses

H. Feldmann[1] and M.P. Kiley[2]

1	Classification	1
1.1	Taxonomic Classification	1
1.2	Biohazard Classification	4
2	Structure	5
2.1	Morphology and Structure of Virion Particles	5
2.2	Genome Structure	6
2.3	Polypeptides	7
2.3.1	Nucleoprotein	8
2.3.2	Virion Structural Protein 35 – Polymerase Cofactor	9
2.3.3	Virion Structural Protein 40 – Matrix Protein	10
2.3.4	Glycoprotein	10
2.3.5	Virion Structural Protein 30 – Minor Nucleoprotein?	11
2.3.6	Virion Structural Protein 24 – Membrane-Associated Protein	11
2.3.7	Large Protein – RNA-Dependent RNA Polymerase	12
2.3.8	Nonstructural Protein	12
3	Replication Cycle	13
3.1	Virus Growth in Cell Cultures	13
3.2	Virus Entry	14
3.3	Transcription	14
3.4	Translation, Processing, and Transport of Viral Proteins	16
3.5	Replication	17
3.6	Virus Assembly and Exit	17
4	Final Remarks	18
	References	18

1 Classification

1.1 Taxonomic Classification

When Marburg virus (MBGV) was first visualized by electron microscopy it was clear that morphologically it was very similar to the rhabdoviruses. This led to initial proposals to classify filoviruses in the family Rhabdoviridae. In 1982, the

---

[1]Institut für Virologie, Philipps-Universität, Robert-Koch-Str. 17, 35037 Marburg, Germany
[2]National Program Staff, Agricultural Research Service, United States Department of Agriculture, Beltsville, MD 21113, USA

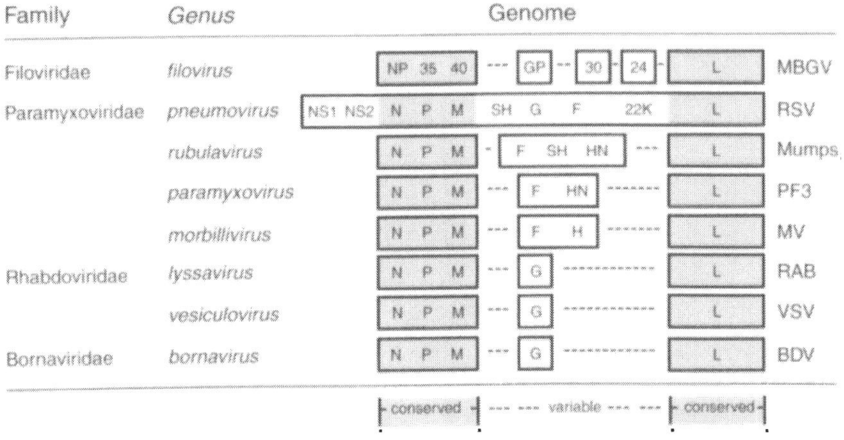

Family	Genus	Genome	
Filoviridae	filovirus	NP 35 40 --- GP -- 30 - 24 - L	MBGV
Paramyxoviridae	pneumovirus	NS1 NS2 N P M SH G F 22K L	RSV
	rubulavirus	N P M - F SH HN --- L	Mumps
	paramyxovirus	N P M --- F HN ------- L	PF3
	morbillivirus	N P M --- F H ------- L	MV
Rhabdoviridae	lyssavirus	N P M --- G ----------- L	RAB
	vesiculovirus	N P M --- G ----------- L	VSV
Bornaviridae	bornavirus	N P M --- G ----------- L	BDV

⊢ conserved ⊣ --- --- variable --- --- ⊢ conserved ⊣

Fig. 1. Comparison of genome organizations of nonsegmented negative-strand RNA viruses (Mononegavirales). Compared are genomes of viruses belonging to different genera of the four families Bornaviridae, Filoviridae, Paramyxoviridae, and Rhabdoviridae (Order Mononegavirales). Conserved and variable regions are identified. *RSV*, human respiratory syncytial virus; *MBGV*, Marburg virus; *Mumps*, mumps virus; *PF3*, human parainfluenza 3 virus; *MV*, measles virus; *RAB*, rabies virus; *VSV*, vesicular stomatitis virus; *BDV*, Borna disease virus; *N*, NP, (*BDV, p38*), nucleoprotein gene; *P*, (*BDV, p24*), phosphoprotein gene; *35*, virion structural protein (VP) 35 gene (phosphoprotein gene?); *M* (*BDV, gp18*), matrix protein gene; *40*, VP40 gene (matrix protein gene); *G*, GP (*BDV, gp57*), *F, H, HN*, glycosylated membrane protein genes; *SH*, small hydrophobic protein gene; *22 K*, non-glycosylated membrane protein gene; *30*, VP30 gene (unknown function); *24*, VP24 gene, (unknown function); *L* (*BDV, p190*), polymerase gene; *NS1, NS2*, genes of unknown function

family Filoviridae, with a single genus, *Filovirus*, was suggested on the basis of unique morphologic, morphogenetic, physicochemical, and biological features of its members (KILEY et al. 1982). Today all nonsegmented negative-strand (NNS) RNA viruses are grouped in the order Mononegavirales bearing four distinct families, Bornaviridae, Filoviridae, Paramyxoviridae, and Rhabdoviridae (ICTV 1991, 1996). All Mononegavirales share a similar genome organization with conserved regions at both ends encoding the core proteins and the RNA-dependent RNA polymerase (L). Between these conserved areas is a more variable region that generally contains genes encoding envelope and various membrane-associated proteins. The only exception is the genome of filoviruses, which contains a second minor nucleoprotein (VP30) in this region (Fig. 1). Filovirus genomes are more complex than those of lyssaviruses and vesiculoviruses and their genome organization aligns more closely with members of the genera *Paramyxovirus* and *Morbillivirus*. This relationship has been confirmed by comparison of the deduced amino acid sequences of the nucleoproteins (NPs) and L proteins of several NNS RNA viruses (FELDMANN et al. 1993; SANCHEZ et al. 1993).

The genus *Filovirus* can be separated into two distinct species, Marburg and Ebola, which differ in their glycoprotein (GP) genes by at least 55% at the nucleotide and 67% at the amino acid levels (SANCHEZ et al. 1996). MBGV strains seem to be more homogeneous and different subspecies (genotypes) have not been

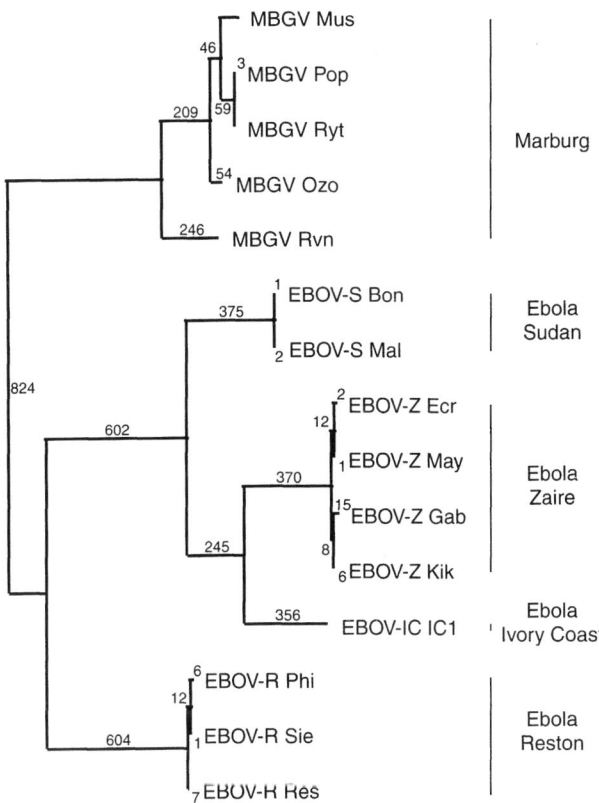

Fig. 2. Phylogenetic relationship of filoviruses. The phylogenetic relationship of all molecular characterized filoviral GP gene nucleotide sequences is shown. The sequences were analyzed by the maximum parsimony method using the PAUP program run on a MacIntosh computer. Bootstrap confidence limits of 100 replicates were always >90% and, thus, not indicated at appropriate branch points. The horizontal distances represent the number of nucleotide step differences (indicated adjacent to the lines). Vertical differences are for graphic representation only. *MBGV*, Marburg virus; *Mus*, strain Musoke (Kenya 1980); *Pop*, strain Popp (Uganda/Germany 1967); *Ryt*, strain Ratayczak (Uganda/Germany 1967); *Ozo*, strain Ozolin (Zimbabwe 1975); *Rvn*, strain Ravn (Kenya 1987); *EBOV*, Ebola virus; *S*, subspecies Sudan; *Bon*, strain Boniface (Sudan 1976); *Mal*, strain Maleo (Sudan 1979); *Z*, subspecies Zaire; *Ecr*, strain Eckron (Zaire 1976); *May*, strain Mayinga (Zaire 1976); *Gab*, strain Gabon (1994); *Kik*, strain Kikwit (Zaire 1995); *IC*, subspecies Ivory Coast; *IC1*, strain Cote d'Ivoire (1994); *R*, subspecies Reston; *Phi*, strain Philippines (1992); *Sie*, strain Siena, Italy (1992); *Res*, strain Reston, Virginia, USA (1989/90)

described (KILEY et al. 1988; FELDMANN et al. 1994). Recent phylogenetic analyses, however, based on the nucleotide sequences of the second and the fourth genes of five different MBGV strains revealed at least two different genetic lineages for the Marburg species with the Ravn strain being distinct from the others (Fig. 2) (JOHNSON et al. 1996; SANCHEZ et al. 1998). Significant differences within the Ebola virus (EBOV) species were first described based on peptide and oligonucleotide mapping studies (BUCHMEIER et al. 1983; COX et al. 1983), which have since been confirmed by nucleotide sequence analyses. Currently, EBOVs are subdivided into four subspecies, Ivory Coast (IC), Reston (R), Sudan (S), and Zaire (Z), which all present a monophyletic lineage (Fig. 2). Each subspecies differs from heterologous types by 37%–41% and 34%–43% at the nucleotide (GP gene) and amino acid (GP) level, respectively (FELDMANN and KLENK 1996; SANCHEZ et al. 1996). The molecular differences seen between the species Ebola and Marburg are so extensive that a taxonomic reevaluation of the genus *Filovirus* may be necessary.

Nucleotide sequence comparison between MBGV and EBOV shows only scattered similarities, which is in contrast to strong similarities seen between amino acid sequences of the structural proteins (FELDMANN et al. 1993; SANCHEZ et al. 1993). Despite this amino acid similarity, there is no indication that there is any significant serological (antigenic) cross-reactivity between EBOV and MBGV, but the subspecies of EBOV share common epitopes (RICHMAN et al. 1983; FELDMANN et al. 1994). This finding indicates that the nucleotide sequences of these agents may have diverged at some point in the distant past, but the structural proteins have maintained similar structures and functions. The divergence seen between the two species of filoviruses and among the subspecies of EBOV might also imply that these viruses have a common ancient origin and have slowly co-evolved with their as-yet-unknown but predicted natural animal hosts (SANCHEZ et al. 1996). In addition, the surprisingly high conservation of the GP genes of filovirus strains within each EBOV subspecies, especially those of subspecies Zaire (1976 and 1995 from former Zaire; 1994–1997 from Gabon) (GEORGES-COURBOT et al. 1997; VOLCHKOV et al. 1997) and those of most of the MBGV strains (SANCHEZ et al. 1998) may indicate that the different viruses have evolved into specific niches and may reflect a similar divergence in the natural hosts. The phylogenetic relationship of EBOV-R strains (Fig. 2), filoviruses isolated from Philippine primates, to the African subspecies may indicate that EBOV-R may indeed be indigenous to Asia and not be introduced to this region from Africa in the recent past.

1.2 Biohazard Classification

All filoviruses are classified as biosafety level 4 (BSL 4) agents based on their high mortality rate, potential transmissibility, and absence of either effective vaccines or chemotherapy. A maximum containment (BSL 4) facility is required for all laboratory work with infectious material (CENTERS FOR DISEASE CONTROL AND PREVENTION 1993). Detailed information has been published regarding the management of patients with suspected filoviral hemorrhagic fever and ways to minimize spread of virus during outbreaks, especially in Africa (CENTERS FOR DISEASE CONTROL AND PREVENTION 1988; WORLD HEALTH ORGANIZATION 1995), and to quarantine and properly handle imported monkeys (CENTERS FOR DISEASE CONTROL AND PREVENTION 1990).

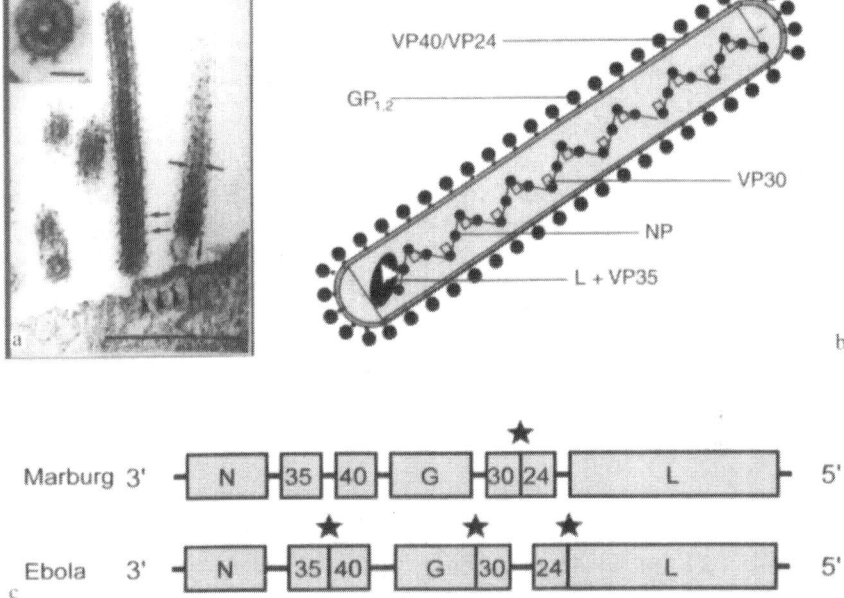

Fig. 3a–c. Characteristics of filoviral particles. **a** An electron micrograph (ultrathin section) showing budding of Marburg virus particles from the plasma membrane of infected primary cultures of human endothelial cells. Particles consist of a nucleocapsid surrounded by a membrane in which spikes are inserted (*arrows*). The nucleocapsid contains a central channel (*inset*). The plasma membrane of infected cells is often thickened at locations where budding occurs (*arrowheads*). *Bar*, 0.5 μm; bar inset, 50 nm. **b** Four proteins are involved in nucleocapsid formation; polymerase or large (L) protein (*ellipse*), nucleoprotein (NP) (*black circles*), virion structural protein (VP) 30 (*gray squares*) and VP35 (*white triangles* triangle. The glycoprotein; $GP_{1,2}$ is a transmembrane protein and anchored with the COOH-terminal part in the virion membrane (*black circles on outside of ellipse*). Homotrimers of GP form the spikes on the virion surface (*arrows* in **a**). VP40 and VP24 are membrane-associated proteins. **c** Genome organization of filoviruses. Filoviral genomes consist of a single, negative-stranded, linear RNA molecule. Differences in organization between viruses of the Marburg and Ebola species are indicated. *Asterisk*, position of gene overlap; *Ebola*, Ebola virus, subspecies Zaire, strain Mayinga; *G*, glycoprotein gene; *L*, polymerase (L) gene; *Marburg*, Marburg virus, strain Musoke; *N*, nucleoprotein gene; *24/30/35/40*, virion structural protein (VP) genes

2 Structure

2.1 Morphology and Structure of Virion Particles

Filoviral particles have a M_r of approximately $3-6 \times 10^8$ and a density in potassium tartrate of 1.14 g/cm^3 (ELLIOTT et al. 1985; KILEY et al. 1988). Virions are bacilliform in shape, but particles can also appear as branched, circular, U or 6-shaped, and long filamentous forms. Cell culture-derived MBGV particles show a much higher proportion of circular forms than EBOV subspecies which, in contrast, are characterized by predominantly filamentous forms. This morphology is unusual for viruses and has been important in the classification and nomenclature (filo-, Latin for thread) (Fig. 3). Filoviral virions show a uniform diameter of approximately 80 nm, but vary greatly in length. MBGV virions recovered from culture fluids of different cell types are uniformly shorter in mean unit length (795–828 nm) than EBOVs which vary from 974–1063 nm for EBOV-S and 990–1086 nm for EBOV-Z to 1026–1083 nm for EBOV-R and cannot be distinguished by particle length (GEISBERT and JAHRLING 1995). Negatively contrasted particles, regardless of serotype or host cell, contain an electron-dense central axis (19–25 nm in diameter) surrounded by an outer helical layer (45–50 nm in diameter) with cross-striations of 5 nm intervals. This central core is formed by the ribonucleoprotein (RNP) complex, which is surrounded by a lipid envelope derived from the host cell plasma membrane. Spikes of approximately 7 nm in diameter and spaced at about 5–10 nm intervals are seen as globular structures on the surface of virions (SIEGERT et al. 1967; PETERS et al. 1971; MURPHY et al. 1978; KILEY et al. 1982; SCHNITTLER et al. 1993; GEISBERT and JAHRLING 1995).

2.2 Genome Structure

Genomes of filoviruses consist of a single negative-stranded linear RNA molecule that has a M_r of 4.2×10^6 and constitutes 1.1% of the virion mass (REGNERY et al. 1980; KILEY et al. 1982). The RNA is noninfectious, does not contain a poly(A) tail, and upon entry into the cytoplasm of host cells is transcribed to generate polyadenylated subgenomic mRNA species (KILEY et al. 1982; FELDMANN et al. 1992; SANCHEZ et al. 1993). The entire nucleic acid sequences of two different MBGV strains (FELDMANN et al. 1992; BUKREYEV et al. 1995) and the Mayinga strain of EBOV-Z (SANCHEZ et al. 1993; VOLCHKOV et al. 1993) have been elucidated. Filovirus genomes have a length of approximately 19 kb (MBGV 19.1 kb; EBOV, 18.9 kb) and are significantly larger than those of other members of the order Mononegavirales. Analyses of the open reading frames (ORFs) on the antigenomic sequence revealed the following characteristic, linear gene order for filoviruses: 3′ leader-NP-VP35-VP40-GP-VP30-VP24-L-5′ trailer (Fig. 3).

Genes are delineated by transcriptional signals at their 3′ and 5′ ends that have been identified by their conservation (FELDMANN et al. 1992; SANCHEZ et al. 1993; VOLCHKOV et al. 1993; BUKREYEV et al. 1995) and by sequence analysis of mRNA species (SANCHEZ et al. 1989, 1992; MÜHLBERGER et al. 1996). Transcriptional start and stop signals are conserved among filoviruses, and the sequences 3′-CUNCNUNUAAUU-5′ and 3′-UAAUUCUUUUU-5′ represent the consensus motifs, respectively (Fig. 4). Filoviral genes are usually separated from each other by intergenic regions that vary in length and nucleotide composition. However, some genes overlap, especially those of EBOV, and the positions and numbers of overlaps vary among filoviruses. Viruses belonging to the EBOV-Z subspecies possess three overlaps, between the VP35 and VP40, GP and VP30 and VP24 and L genes, whereas MBGV strains have only a single overlap involving the VP30 and VP24 genes (Fig. 3). The length of the overlaps is limited to five highly conserved nu-

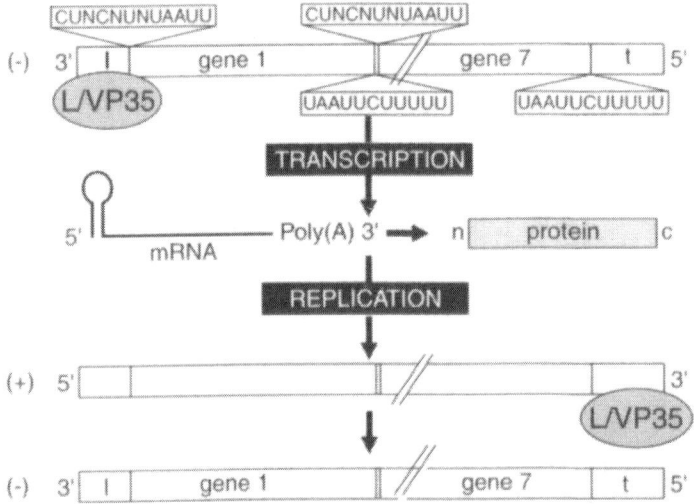

Fig. 4. Model of transcription and replication of filoviruses. Each gene on the linear arranged nonsegmented negative ( )-sense genome is flanked by conserved transcriptional start (3'-CUNCNUNUAAUU-5'; indicated *above*) and termination signals (3'-UAAUUCUUUUU-5'; indicated *beneath*; exception MBGV VP40 and EBOV L genes). Transcription starts at the 3' end of the ( )-sense genome and leads to polyadenylated messenger RNA species. For replication a full-length positive (+)-sense antigenome is synthesized which serves as the template for the synthesis of progeny ( )-sense RNA. *c*, COOH-terminal end of proteins; *l*, 3' untranslated region (leader); *L*, viral RNA-dependent RNA polymerase; *n*, NH₂-terminal end of proteins; *Poly(A)*, polyadenylation of messenger RNA species; *t*, 5' untranslated region (trailer); VP35, virion structural protein 35 KDa

cleotides within the transcriptional signals (3'-UAAUU-5') that are found at the internal ends of the conserved sequences (FELDMANN et al. 1992; SANCHEZ et al. 1993) (Fig. 4). Most genes tend to possess long noncoding sequences at their 3' and/or 5' ends which contribute to the increased length of the genome. Extragenic sequences are found at the 3'-leader and 5'-trailer ends of the genome. The 3'-leader sequences show a high content of adenine and uridine residues, with approximately twice as much uridine as adenine. An extremely high degree of homology between MBGV and EBOV is found in a run of 12 identical nucleotides at the very 3' end and a box of five nucleotides (3'-UAAAA-5') in the middle of the leader sequence. Those leader and trailer sequences are complementary to each other at the extreme ends (KILEY et al. 1986; FELDMANN et al. 1992), a feature that is shared by many NNS RNA viruses.

2.3 Polypeptides

Filoviral particles consist of seven structural polypeptides with presumed identical functions for the different viruses. The electrophoretic mobility patterns are species-specific with obvious differences in the migration of the GP and the virion structural proteins (VPs) 40, 35, and 30. Minor differences also distinguish among the

EBOV subspecies (FELDMANN et al. 1993, 1994). Four proteins are associated with the viral genomic RNA in the RNP complex: NP, VP30, VP35, and the L protein (Fig. 3). The three remaining structural proteins are membrane-associated; GP shows a type I transmembrane protein profile (WILL et al. 1993), while VP24 and VP40 are probably located at the inner side of the membrane. Metabolic labeling using different [^{3}H]-carbohydrates demonstrated the presence of only one glycosylated protein ($GP/GP_{1/2}$) in mature particles (FELDMANN et al. 1991; GEYER et al. 1992; WILL et al. 1993). Labeling with [^{32}P]-orthophosphate revealed two phosphorylated proteins, the major (NP) and minor (VP30) phosphoproteins of virion particles (ELLIOTT et al. 1985, 1993; BECKER et al. 1994). A nonstructural protein (sGP) has been described for all EBOVs (VOLCHKOV et al. 1995; SANCHEZ et al. 1996).

2.3.1 Nucleoprotein

The nucleoprotein (NP) is encoded by gene 1 at the extreme 3' end of the linear unsegmented RNA genome. NPs differ slightly in their electrophoretic mobility patterns ranging from 95 kDa for MBGV to 105 kDa for EBOV strains. The M_r calculated from the deduced amino acid sequences of the corresponding genes of MBGV (695 amino acids) and EBOV (739 amino acids) are 78 kDa and 83 kDa, respectively, and the differences in lengths are related to the less conserved COOH-termini of the protein (SANCHEZ et al. 1989, 1992; FELDMANN et al. 1992; BUKREYEV et al. 1995). Thus, filovirus NPs possess an unusually high M_r compared with other NNS RNA virus nucleoproteins, which range from 42 to 62 kDa. This suggests additional functions for the filovirus NP located in its unique COOH-terminus. The NP is the major structural phosphoprotein (ELLIOTT et al. 1985, 1993; KILEY et al. 1988; BECKER et al. 1994) and only the phosphorylated form of the protein is incorporated into virions as demonstrated for MBGV (BECKER et al. 1994). This implies that phosphorylation is needed to form stable virion RNP complexes. Sequence comparison of NPs of MBGV and EBOV show a high degree of homology within the first 400 predicted amino acids. The alignment shows that the region from position 130 to 392 (MBGV sequence) has a very strong similarity and is highlighted by a run of 34 identical amino acids at position 296–329. The fact that two of the three cysteine residues of NP are conserved may indicate their role in proper folding of the molecule. A small region in the middle of the MBGV and EBOV NP sequences was found to contain a significant amino acid homology with paramyxoviruses and to a lesser extent with rhabdoviruses (SANCHEZ et al. 1992). The NP proteins of filoviruses and many other NNS RNA viruses also have the hydrophobicity of their NH_2-termini in common. A role of this region in either protein folding and/or RNA binding has been postulated (BARR et al. 1991; MORGAN 1991). The less conserved COOH-terminal half of filoviral NPs which is hydrophilic and very acidic may function in the assembly process by interacting with the matrix proteins or the presumed second proposed nucleoprotein VP30 (ELLIOTT et al. 1985; KILEY et al. 1988; SANCHEZ et al. 1992). Similar functions have been discussed before for the variable COOH-termini of paramyxovirus nucleoproteins (BARR et al. 1991; PEEPLES 1991). NP is the major component of the RNP complex and tightly bound within the complex. Although RNA binding has not yet been demonstrated, there is little doubt that this protein is the functional analogue of the nucleoproteins of paramyxoviruses and rhabdoviruses (Fig. 3, Table 1).

2.3.2 Virion Structural Protein 35 – Polymerase Cofactor

Virion structural protein 35 (VP35) is encoded by gene 2. It has a length of 329 amino acids with MBGV (BUKREYEV et al. 1993a); VP35 of EBOV is 351 or 340 amino acids long (SANCHEZ et al. 1993; BUKREYEV et al. 1993a, respectively). RNP complex association of the protein is much weaker than is the case with NP and VP30 as demonstrated by nonionic detergent treatment of virion particles (ELLIOTT et al. 1985; KILEY et al. 1988). VP35 of EBO virions appears not to be phosphorylated (ELLIOTT et al. 1985, 1993). Expression studies of MBGV VP35 in insect cells (SF9 cells), however, revealed weak phosphorylation of this protein (H. Feldmann, unpublished data). Thus, VP35 may exist in a phosphorylated and an unphosphorylated form as has been demonstrated for NP. Hydropathy plots of MBGV and EBOV VP35 showed similar profiles and a prominent common hydrophilic domain in close proximity to the NH_2-terminus (MBGV, positions 28–42; EBOV, 57–76). This region may be involved in template binding which is supported by the fact that VP35 binds nonspecifically to nucleic acids (H. Feldmann, unpublished data). In spite of the inconsistent data on phosphorylation among filoviruses and the lack of sequence homology, the genome position of the corresponding gene combined with the association in the RNP complex suggest that VP35 is functionally analogous to the P proteins of paramyxoviruses and rhabdoviruses. Further studies will be required to determine if "P" protein would be an appropriate designation for this protein (Fig. 3, Table 1).

Table 1. Filoviral proteins and their proposed functions

Designation	Virus species	Encoding gene	Localization	Proposed function
NP	MBG/EBO	1	Ribonucleoprotein complex	Encapsidation
VP35	MBG/EBO	2	Ribonucleoprotein complex	polymerase cofactor
VP40	MBG/EBO	3	Membrane-associated	Matrix protein
GP	MBG/EBO	4[a]	Transmembrane protein (type I)	Receptor binding, fusion
sGP	EBO	4	Nonstructural, secreted	Immune modulation
VP30	MBG/EBO	5	Ribonucleoprotein complex	Encapsidation
VP24	MBG/EBO	6	Membrane-associated	Unknown
L	MBG/EBO	7	Ribonucleoprotein complex	RNA-dependent

NP, nucleoprotein; VP, virion structural protein; GP, glycoprotein; L, large protein (polymerase); sGP, small glycoprotein; MBG, species Marburg of filoviruses; EBO, species Ebola of filoviruses.
[a]Expressed by RNA editing.

2.3.3 Virion Structural Protein 40 – Matrix Protein

Virion structural protein 40 (VP40) of filoviruses is encoded by gene 3. It is 303 and 326 amino acids long with MBGV and EBOV, respectively (BUKREYEV et al. 1993a; SANCHEZ et al. 1993). Differences in the electrophoretic mobilities of VP40 can be used to distinguish between MBGV and EBOV strains and even among the EBOV subspecies (FELDMANN et al. 1994). Nitrocellulose-bound VP40 binds in a radio-overlay protein assay nonspecifically to nucleic acids (H. Feldmann, unpublished data). VP40 is not associated with the RNP complex and behaves like a membrane-associated protein following nonionic detergent treatment of virion particles (ELLIOTT et al. 1985; KILEY et al. 1988). This finding, together with a predominantly hydrophobic profile, the abundance in virion particles, and the genome localization of the corresponding gene, suggest that VP40 is the analogue of the matrix proteins of NNS RNA viruses (Fig. 3, Table 1).

2.3.4 Glycoprotein

Glycoprotein (GP), encoded by gene 4 of the genome, is the only glycosylated structural protein of virions. GP of MBGV and EBOV is 681 and 676 amino acids in length, respectively (VOLCHKOV et al. 1992; BUKREYEV et al. 1993b; SANCHEZ et al. 1993; WILL et al. 1993). Filoviral GPs ($GP_{1/2}$) are type I transmembrane proteins anchored in the membrane via a COOH-terminal hydrophobic domain. It has been shown with EBOV that GP undergoes posttranslational proteolytic cleavage into the NH_2-terminal fragment GP_1 and the COOH-terminal fragment GP_2 which are present as a disulfide-linked complex ($GP_{1/2}$) in the mature spike (VOLCHKOV et al. 1998a) (for details see elsewhere in this volume). GP contains N- and O-glycans that account for up to 50% of the molecular weight of the mature protein (GEYER et al. 1992; WILL et al. 1993). Oligosaccharide side chains differ in their terminal sialylation patterns which seem to be strain-, as well as cell line-, dependent (FELDMANN et al. 1991, 1994; GEYER et al. 1992; WILL et al. 1993). Detailed structural analyses of filoviral carbohydrates are available for MBGV only. The structures include oligomannosidic and hybrid type N-glycans as well as bi-, tri-, and tetra-antennary complex species, and high amounts of neutral mucin-type O-glycans (GEYER et al. 1992). Comparison of filoviral GP sequences showed conservation at the NH_2- and COOH-terminal ends of the proteins in which the two hydrophobic domains (signal peptide, membrane anchor) and most of the highly conserved cysteine residues are located. The middle region is variable, extremely hydrophilic, and carries the bulk of the glycosylation sites for N- (EBOV 17 sites; MBGV 22 sites) and O-glycans. Recently it has been shown that the two cysteine residues at position 671 and 673 of MBGV GP are acylated (FUNKE et al. 1995). Acylation at the border between membrane anchor region and cytoplasmic tail has been shown for many viral type I transmembrane proteins. The special arrangement of all cysteine residues in the molecule favors an intramolecular cysteine bridge formation between the two cleavage products of the molecule, GP_1 and GP_2, resulting in a stem region consisting of GP_1 and GP_2 with a crown-like GP_1 domain on the top carrying the mass of the carbohydrate side chains (VOLCHKOV et al. 1998a). For MBGV it has been shown that mature GP is inserted in the membrane as a homotrimer (FELDMANN et al. 1991).

In general, filoviral GPs lack significant homologies with envelope proteins of other NNS RNA viruses. However, an extended region of 160 amino acids including the membrane anchor, cleavage site, and a putative fusion domain shows a significant homology to envelope proteins of several retroviruses (VOLCHKOV et al. 1992; BUKREYEV et al. 1993b; SANCHEZ et al. 1993; Will et al. 1993). A domain of 26 amino acids within this region aligns with an immunosuppressive domain in retroviruses that has been assumed to be involved in inhibition of blastogenesis of lymphocytes, decrease in monocyte chemotaxis and macrophage infiltration, inhibition of human natural killer cell activity, and block of protein kinase C activity (CIANCIALO et al. 1985; HARRIS et al. 1987; KADOTA et al. 1991). The fact that GP is the only surface protein of virions suggests a function in binding to cellular receptors (TAKADA et al. 1997; YANG et al. 1998). A fusion domain has been proposed GP however, experimental data to support such a biological function are missing (GALLAHER 1996). Additionally, GP is also under consideration as the major viral antigen and the main target for the host immune response (Fig. 3, Table 1).

2.3.5 Virion Structural Protein 30 – Minor Nucleoprotein?

Virion structural protein 30 (VP30) is encoded by gene 5 and intimately associated with the RNP complex (ELLIOTT et al. 1985; KILEY et al. 1988; FELDMANN et al. 1992). The protein has a length of 260 and 281 amino acids with EBOV and MBGV, respectively (SANCHEZ et al. 1993; BUKREYEV et al. 1995). The protein binds RNA under denaturing conditions in an RNA protein overlay assay and forms complexes with the NP. Complexes of coexpressed recombinant VP30 and NP can be precipitated using either VP30- or NP-specific antibodies (H. Feldmann, unpublished data). VP30 of EBOV has been identified as the minor phosphoprotein of virions (ELLIOTT et al. 1985, 1993). The available data indicate that VP30 may work as a functional unit in encapsidation of the RNA genome (KILEY et al. 1988) which could be achieved by either binding to NP and/or binding to genomic RNA. It could also play a role as an additional cofactor of the transcriptase/replicase complex (Fig. 3, Table 1).

2.3.6 Virion Structural Protein 24 – Membrane-Associated Protein

Virion structural protein 24 (VP24) is encoded by gene 6 of filoviruses. It is 253 and 251 amino acids in length with MBGV and EBOV, respectively (FELDMANN et al. 1992; SANCHEZ et al. 1993; BUKREYEV et al. 1995). The protein is membrane-associated. Unlike VP40, it is not completely removed from the RNP complex under isotonic conditions (ELLIOTT et al. 1985; KILEY et al. 1988). VP24 presumably serves as a second matrix protein and may bind to the cytoplasmic tail of GP and/

or may link the membrane proteins (VP40 and/or GP) to the RNP complex. There are minor differences in the SDS-PAGE migration profile of this protein among filovirus strains. Such changes have also been observed between EBOV wild type (subspecies Zaire, strain Mayinga) and a highly pathogenic variant isolated after several passages from guinea pigs (VOLCHKOV et al. 1994). To what extent these changes contribute to the higher pathogenic potential of the variant in guinea pigs is currently unknown (Fig. 3, Table 1).

2.3.7 Large Protein – RNA-Dependent RNA Polymerase

The large (L) protein is encoded at the 5' end of the linear genome and has a predicted M_r of 267 kDa (2331 amino acids) for the Musoke (MÜHLBERGER et al. 1992) and Popp strains (BUKREYEV et al. 1995) of MBGV. Computer-assisted comparison revealed significant homologies to L proteins of other NNS RNA viruses. Homologies are mainly located in the NH_2-terminal half of the protein and concentrated within three common domains, named boxes A, B, and C (BARIK et al. 1990; MÜHLBERGER et al. 1992). Other common features are a high content of leucine and isoleucine residues, a large positive net charge, clusters of basic amino acids, putative ATP binding sites, two neighboring cysteine residues located in the COOH-terminal half of the protein, and the genome localization of the encoding gene. A highly conserved peptide motif GDNQ, located at the COOH-terminal end of domain B (positions 744–747) flanked by hydrophobic amino acid residues, seems to be correlated with enzymatic functions of the protein. Mutations in this domain, which is present in most NNS RNA virus L proteins, abolished activity as shown for other NNS RNA virus L proteins (SLEAT and BANERJEE 1993). Furthermore, an LDD motif is present at position 1095–1097. Similar motifs with alterations in the first amino acid have been described and discussed as active sites for some RNA-dependent RNA polymerases of plant, animal, and bacterial viruses (KAMER and ARGOS 1984; CHIU et al. 1985; WAIN-HOBSON et al. 1985; KEMDIRIN et al. 1986). The L protein is regarded as an RNA-dependent RNA polymerase (for details see elsewhere in this volume) (Fig. 3, Table 1).

2.3.8 Nonstructural Protein

A nonstructural glycoprotein has recently been discovered with EBOV (VOLCHKOV et al. 1995; SANCHEZ et al. 1996) (for details see elsewhere in this volume). This protein, designated sGP, is the primary expression product of gene 4 of EBOV and is translated from unedited viral transcripts (see Sect. 3.3; Table 1). sGP shares ~300 NH_2-terminal amino acids with GP, but has a different COOH-terminus (~70 amino acids) which contains many charged residues as well as conserved cysteines. The protein is directed into the endoplasmic reticulum and becomes N- and probably O-glycosylated. sGP is secreted into culture medium in high quantities. The function of this nonstructural protein is unknown, but it could modulate the host immune response (YANG et al. 1998). A similar protein is not found with MBGV (Table 1).

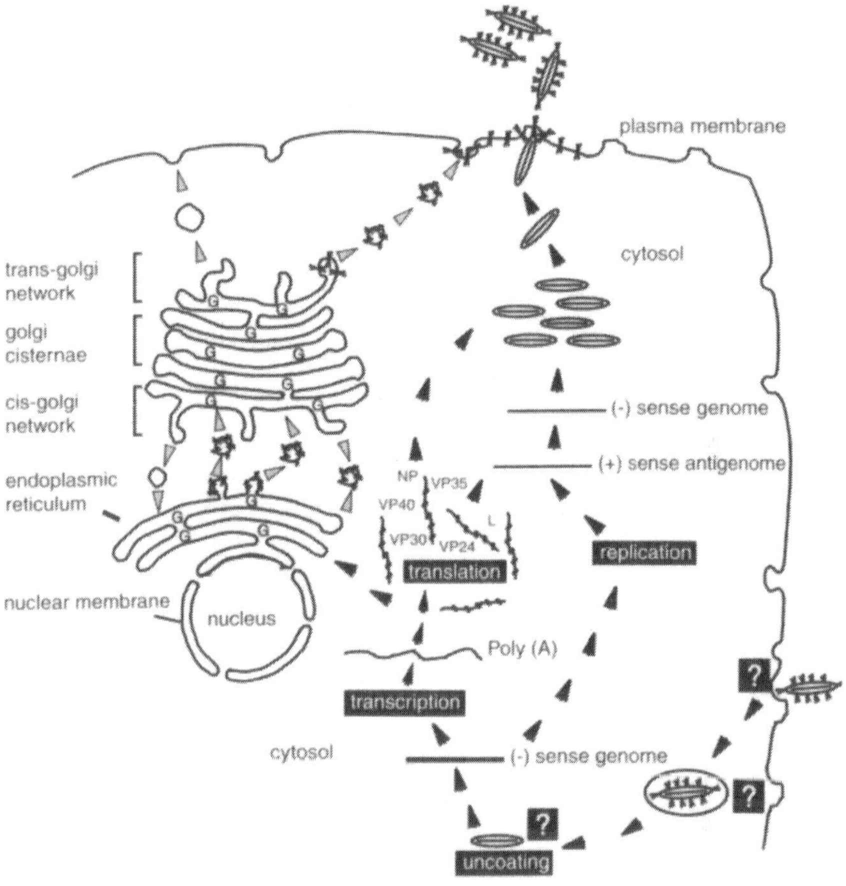

Fig. 5. Replication cycle of filoviruses. Filoviruses conclusively replicate in the cytoplasm of an infected cell. Virus entry is probably receptor-mediated followed by fusion and uncoating, two steps which have not yet been examined. Transcription of polyadenylated messenger RNAs and replication of the negative (−)-sense genome via a full-length positive (+)-sense antigenome are described in more detail in Sects. 3.3–3.5 and are illustrated in Fig. 4. Maturation of virions seems to take place at or near the plasma membranes where budding of particles occurs. *G*, glycoprotein; *L*, viral RNA-dependent RNA polymerase; *NP*, nucleoprotein; *VP*, virion structural protein; *24, 30, 35, 40*, molecular weight in kDa

3 Replication Cycle

3.1 Virus Growth in Cell Cultures

The Vero cell line, especially the E6 clone, is most widely used for virus isolation and propagation. Primary virus isolation has also been successful in MA-104 and SW13 cells (McCormick et al. 1983; Jahrling et al. 1990). In addition, a variety of other cells have been tested as substrates for filovirus replication (van der Groen et al. 1978; McCormick et al. 1983; Peters et al. 1992; Geisbert and Jahrling 1995). These include a recently developed human microvascular endothelial cell line (HMEC-1), primary cultures of human umbilical cord vein endothelial cells (HUVEC), and human peripheral blood monocytes/macrophages (Schnittler et al. 1993; Feldmann and Klenk 1996; Feldmann et al. 1996).

MBGV and EBOV (subspecies Zaire) cause lytic infections in cell culture. The Sudan and Reston subspecies of EBOV replicate more slowly upon primary isolation, and the cytopathogenic effect is not as prominent as with the Zaire subspecies. The course of infection in tissue culture can be monitored by an indirect immunofluorescence assay (IFA) or by plaque assay. In cases of little or no cytopathogenic effect, reverse transcriptase-polymerase chain reaction (RT-PCR) on viral RNA isolated from infected cells and tissue culture supernatants can be helpful for quantification (Schnittler et al. 1993; Sanchez and Feldmann 1996).

EBOV subgenomic RNA synthesis in tissue culture is detectable by at least 6–7 h post-infection, reaches a maximum by 18 h, and declines thereafter; cytopathogenic effects are not seen before 48 h post-infection. The first mRNA to be detected is NP-specific which reaches levels sufficient to produce protein by 7 h post-infection. All proteins are detectable by in vitro translation of polyadenylated RNA isolated 18 h post-infection, thereafter the yield of translation products decreases (Sanchez and Kiley 1987). RT-PCR assays on genomic RNA of MBGV particles from supernatants of infected cells indicated that the replication cycle is approximately 12 h (Schnittler et al. 1993).

3.2 Virus Entry

Cell entry is presumably mediated by GP (Fig. 5). The asialoglycoprotein receptor (ASGP-R) of hepatocytes has been, postulated shown to serve as a receptor for MBGV (Becker et al. 1995). However, since ASGP-R is not expressed on many cells that support the growth of filoviruses, other receptors have to be postulated. The next step in virus entry presumably involves fusion. That proteolytic cleavage of GP, as observed with EBOV, is necessary for fusion activity is likely, but has not been definitely shown yet. It is also not known whether the fusion process occurs directly at the plasma membrane or in endocytotic vesicles. Syncytia formations of infected cells at neutral as well as low pH conditions have never been reported. Ultrastructural studies on infected cells early post-infection showed that filoviruses appear to be closely associated with coated pits along the plasma membrane indicating endocytosis as a possible mechanism for entry (Geisbert and Jahrling 1995). Penetration by receptor endocytosis is supported by studies which employed lysosomotropic agents (Mariyankova et al. 1993). The mechanism of uncoating is still unknown (Fig. 5).

3.3 Transcription

Filoviral transcription and replication take place in the cytoplasm of infected cells (Fig. 5), and the mechanisms resemble those of many other NNS RNA viruses.

Transcription starts at the extreme 3' end and probably leads to the synthesis of a short (+)-leader sequence that is terminated when the first transcription start site is encountered (Fig. 4). The transcription start signals of filovirus genes show the consensus sequence 3'-CUNCNUNUAAUU-5' which are located at the 3' end of each gene. The sequence 3'-UAAUUCUUUUU(U)-5' at the 5' end of each gene serves as a transcription stop and polyadenylation signal (exception MBGV VP40 and EBOV L genes). Both signals carry the pentamer 3'-UAAUU-5', a unique feature among NNS RNA viruses (Fig. 4). The function of the pentamer is unknown, but it could serve as the recognition site for positioning the polymerase complex. The surrounding semiconserved sequences may then mediate the exact initiation of transcription and termination/polyadenylation events (FELDMANN et al. 1993). The seven genes encoding the structural proteins are subsequently transcribed to produce seven monocistronic polyadenylated mRNA species (Fig. 4). There is no evidence for larger amounts of bi- or multicistronic subgenomic RNA species (SANCHEZ and KILEY 1987; FELDMANN et al. 1992; SANCHEZ et al. 1993).

Analyses of MBGV mRNA species have shown that the 5' ends of the transcripts are identical with the transcription start signals (Fig. 4). Thus, mRNA synthesis starts precisely at the first nucleotide of the transcription start signal (MÜHLBERGER et al. 1996). Computer-assisted analyses of the 5' ends of filoviral mRNA species predicted stable secondary structures with conserved nucleotides located in the stem regions of the hairpins. Nucleotide substitutions at conserved positions are usually accompanied by compensatory mutations of the base-pairing nucleotide. This indicates the significance of the secondary structures at the 5' ends of the transcripts which might play a role in transcript stability and ribosome binding (SANCHEZ et al. 1993; MÜHLBERGER et al. 1996). Today it is not clear if filoviral transcripts carry cap structures at the 5' end, but indirect evidence comes from experiments with S-adenosylhomocysteine hydrolase (SAH) inhibitors that are capable of reducing filovirus replication in vitro and in vivo (HUGGINS et al. 1996). The 3' ends of the transcripts carry a poly(A) tail probably generated by a stuttering mechanism of the viral polymerase at a run of five or six uridine residues located at the 5' end of all transcription stop signals (FELDMANN et al. 1992; VOLCHKOV et al. 1992; SANCHEZ et al. 1993; MÜHLBERGER et al. 1996). Filoviral transcripts contain unusually long untranslated regions especially at the 3' ends. The correct AUG start codons, opening the ORFs encoding the viral proteins, are not always those that are located closely to the transcription start signals nor are they always in the proposed favorable context to serve as an initiation codon for eukaryotic ribosomes (A/G-AUGG/A) (KOZAK 1986). Thus, the long 5' noncoding regions of the transcripts could be involved in the regulation of translation by directing downstream AUG start codons to the ribosomes for the initiation of protein synthesis.

The role of gene overlaps in regulation of transcription is unknown (Fig. 4). SANCHEZ et al. (1993) proposed that, following mRNA synthesis, transcription is reinitiated by reposition of the polymerase at the downstream start site. This "back up" mechanism is supported by the finding that attenuation of filovirus genes with start sites in overlaps does not occur to any higher degree, as has been noted for a much larger overlap found in the respiratory syncytial virus genome (COLLINS 1991). Alternatively, the polymerase may occasionally terminate transcription without polyadenylation at the overlap and initiate transcription of the downstream gene, but there is no evidence for detectable levels of transcripts lacking poly(A) tails.

The organization and transcription of the GP gene of EBOV are unusual (for details see elsewhere in this volume). Full-length GP is expressed by transcriptional editing of a single nontemplated adenosine residue at a run of seven uridine residues on the genomic RNA (VOLCHKOV et al. 1995; SANCHEZ et al. 1996). The primary gene product is a small nonstructural glycoprotein (sGP) that is secreted from infected cells. In addition, virus variants have been selected after passaging in tissue culture and animals that express full-length GP from a single ORF. Those variants acquired a mutation that added a single uridine nucleotide at the editing site connecting the former two ORFs to a single larger ORF that encodes the full-length GP (SANCHEZ et al. 1993; VOLCHKOV et al. 1994). The MBGV GP is expressed in a single frame and the gene does not contain sequences favoring mechanisms such as editing or frameshifting. A second overlapping ORF has been described for the GP gene of the Musoke strain (WILL et al. 1993). Sequence analyses of all known MBGV strains revealed that this ORF is not conserved and, thus, unlikely to be used for expression of a protein (BUKREYEV et al. 1993b; SANCHEZ et al. 1998).

3.4 Translation, Processing, and Transport of Viral Proteins

Translation of viral proteins is mediated by the cellular machinery (Fig. 5). GP and EBOV sGP are translated from distinct GP gene-specific transcripts at membrane-bound ribosomes and are directed by an NH_2-terminal hydrophobic domain into the endoplasmic reticulum (ER) where they enter the exocytotic transport route to the cell surface. Studies on MBGV have shown that GP undergoes during transport a complex sequence of processing events involving removal of the signal peptide (WILL et al. 1993), N-glycosylation (FELDMANN et al. 1991; GEYER et al. 1992), and trimerization (FELDMANN et al. 1991), all of which occur in the ER, followed by acylation in a pre-Golgi compartment (FUNKE et al. 1995), and by O-glycosylation and maturation of N-glycans in the Golgi apparatus (FELDMANN et al. 1991; GEYER et al. 1992; WILL et al. 1993; BECKER et al. 1996). Depending on the host cell, there are wide variations in the amount of neuraminic acid present on MBGV GP (FELDMANN et al. 1994) (for details see elsewhere in this volume). Recent studies have shown that processing of GP involves proteolytic cleavage in the trans-Golgi network by a proprotein convertase, most likely furin, into a larger NH_2-terminal GP_1 and a smaller COOH-terminal GP_2 subunit. Mature GP on viral particles ($GP_{1/2}$) consists of disulfide-linked GP_1 and GP_2 subunits (VOLCHKOV et al. 1998a). GP_1 is partly shed after release of its disulfide linkage to the smaller transmembrane subunit GP_2 (VOLCHKOV et al. 1998b) (for details see elsewhere in this volume).

All other viral proteins are translated at free ribosomes in the cytosol of infected cells (Fig. 5). For NP and VP30 phosphorylation has been demonstrated (ELLIOTT et al. 1985, 1993; BECKER et al. 1994). The RNA-dependent RNA polymerase is thought to possess kinase activity (MÜHLBERGER et al. 1992), and recently first insight has been obtained in the mechanisms involved in RNP complex formation (for details see elsewhere in this volume).

3.5 Replication

The switch mechanism between transcription and replication is unknown (Figs. 4, 5), but as with many other NNS RNA viruses, synthesis of the NP could be a key factor. Encapsidation and polymerase complex entry sites are probably located on the leader sequence, and the fact that the ends of the genome are complementary suggests a single identical encapsidation site on genome and antigenome; this would function for both transcription as well as replication (FELDMANN et al. 1992). Replication involves a full-length positive (+)-strand antigenome which serves as the template for synthesis of negative (−)-strand genomic molecules. Encapsidated genomic RNA forms RNP complexes that go into the formation of new infectious virions (Figs. 4, 5).

3.6 Virus Assembly and Exit

The first obvious morphogenic event during filovirus infections is the appearance of amorphous viral material that gradually increases in electron density and progressively accumulates to larger inclusions in the cytoplasm. Early forms of inclusions are similar in structure for all filoviruses. Following maturation preformed RNP complexes are already found in intermediate inclusions of EBOV-infected cells. As inclusions mature and progress to the plasma membrane, the preformed RNP complexes increase in number. As MBGV inclusions mature, however, the dense matrices accumulate beneath the plasma membrane to assemble as RNP complexes (GEISBERT and JAHRLING 1995). Immune electron microscopic studies indicate that all filoviral inclusions likely contain at least VP40 and NP that may interact due to the net positive and negative charge of the matrix protein (VP40) and NP, respectively (ELLIOTT et al. 1993; GEISBERT and JAHRLING 1995). Virions usually bud at the plasma membrane, and the budding process is probably mediated at membrane locations where GP is incorporated. During the budding process the cytoplasmic tail of GP may interact with VP40 and/or VP24 and, thus, mediates the linkage to the RNP complex. Particles mature preferentially in a vertical mode, but budding via the longitudinal axis has also been observed (SCHNITTLER et al. 1993; GEISBERT and JAHRLING 1995). In macrophages budding has also been observed at intracytoplasmic membranes surrounding vacuoles which form during infection (FELDMANN et al. 1996).

4 Final Remarks

Several steps in the replication cycle of filoviruses as described in this chapter are based on comparison with other NNS RNA viruses rather than solid experimental data on filoviruses themselves. In order to identify functional elements on the RNA genome and to study the involvement of viral proteins in transcription and replication recombinant systems such as minireplicons or a full-length genomic clone are needed. These systems are available for some other RNA viruses such as rabies (SCHNELL et al. 1994), VSV (LAWSON et al. 1995), measles (RADECKE et al. 1995), and influenza A (PALESE et al. 1996) viruses. Data on a minireplicon system for MBGV are presented elsewhere in this volume. Other systems are currently being investigated at other laboratories.

Acknowledgements. The authors are grateful for the discussions and suggestions of many friends and colleagues in the field, especially Anthony Sanchez. We further acknowledge the support over the years by grants of the Deutsche Forschungsgemeinschaft (Forschergruppe Kl 238/1-1, SFB 286 and 535, grant Fe 286/4-1), and the Kempkes-Stiftung (grant 21/95), given to the Institut für Virologie, Philipps Universität, Marburg, Germany, and by grants of USAMRIID, given to the Centers for Disease Control and Prevention, Special Pathogens Branch. H.F. held a fellowship of the National Research Council (NRC) at the Centers for Disease Control and Prevention, Special Pathogens Branch (1992-1994).

References

Barik SE, Rud W, Luk D, Banerjee AK, Yong Kang C (1990) Nucleotide sequence analysis of the L gene of vesicular stomatitis virus (New Jersey serotype): identification of conserved domains in L proteins of nonsegmented negative-strand RNA viruses. Virology 175:332–337

Barr J, Chamers P, Pringle CR, Easton AJ (1991) Sequence of the major nucleocapsid gene of pneumonia virus of mice: sequence comparison suggests structural homology between nucleocapsid proteins of pneumoviruses, paramyxoviruses, rhabdoviruses and filoviruses. J Gen Virol 72:677–685

Becker S, Huppertz S, Klenk HD, Feldmann H (1994) The nucleoprotein of Marburg virus is phosphorylated. J Gen Virol 75:809–818

Becker S, Spiess M, Klenk HD (1995) The asialoglycoprotein receptor is a potential liver-specific receptor for Marburg virus. J Gen Virol 76:393–399

Becker S, Klenk HD, Mühlberger E (1996) Intracellular transport and processing of the Marburg virus surface protein in vertebrate and insect cells. Virology 225:145–155

Buchmeier MJ, DeFries R, McCormick JB, Kiley MP (1983) Comparative analysis of the structural polypeptides of Ebola virus from Sudan and Zaire. J Infect Dis 147:276–281

Bukreyev AA, Volchkov VE, Blinov VM, Netesov SV (1993a) The VP35 and VP40 proteins of filoviruses: homology between Marburg and Ebola viruses. FEBS Lett 322:41–46

Bukreyev AA, Volchkov VE, Blinov VM, Netesov SV (1993b) The GP protein of Marburg virus contains the region similar to the 'immunosuppressive domain' of oncogenic retroviruses P15 E proteins. FEBS Lett 323:183–187

Bukreyev AA, Volchkov VE, Blinov VM, Dryga SA, Netesov SV (1995) The nucleotide sequence of the Popp (1967) strain of Marburg virus: a comparison with the Musoke (1980) strain. Arch Virol 140:1589–1600

Centers for Disease Control and Prevention (1988) Management of patients with suspected viral hemorrhagic fever. MMWR 37, Suppl 3:1–16

Centers for Disease Control and Prevention (1990) Update: Ebola-related filovirus infection in nonhuman primates and interim guidelines for handling nonhuman primates during transit and quarantine. MMWR 39:22–24, 29–30

Centers for Disease Control and Prevention (1993) Biosafety in microbiology and biomedical laboratories. US Department of Health and Human Services (HHS), publication No. (CDC) 93-8395, US Government Printing Office, Washington DC

Chiu IM, Yaniv A, Dahlberg JE, Gazit A, Skunitz SF, Tronick SR, Aaronson SA (1985) Nucleotide sequence evidence for relationship of AIDS retrovirus to lentiviruses. Nature (London) 317:366-368

Cianciolo GJ, Copeland TJ, Oroszlan S, Snyderman R (1985) Inhibition of lymphocyte proliferation by a synthetic peptide homologous to retroviral envelope proteins. Science 230:453-455

Collins PL (1991) The molecular biology of human respiratory syncytial virus (RSV) of the genus Pneumovirus. In: Kingsbury DW (ed) The paramyxoviruses. Plenum, New York, pp 103-162

Cox NJ, McCormick JB, Johnson KM, Kiley MP (1983) Evidence for two subspecies of Ebola virus based on oligonucleotide mapping of RNA. J Infect Dis 147:272-275

Elliott LH, Kiley MP, McCormick JB (1985) Descriptive analysis of Ebola virus proteins. Virology 147:169-176

Elliott LH, Sanchez A, Holloway BP, Kiley MP, McCormick JB (1993) Ebola protein analysis for the determination of genetic organization. Arch Virol 133:423-436

Feldmann H, Will C, Schikore M, Slenczka W, Klenk HD (1991) Glycosylation and oligomerization of the spike protein of Marburg virus. Virology 182:353-356

Feldmann H, Mühlberger E, Randolf A, Will C, Kiley MP, Sanchez A, Klenk HD (1992) Marburg virus, a filovirus: messenger RNAs, gene order, and regulatory elements of the replication cycle. Virus Res 24:1-19

Feldmann H, Klenk HD, Sanchez A (1993) Molecular biology and evolution of filoviruses. Arch Virol, Suppl. 7:81-100

Feldmann H, Nichol ST, Klenk HD, Peters CJ, Sanchez A (1994) Characterization of filoviruses based on differences in structure and antigenicity of the virion glycoprotein. Virology 199:469-473

Feldmann H, Bugany H, Mahner F, Klenk HD, Drenckhahn D, Schnittler HJ (1996) Filovirus-induced endothelial leakage triggered by infected monocytes/macrophages. J Virol 70:2208-2214

Feldmann H, Klenk HD (1996) Filoviruses: Marburg and Ebola. In: Maramorosch K, Murphy FA, Shatkin AJ (eds) Advances in virus research, vol. 47, Academic, San Diego, pp 1-52

Funke C, Becker S, Dartsch H, Klenk HD, Mühlberger E (1995) Acylation of the Marburg virus glycoprotein. Virology 208:289-297

Gallaher WR (1996) Similar structural models of the transmembrane proteins of Ebola and avian sarcoma viruses. Cell (Letter) 85:477-478

Geisbert TW, Jahrling PB (1995) Differentiation of filoviruses by electron microscopy. Virus Res 39:129-150

Georges-Courbut MC, Sanchez A, Lu CY, Baize S, Leroy E, Lansout-Soukate J, Tevy- Benissan C, Georges A, Trappier SG, Zaki SR, Swanepoel R, Leman PA, Rollin PE, Peters CJ, Nichol ST, Ksiazek TG (1997) Isolation and phylogenetic characterization of Ebola virus causing different outbreaks in Gabon. Emerging Infectious Diseases 3:59-62

Geyer H, Will C, Feldmann H, Klenk HD, Geyer R (1992) Carbohydrate structure of Marburg virus glycoprotein. Glycobiology 2:299-312

Harris DT, Cianciolo GJ, Snyderman R, Argov S, Koren HR (1987) Inhibition of human natural killer cell activity by a synthetic peptide homologous to a conserved region in the retroviral protein, p15 E. J Immunol 138:889-894

Huggins J, Tseng C, Laughlin C, Bray M (1996) Antiviral drug therapy of filovirus infection. International Colloquium on Ebola Virus Research, Antwerp, Belgium

ICTV (1991) The order Mononegavirales. Paramyxovirus Study Group of the Vertebrate Subcommittee. Virology Division News. Arch Virol 117:137-140

ICTV (1996) International Committee on Taxonomy of Viruses, Virology Division; Xth International Congress of Virology, Jerusalem, Israel

Jahrling PB, Geisbert TW, Galgard DW, Johnson ED, Ksiazek TG, Hall WC, Peters CJ (1990) Preliminary report: isolation of Ebola virus from monkeys imported to USA. Lancet 335:502-505

Johnson ED, Johnson BK, Silverstein D, Tukei P, Geisbert TW, Sanchez AN, Jahrling PB (1996) Characterization of a new Marburg virus isolated from a 1987 fatal case in Kenya. Arch Virol, Suppl 11:101-114

Kadota J, Cianciolo GJ, Snyderman R (1991) A synthetic peptide homologous to retroviral transmembrane envelope proteins depressed protein kinase C-mediated lymphocyte proliferation and directly inactivated protein kinase C: a potential mechanism for immunosuppression. Microbiol Immunol 35:443-459

Kamer G, Argos P (1984) Primary structural comparison of RNA-dependent polymerases from plant, animal and bacterial viruses. Nucleic Acids Res 12:7269–7282

Kemdirin S, Palefsky J, Briedis DJ (1986) Influenza B virus PB1 protein: nucleotide sequence of the genome RNA segment predicts a high degree of structural homology with the corresponding influenza A virus polymerase proteins. Virology 152:126–135

Kiley MP, Bowen ETW, Eddy GA, Isaäcson M, Johnson KM, McCormick JB, Murphy FA, Pattyn SR, Peters D, Prozesky OW, Regnery RL, Simpson DIH, Slenczka W, Sureau P, van der Groen G, Webb PA, Wulff H (1982) Filoviridae: a taxonomic home for Marburg and Ebola viruses? Intervirology 18:24–32

Kiley MP, Wilusz J, McCormick JB, Keene JD (1986) Conservation of the 3' terminal nucleotide sequence of Ebola and Marburg viruses. Virology 149:251–254

Kiley MP, Cox NJ, Elliott LH, Sanchez A, DeFries R, Buchmeier MJ, Richman DD, McCormick JB (1988) Physicochemical properties of Marburg virus: evidence for three distinct virus strains and their relationship to Ebola virus. J Gen Virol 69:1957–1967

Kozak M (1986) Point mutations define a sequence flanking the AUG initiator codon that modulates translation by eukaryotic ribosomes. Cell 44:283–292

Lawson ND, Stillman EA, Whitt MA, Rose JK (1995) Recombinant vesicular stomatitis virus from DNA. Proc Natl Acad Sci USA 92:4477–4481

Mariyankova RF, Giushakowa SE, Pyzhik EV, Lukashevich IS (1993) Marburg virus penetration into eukaryotic cells. Vopr Virusol 2:74–76

McCormick JB, Bauer SP, Elliott LH, Webb PA, Johnson KM (1983) Biological differences between strains of Ebola virus from Zaire and Sudan. J Infect Dis 147:264–267

Morgan EM (1991) Evolutionary relationships of paramyxoviruses nucleocapsid-associated proteins. In: Kingsbury DW (ed) The paramyxoviruses Plenum, New York, pp 163–179

Mühlberger E, Sanchez A, Randolf A, Will C, Kiley MP, Klenk HD, Feldmann H (1992) The nucleotide sequence of the L gene of Marburg virus, a filovirus: homologies with paramyxoviruses and rhabdoviruses. Virology 187:534–547

Mühlberger E, Trommer S, Funke C, Volchkov V, Klenk HD, Becker S (1996) Termini of all mRNA species of Marburg virus: sequence and secondary structure. Virology 223:376–380

Murphy FA, van der Groen G, Whitfield SG, Lange JV (1978) Ebola and Marburg virus morphology and taxonomy. In: Pattyn SR (ed) Ebola virus hemorrhagic fever, 1st edn. Elsevier/North-Holland, Amsterdam, pp 61–84

Palese P, Zheng H, Engelhardt O, Pleschka S, Garcia-Sastre A (1996) Negative-strand RNA viruses: genetic engineering and applications. Proc Natl Acad Sci USA 93:11354–1358

Peeples ME (1991) Paramyxovirus M proteins: pulling it all together and putting it on the road. In: Kingsbury DW (ed) The paramyxoviruses. Plenum, New York, pp 427–456

Peters CJ, Jahrling PB, Ksiazek TG, Lupton H (1992) Filovirus contamination of cell cultures. Develop Biol Standard 76:267–274

Peters D, Müller G, Slenczka W (1971) Morphology, development, and classification of Marburg virus. In: Martini GA, Siegert R (eds) Marburg virus disease, 1st edn. Springer, Heidelberg Berlin New York, pp 68–83

Radecke F, Spielhofer P, Schneider H, Kaelin K, Huber M, Dotsch C, Christiansen G, Billeter MA (1995) Rescue of measles viruses from cloned DNA. EMBO J 14:5773–5784

Regnery RL, Johnson KM, Kiley MP (1980) Virion nucleic acid of Ebola virus. J Virol 36:465–469

Richman DD, Cleveland PH, McCormick JB, Johnson KM (1983) Antigenic analysis of strains of Ebola viruses: identification of two Ebola virus subspecies. J Infect Dis 147:268–271

Sanchez A, Kiley MP (1987) Identification and analysis of Ebola virus messenger RNAs. Virology 157:414–420

Sanchez A, Kiley MP, Holloway BP, McCormick JB, Auperin DD (1989) The nucleoprotein gene of Ebola virus: cloning, sequencing, and in vitro expression. Virology 170:81–91

Sanchez A, Kiley MP, Klenk HD, Feldmann H (1992) Sequence analysis of the Marburg virus nucleoprotein gene: comparison to Ebola virus and other nonsegmented negative-strand RNA viruses. J Gen Virol 73:347–357

Sanchez A, Kiley MP, Holloway BP, Auperin DD (1993) Sequence analysis of the Ebola virus genome: organization, genetic elements, and comparison with the genome of Marburg virus. Virus Res 29:215–240

Sanchez A, Feldmann H (1996) Detection of Marburg and Ebola virus infections by polymerase chain reaction assays. In: Becker Y, Darai G (eds) Frontiers in virology – diagnosis of human viruses by polymerase chain reaction echnology, 2. edn. Springer, Berlin Heidelberg New York, pp 411–418

Sanchez A, Trappier SG, Mahy BWJ, Peters CJ, Nichol ST (1996) The virion glycoprotein of Ebola viruses are encoded in two reading frames and are expressed through transcriptional editing. Proc Natl Acad Sci USA 93:3602–3607

Sanchez A, Trappier SG, Ströher U, Nichol ST, Bowen MD, Feldmann H (1998) Variation in the glycoprotein and VP35 genes of Marburg virus strains. Virology 240:138–146

Schnell MJ, Mebatsin T, Conzelmann KK (1994) Infectious rabies viruses from cloned cDNA. EMBO J 13:4195–4203

Schnittler HJ, Mahner F, Drenckhahn D, Klenk HD, Feldmann H (1993) Replication of Marburg virus in human endothelial cells. A possible mechanism for the development of viral hemorrhagic disease. J Clin Invest 91:1301–1309

Siegert R, Shu HL, Slenczka W, Peters D, Müller G (1967) Zur Äthiologie einer unbekannten von Affen ausgegangenen Infektionskrankheit. Dtsch Med Wochenschr 92:2341–2343

Sleat DE, Banerjee AK (1993) Transcriptional activity and mutational analysis of recombinant vesicular stomatitis virus RNA polymerase. J Virol 67:1334–1339

Takada A, Robinson C, Goto H, Sanchez A, Murti G, Whitt M, Kawaoka Y (1997) A system for functional analysis of Ebola virus glycoprotein. Proc Natl Acd Sci USA 94:14764–14769

van der Groen G, Johnson KM, Webb FA, Wulff H, Lange J (1978) Results of Ebola antibody survey in various population groups. In: Pattyn SR (ed) Ebola virus hemorrhagic fever, 1. edn. Elsevier/North-Holland, Amsterdam, pp 203–208

Volchkov VE, Blinov VM, Netesov SV (1992) The envelope glycoprotein of Ebola virus contains an immunosuppressive-like domain similar to oncogenic retroviruses. FEBS Lett 305:181–184

Volchkov VE, Blinov VM, Kotov AN, Chepurnov AA, Netesov SV (1993) The full-length nucleotide sequence of the Ebola virus. IXth International Congress of Virology. Glasgow, Scotland, p 299

Volchkov VE, Chepurnov A, Dryga S, Becker S, Blinov V, Kotov A, Ternovoj V, Klenk HD, Netesov SV (1994) Molecular characterization of a pathogenicity variant of Ebola virus. Ninth International Conference on Negative Strand RNA Viruses. Estoril, Portugal, p 176

Volchkov VE, Becker S, Volchkova VA, Ternovoj VA, Kotov AN, Netesov SV, Klenk HD (1995). GP mRNA of Ebola virus is edited by the Ebola virus polymerase and by T7 and vaccinia virus polymerases. Virology 214:421–430

Volchkov VE, Volchkova VA, Eckel C, Klenk HD, Bouloy M, LeGuenno B, Feldmann H (1997) Emerging of subspecies Zaire Ebola virus in Gabon. Virology 232:139–144

Volchkov VE, Feldmann H, Volchkova VA, Klenk HD (1998a) Processing of the Ebola virus glycoprotein by the proprotein convertase furin. Proc Natl Acad Sci USA 95:5762–5767

Volchkov VE, Volchkova VA, Slenczka W, Klenk HD, Feldmann H (1998b) Release of viral glycoproteins during Ebola virus infection. Virology 245:110–119

Wain-Hobson S, Sonigo P, Danos O, Cole S, Alizon M (1985) Nucleotide sequence of the AIDS virus, LAV. Cell 40:9–17

Will C, Mühlberger E, Linder D, Slenczka W, Klenk HD, Feldmann H (1993) Marburg virus gene four encodes for the virion membrane protein, a type I transmembrane glycoprotein. J Virol 67:1203–1210

World Health Organization (1995) Viral haemorrhagic fever – management of suspected cases. Weekly Epidemiol Rec 70:249–256

Yang Z, Delgado R, Xu L, Todd RF, Nabel EG, Sanchez A, Nabel GJ (1998) Distinct cellular interactions of secreted and transmemebrane Ebola virus glycoproteins. Science 279:1034–1037

Co- and Posttranslational Modifications and Functions of Marburg Virus Proteins

S. BECKER and E. MÜHLBERGER

1	Introduction	23
2	Structural Proteins of Filoviruses	24
3	Structural and Functional Analysis of Recombinant Marburg Virus Proteins	26
3.1	Posttranslational Modifications and Function of Marburg Virus Nucleocapsid Proteins	26
3.1.1	Phosphorylation of MBGV Nucleocapsid Proteins	26
3.1.2	Function of MBGV Nucleocapsid Proteins	27
3.2	Posttranslational Modification and Transport of the Surface Protein of Marburg Virus	29
References		32

1 Introduction

During the first two decades following the isolation of the prototype of filoviruses, Marburg virus (MBGV; for review see MARTINI and SIEGERT 1971), research was hindered by the exceptionally high pathogenicity of the agent both for humans and for nonhuman primates. With the development of recombinant DNA techniques and the subsequent determination of the nucleotide sequences of MBGV [EMBL Nucleotide Sequence Database, accession numbers Z12132 (MBGV Musoke), X64405, X64406, X68493, X68494, X68495, and Z29337 (MBGV Popp)] and Ebola virus (EBOV) genomes (EMBL Data Libraries accession number L11365) it became possible to employ recombinant proteins for filovirus research that remarkably reduced the risk to the investigator. Moreover, recombinant expression of viral proteins is essential in order to elucidate the function of the various filoviral proteins and to study structure/function relationships, e.g., of proteins involved in the transcription/replication process.

Institut für Virologie, Philipps-Universität, Robert-Koch-Str. 17, 35037 Marburg, Germany

2 Structural Proteins of Filoviruses

In Fig. 1, A MBGV particle is schematically depicted to show localization of the various structural proteins. The envelope of the virion carrying protruding spikes is derived from host cell membranes. Spikes are homotrimers of the surface protein (GP, FELDMANN et al. 1991). The space between envelope and nucleocapsid contains two putative matrix proteins (VP24 and VP40). The nucleocapsid complex is composed of four proteins, nucleoprotein (NP), VP35, VP30, and L protein (L) (KILEY et al. 1988). L is present in too small amounts to be visible on the autoradiogram. The predicted and apparent molecular weights and the number of amino acids of MBGV structural proteins are shown in Table 1. The apparent molecular weights of NP, GP, and L differ from the molecular weights predicted from the nucleotide sequences. A similar pattern of seven structural proteins is found with EBOV (KILEY et al. 1980; ELLIOTT et al. 1985).

Within the Order Mononegavirales genome organization is highly conserved. The localization of a particular gene on the linear genome therefore gives some information on the function of the respective gene product (FELDMANN et al. 1992). The product of the first gene of nonsegmented negative sense (NNS) RNA viruses (N protein in Paramyxo- and Rhabdoviridae) encapsidates the RNA. Products of

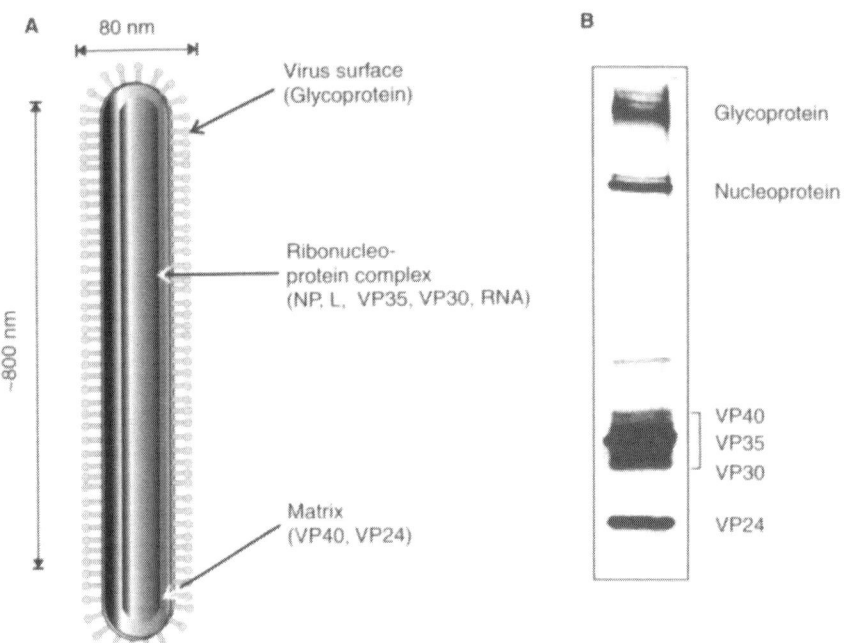

Fig. 1. A Marburg virus (MBGV) particle; B SDS profile of viral structural proteins (VP)

Table 1. Apparent (10% SDS-PAGE) and predicted (computer-aided) molecular weight of the structural proteins of MBGV and EBOV

Protein	Apparent molecular weight (kDa)	Number of amino acids	Predicted weight (kDa)
NP	92, 94 (92 only intracellular)	695	77.7
VP35	35	329	36.1
VP40	38	303	33.8
GP	90 (unglycosylated, only intracellular) 170–200 (glycosylated)	681	74.4 (unglycosylated)
VP30	32	277	31.5
VP24	28	254	28.6
L	220	2331	267.1

NP, nucleoprotein; VP24, VP30, VP35, VP40, viral structural proteins designated according to their counterparts in EBOV; GP, surface protein; L, large protein (polymerase).
SDS-PAGE was performed with authentic viral structural proteins and recombinant proteins expressed with the vaccinia virus T7 system.

the second (P protein) and last (L protein) gene are the components of the viral RNA-dependent RNA polymerase. Genes in the middle of the genome code for surface and matrix proteins. Therefore the following functions for filoviral structural proteins are suggested: The first gene of MBGV and EBOV codes for NP, the major nucleocapsid component presumably acting as RNA-encapsidating protein. It is 695 amino acids in length and has an apparent M_r of 94 kDa, as can be deduced from SDS-PAGE (KILEY et al. 1988; SANCHEZ et al. 1989; SANCHEZ et al. 1992).

L (encoded by the last gene) is present in nucleocapsids only in catalytic amounts corresponding to its presumed function as RNA-dependent RNA polymerase (MÜHLBERGER et al. 1992). The open reading frame (ORF) of the L gene codes for 2330 amino acids leading to a predicted M_r of 267 kDa. Since none of the antisera directed against MBGV recognized L, bacterial expression of the NH_2-terminal part of the protein was performed to raise antibodies against the protein. The antiserum obtained recognized a 220 kDa protein in purified virions. Additionally, a Flag-tagged L (L-Flag) was constructed and expressed using the vaccinia virus T7 expression system. Anti-Flag antibodies also detected a 220 kDa protein band in lysates of L-Flag-expressing HeLa cells indicating that L migrates in SDS-PAGE considerably faster than expected (BECKER et al., in press).

Regarding the two remaining nucleocapsid proteins, VP35 (product of the second gene) and/or VP30 (encoded by the fifth gene) might serve as accessory proteins in analogy to the P protein of other NNS RNA viruses. Recombinant expression showed VP30 migrating as a double band at 30–32 kDa. In preparations of viral structural proteins VP30 was found only in small amounts. VP35 expressed using the vaccinia virus T7 system migrated as a single 35 kDa protein band. Both recombinant VP30 and VP35 were recognized by antibodies directed against the nucleocapsid complex of MBGV.

The position of the VP40 gene (third gene) and the localization of the protein inside the virion between membrane and nucleocapsid indicates a putative matrix protein (Becker et al, in press; FELDMANN et al. 1992; ELLIOTT et al. 1985). GP (fourth gene), the only surface protein, is presumed to mediate virus binding to target cells (FELDMANN et al. 1991; WILL et al. 1993; BECKER et al. 1995). VP24 (sixth gene) is thought to be membrane associated (ELLIOTT et al. 1985), but its function remains obscure.

3 Structural and Functional Analysis of Recombinant Marburg Virus Proteins

Early investigations on recombinant expression of filoviral structural proteins were performed by SANCHEZ et al. (1989, 1992, 1993), ELLIOTT et al. (1993), and FELDMANN et al. (1992). Except L, all proteins of EBOV and MBGV were in vitro translated from cDNA clones. In the following years, filoviral proteins were expressed in bacterial, mammalian, and insect cell expression systems to study the structure and function of the viral components (BECKER et al. 1994, 1996; FUNKE et al. 1995; VOLCHKOV et al. 1995; SANCHEZ et al. 1996). So far, investigations of recombinant filoviral proteins have focused on the proteins presumed to drive the process of replication and transcription of MBGV (NP, VP35, VP30, and L protein) and on the surface proteins of MBGV and EBOV (GP).

3.1 Posttranslational Modifications and Function of Marburg Virus Nucleocapsid Proteins

3.1.1 Phosphorylation of MBGV Nucleocapsid Proteins

Proteins of the nucleocapsid complex were expressed using the vaccinia virus T7 expression system (FUERST et al. 1986) or the baculovirus expression system to study posttranslational modifications. In both systems recombinant NP migrated on SDS-PAGE as a double band which could not be detected in viral structural proteins, where only one sharp protein band was detected (BECKER et al. 1994). Labeling of the recombinant protein with $^{32}P_i$ revealed only the upper band, migrating at approximately 94 kDa, to be phosphorylated. Analysis of the authentic NP in preparations of purified virions showed that only the phosphorylated form of NP was incorporated into the virion. Since MBGV-infected cells also contained both forms of NP, it was assumed that phosphorylation of NP is a prerequisite for incorporating the protein into the nucleocapsid complex. Whether phosphorylation of NP plays a role in protein-protein or protein-RNA interaction is not known. Investigations concerning phosphorylation of nucleoproteins of other Mononegavirales revealed a complex situation. Some of the viruses contain phosphorylated nucleoproteins, such as rabies virus (SOKOL and KOPROWSKI 1975), mumps virus (NARUSE et al. 1981), and measles virus (ROBBINS et al. 1980). Examples of NNS RNA viruses whose nucleoproteins are not phosphorylated are respiratory syncytial virus (LAMBERT et al. 1988)

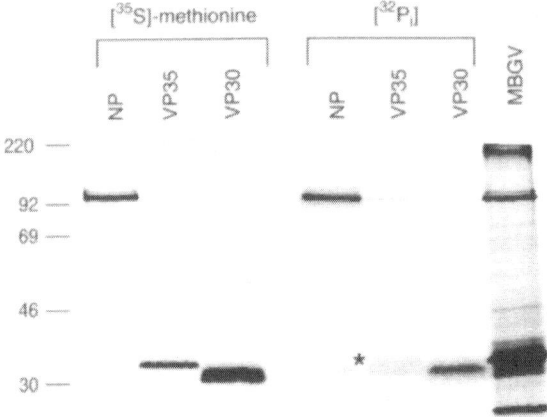

Fig. 2. Marburg virus (MBGV)-infected C1008 cells were labeled with [^{35}S]methionine or [^{32}P$_i$]. Virions were purified from the culture medium as described by MÜHLBERGER et al. 1992 and structural proteins were immunoprecipitated with monospecific antibodies. Proteins were separated by SDS-PAGE and dried gels subjected to autoradiography. *Asterisk* marks position of [^{32}P$_i$]-labeled VP35. MBGV: MBGV structural proteins. [^{35}S] methionine labeled

and vesicular stomatitis virus (SOKOL and KOPROWSKI 1975). Little is known about how phosphorylation influences the function of nucleoproteins, but GOMBART et al. (1995) found the nucleoprotein of measles virus to be incorporated into nucleocapsids dependent on its phosphorylation state.

Analysis of ^{32}P$_i$-labeled viral structural proteins showed that besides NP at least one additional MBGV protein in the range of 30–35 kDa was phosphorylated. Precipitation of ^{32}P$_i$-labeled MBGV structural proteins with monospecific antibodies revealed that VP30 as well as VP35 was phosphorylated (Fig. 2). However, the signal of VP35 was very weak compared with that of VP30 or NP signal. In addition to the specific VP35 band (see asterisk in Fig. 2) contaminants of phosphorylated VP30 and NP occurred. Investigations regarding phosphorylation of EBOV proteins were performed with viral structural proteins (ELLIOTT et al. 1985, 1993). Metabolic labeling of viral particles with ^{32}P$_i$ showed that NP and VP30 were phosphorylated. As with MBGV, NP was strongly labeled whereas VP30 gave a weaker signal. Phosphorylation of EBOV VP35 was not detected.

3.1.2 Function of MBGV Nucleocapsid Proteins

As mentioned above, VP35 and/or VP30 are presumed to act as P protein equivalent. P proteins of NNS RNA viruses are encoded by the second gene of the respective viruses. They are phosphorylated (VIDAL et al. 1988; TAKACS et al. 1992) and essential for transcription and replication (BARIK and BANERJEE 1992; BARIK et al. 1995; CURRAN 1996; SPADAFORA et al. 1996). The fact that VP35 is also encoded by the second gene points to its putative function as P protein. On the

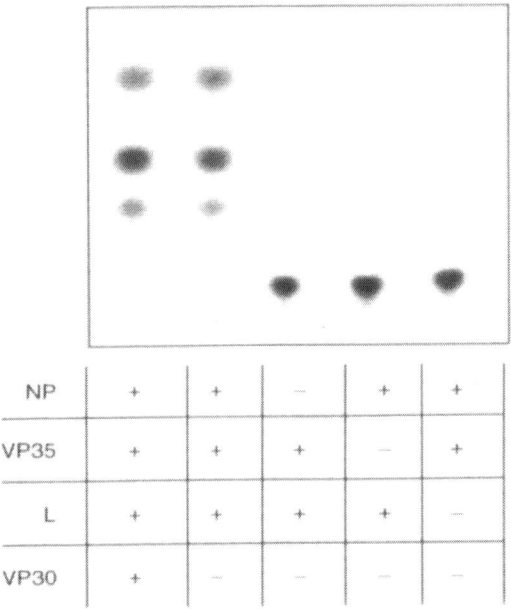

Fig. 3. Nucleocapsid proteins essential for Marburg virus (MBGV)-specific replication. HeLa cells were infected with vaccinia virus strain MVA-T7 (SUTTER et al. 1995) and transfected with recombinant pTM1 plasmids containing the genes for nucleoprotein (NP), and the nucleocapsid proteins VP35, VP30, or L. At 3.5 h after the first transfection, the same cells were transfected with a MBGV-specific negative sense minigenomic RNA encoding the CAT gene. At 3 days postinfection cells were harvested and CAT gene expression was determined

other hand localization of VP30 in the nucleocapsid in combination with its strong phosphorylation suggests that VP30 might serve as a P homologue. VP30 (fifth gene) has no counterpart in other members of Mononegavirales. Most viruses of the Order Mononegavirales carry at this particular position of the genome genes coding for membrane or matrix proteins (FELDMANN et al. 1992). However, COLLINS et al. (1996) recently published that the product of the M2 gene, which is located in the direct vicinity of the L gene of respiratory syncytial virus (RSV), encodes a viral structural protein colocalized with N and P proteins in infected cells. M2, also called 22 kDa protein, appears to be essential for proper elongation of viral mRNAs in a reconstituted, cDNA-expressed minigenome system. Thus, another member of the Mononegavirales possesses a more complex form of genome organization concerning localization of genes coding for nucleocapsid and membrane proteins than the prototypes vesicular stomatitis or Sendai virus. Interestingly, comparative sequence analysis of parts of the genomes showed the highest homology between MBGV and RSV (MÜHLBERGER et al. 1992).

To address the question which of the four MBGV nucleocapsid proteins were essential for transcription and replication an artificial replication system based on the vaccinia virus T7 expression system was used as established for other NNS

RNA viruses (CONZELMANN 1996). As replicon a synthetic MBGV-specific minigenome was constructed containing the leader and trailer regions of MBGV genome flanking the CAT gene. When MBGV-infected cells were transfected with the minigenome, the minigenome was replicated and packaged. Using this RNA construct for transfection of vaccinia virus-infected cells expressing the nucleocapsid proteins of MBGV it was shown that NP, VP35, and L were sufficient for reporter gene expression (Fig. 3). These data revealed that indeed VP35 acts as a P equivalent.

3.2 Posttranslational Modification and Transport of the Surface Protein of Marburg Virus

The envelope of MBGV is decorated by the only surface protein, GP (M_r = 170 kDa), which is inserted into the viral membrane as a homotrimer (FELDMANN et al. 1991). The GP gene is 2844 nucleotides in length and encodes a protein of 681 amino acids (WILL et al. 1993). In contrast to EBOV, in which the virion-associated surface protein is only expressed after mRNA editing (VOLCHKOV et al. 1995; SANCHEZ et al. 1996), MBGV GP is encoded by a single ORF. Two strongly hydrophobic regions have been identified in GP, one at the NH_2-terminal, the other in the COOH-terminal region (Fig. 4). Amino acid sequencing revealed that the NH_2-terminal hydrophobic region is not present in the mature protein, indicating that this region serves as signal peptide, which is cleaved after translocation into the ER (WILL et al. 1993). The COOH-terminal hydrophobic domain is probably used as membrane anchor, and the last eight amino acids constitute the cytoplasmic tail. The sequence of GP contains 19 potential N-linked glycosylation sites (Fig. 4) and several clusters of hydroxyamino acids which can serve as O-linked glycosylation sites (WILL et al. 1993). Oligosaccharide analysis revealed that the sugar side chains account for approx. 50% of the molecular weight. Most of the

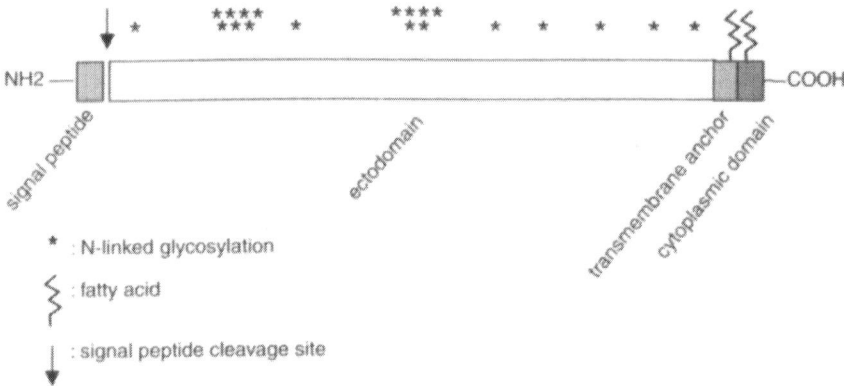

Fig. 4. Marburg virus (MBGV) surface protein

N-linked sugars belong to the complex type oligosaccharides. Furthermore, GP contains a substantial amount of O-linked oligosaccharides, which make up the main part (55 mol%) of the sugar moiety (GEYER et al. 1992). O-glycosylation is reported only from one other viral glycoprotein of the Order Mononegavirales, the respiratory syncytial virus G protein (GRUBER and LEVINE 1985). Oligosaccharides completely lacked terminal sialic acids when MBGV was grown in E6 cells (GEYER et al. 1992). Using other cell lines (Vero, human endothelian cells) GP was shown to be sialylated (FELDMANN et al. 1994). Biochemical analysis of sialylated GP derived from Vero cells revealed that only 10% of the potential sialylation sites were used (Geyer, personal communication). Since GP is the only membrane protein of MBGV, it is assumed to be responsible for virus entry into host cells (BECKER et al. 1995) and to be the major target for the immune response of the infected organism. WILL et al. (1993) and BUKREYEV et al. (1993) reported that the COOH-terminal part of MBGV GP, like EBOV GP (VOLCHKOV et al. 1992), contains a region that shows a high degree of homology to a presumptive immunosuppressive domain observed in some retroviruses.

When GP was expressed in mammalian cells using the vaccinia virus T7 expression system, it was shown to be localized at the plasma membrane, indicating that GP does not need the chaperone-like activity of other viral proteins to be transported correctly. In addition, transport proceeded with similar kinetics as in MBGV-infected cells. The estimated $t_{1/2}$ for gaining resistance to endoglycosydase H was 90 min. Further transport to the plasma membrane was fast. Almost simultaneously with the detection of endo H resistance, GP arrived at the plasma membrane. Although unglycosylated and high-mannose forms of recombinant GP showed an identical size (90 kDa and 140 kDa) as the respective forms of authentic GP in MBGV-infected cells, mature recombinant GP was shown to be 10 kDa smaller (160 kDa) pointing to a slightly altered complex glycosylation of recombinant GP. Lectin analyses, however, did not reveal any differences in the principal features of the glycosylation pattern that is characterized by O-linked oligosaccharides, mainly complex-type N-linked sugars, and a few high-mannose-type oligosaccharides (BECKER et al. 1996; SÄNGER et al., in preparation). Expression of GP using the baculovirus expression system led to a 140 kDa protein. The reduction in size as compared to the authentic protein was caused by an altered glycosylation in insect cells. Combined lectin and glycosidase analysis showed that the 140 kDa protein band was composed of two variants of GP: A large excess of endo H sensitive protein representing the immature form of GP in the endoplasmic reticulum and a small fraction of endo H resistant form which is O-glycosylated and thus transported to the Golgi apparatus. Interestingly, depending on the expression rate, a nonglycosylated form of GP could be detected that was shown to be localized in the cytoplasm (BECKER et al. 1996). The cytoplasmic GP might be the result of an exhausted cellular translocation machinery. The role of N-glycosylation for structural integrity and transport of GP has been investigated with tunicamycin, an inhibitor of N-glycosylation. After tunicamycin treatment neither N- nor O-glycosylation of GP was detectable. Only the nonglycosylated form accumulated leading to the assumption that N-glycosylation of GP is critical for transport of the

Table 2. COOH-terminals of MBGV and EBOV GP

Zaire 76	*WRQWIPAGIGVTGVIIAVIALF* **CIC** KFVF
Zaire 95	*WRQWIPAGIGVTGVIIAVIALF* **CIC** KFVF
MAY 76	*WRQWIPAGIGVTGVIIAVIALF* **CIC** KFVF
SUDMAL	*WRQWIPAGIGITGIIIAIIALL* **CVC** KLLC
RES	*WRQWIPAGIGIIGVIIAIIALL***CIC** KILC
RESPHI	*WRQWIPAGIGIIGVIIAIIALL* **CIC** KILC
RESITAL	*WRQWIPAGIGIIGVIIAIIALL* **CIC** KILC
MBGPOPP	*WGVLTNLGILLLLSIAVLIALS* **CIC** RIFTKYIG
MBGMUS	*WGVLTNLGILLLLSIAVLIALS* **CIC** RIFTKYIG

The highly conserved cysteine residues at the border between transmembrane anchor and cytoplasmic domain are printed in bold letters. The cytoplasmic domain is underlined, transmembrane anchor in italics.
Key Zaire76, Ebola outbreak Zaire 1976; Zaire 95, Ebola outbreak Zaire 1995; MAY 76, Ebola virus Zaire subtype, Mayinga strain; SUDMAL, Ebola, Sudan subtype, Maleo strain; RES, Ebola virus, Reston subtype, Reston outbreak 1989; RESPHI, Ebola virus, Reston subtype, Philippines 1992; RESITAL, Ebola virus, Reston subtype, Italy 1992; MBGVPOPP, Marburg virus, strain Popp, isolated 1967, Marburg; MBGVMUS, Marburg virus, strain Musoke, isolated 1982.
Ebola sequences: SANCHEZ et al. 1996; MBGPOPP sequence, BUKREYEV et al. 1995; MBGMUS, WILL et al. 1993.

protein from endoplasmic reticulum to the Golgi apparatus where O-glycosylation takes place. Similar results were obtained for other glycoproteins where N-glycosylation was shown to be the prerequisite for transport (TATE and BLAKELY 1994; JARVIS et al. 1990).

It has been reported for several viral glycoproteins (SCHMIDT and LAMBRECHT 1985; VEIT et al. 1989, 1991; GAUDIN et al. 1991; COLLINS and MOTTET 1992; YANG and CUMPANS 1996) that in addition to glycosylation and oligomerization, another posttranslational event takes place during the transport from endoplasmic reticulum to the Golgi apparatus (BERGER and SCHMIDT 1985) when fatty acids, mainly palmitic acid, are attached to the region between the transmembrane anchor and the cytoplasmic tail of the molecule. These fatty acids are bound either as oxyesters to serine or threonine residues or via thioester linkage to cysteine residues (for review see TOWLER et al. 1988). Although the physiological meaning of this process is not well understood, there is some evidence that acylation is involved in morphogenesis of viral particles (GAEDIGK-NITSCHKO et al. 1990) and in syncytia formation (BOS et al. 1995). Fatty acid labeling of recombinant MBGV GP in insect cells revealed an acylation of the protein which was also shown for the authentic protein in MBGV-infected cells (FUNKE et al. 1995). Gas chromatographic analysis of the bound fatty acids demonstrated that mainly palmitic acid was attached to the amino acid backbone. However, labeling with radioactive myristic and stearic acid was also possible. Acyl residues were cleaved by mercaptoethanol and hydroxylamine indicating that fatty acids were linked to the protein via thioester linkages. Mutational analysis of the COOH-terminal of GP verified two cysteines (amino acid positions 671 and 673) as fatty acid acceptor sites which can be used alternatively (FUNKE et al. 1995). Cysteine residues at this particular position of the protein are conserved in the family of filoviruses. All surface proteins sequenced so far contained two cysteines directly at the border between transmembrane anchor and cytoplasmic domain (Table 2). Whether acylation of the surface proteins has a function for MBGV morphogenesis and/or invasion of the host cells is still unclear.

References

Barik S, Banerjee A (1992) Phosphorylation by cellular casein kinase II is essential for transcriptional activity of vesicular stomatitis virus phosphoprotein P. Proc Natl Acad Sci USA 89:6570–6574

Barik S, McLean T, Dupuy LC (1995) Phosphorylation of Ser^{232} directly regulates the transcriptional activity of the P protein of human respiratory syncytial virus: Phosphorylation of Ser^{237} may play an accessory role. Virology 213:405–412

Becker S, Hofsäß U, Rinne C, Klenk HD, Mühlberger E. Interactions of Marburg virus nucleocapsid proteins. Virology, in press

Becker S, Huppertz S, Klenk HD, Feldmann H (1994) The nucleoprotein of Marburg virus is phosphorylated. J Gen Virol. 75:809–818

Becker S, Klenk HD, Mühlberger E (1996) Intracellular transport and processing of the Marburg virus surface protein in vertebrate and insect cells. Virology 225:145–155

Becker S, Spiess M, Klenk HD (1995) The asialoglycoprotein receptor is a putative liver specific receptor of Marburg virus. J Gen Virol 76:393–399

Berger M, Schmidt MFG (1985) Protein fatty acyltransferase is located in the rough endoplasmic reticulum. FEBS Letters 187:289–294

Bos EC, Heijnen L, Luytjes W, Spaan WJ (1995) Mutational analysis of the murine coronavirus spike protein: effect on cell-to-cell fusion. Virology 214:453–463

Bukreyev AA, Volchkov VE, Blinov VM, Netesov SV (1993) The GP-protein of Marburg virus contains the region similar to the "immunosuppressive domain" of oncogenic retrovirus P15 E proteins. FEBS letters 323:183–187

Bukreyev AA, Volchkov VE, Blinov VM, Dryga SA, Netesov SV (1995) The complete nucleotide sequence of the Popp (1967) strain of Marburg virus: a comparison with the Musoke strain (1980). Arch Virol 140:1589–1600

Collins PL, Hill MG, Cristina J, Grossfeld H. (1996) Transcription elongation factor of respiratory syncytial virus, a nonsegmented negative-strand RNA virus. Proc Natl Acad Sci USA 93:81–85

Collins PL, Mottet G (1992) Oligomerization and post-translational processing of glycoprotein G of human respiratory syncytial virus: altered O-glycosylation in the presence of brefeldin A. J Gen Virol 73:849–863

Conzelmann KK (1996) Genetic manipulation of non-segmented negative strand RNA viruses. J Gen Virol 77:881–889

Curran J (1996) Reexamination of the Sendai virus P protein domains required for RNA synthesis: A possible supplemental role for the P protein. Virology 221:130–140

Elliott LH, Kiley MP, McCormick JB (1985) Descriptive analysis of Ebola virus proteins. Virology 147:169–176

Elliott LH, Sanchez A, Holloway BP, Kiley MP, McCormick JB (1993) Ebola protein analyses for the determination of genetic organization. Arch Virol 133:423–436

Feldmann H, Will C, Schikore M, Slenczka W, Klenk HD (1991) Glycosylation and oligomerization of the spike protein of Marburg virus. Virology 182:353–356

Feldmann H, Mühlberger E, Randolf A, Will C, Kiley MP, Sanchez A, Klenk HD (1992) Marburg virus, a filovirus: messenger RNAs, gene order, and regulatory elements of the replication cycle. Virus Res 24:1–19

Feldmann H, Nichol ST, Klenk HD, Peters CJ, Sanchez A (1994) Characterization of filoviruses based on differences in structure and antigenicity of the virion glycoprotein. Virology 199:469–473

Fuerst TR, Niles EG, Studier RW, Moss B (1986) Eukaryotic transient-expression system based on recombinant vaccinia virus that synthesizes bacteriophage T7 RNA polymerase. Proc Natl Acad Sci USA 83:8122–8126

Funke C, Becker S, Dartsch H, Klenk HD, Mühlberger E (1995) Acylation of the Marburg virus glycoprotein. Virology 208:289–297

Gaedigk-Nitschko K, Ding M, Levy MA, Schlesinger MJ (1990) Site-directed mutations in the Sindbis virus 6 K protein reveals sites for fatty acylation, and the underacylated protein affects virus release and virion structure. Virology 175:282–291

Gaudin Y, Tuffereau C, Benmansour A, Flamand A (1991) Fatty acylation of rabies virus proteins. Virology 184:441–444

Geyer H, Will C, Feldmann H, Klenk HD, Geyer R (1992) Carbohydrate structure of Marburg virus glycoprotein. Glycobiology 2:299–312

Gombart AF, Hirano A, Wong TC (1995) Nucleoprotein phosphorylated on both serine and threonine is preferentially assembled into the nucleocapsids of measles virus. Virus Res 37:63–73

Gruber C, Levine SJ (1985) Respiratory syncytial virus polypeptides. V. The kinetics of glycoprotein synthesis. J Gen Virol 66:1241–1247

Jarvis DL, Oker-Blom C, Summers MD (1990) Role of glycosylation in the transport of recombinant glycoproteins through the secretory pathway of lepidopteran insect cells. J Cell Biochem 42:181–191

Kiley MP, Cox NJ, Elliott LH, Sanchez A, DeFries R, Buchmeier MJ, Richman DD, McCormick JB (1988) Physicochemical properties of Marburg virus: evidence for three distinct virus strains and their relationship to Ebola virus. J Gen Virol 69:1957–1967

Kiley MP, Regnery RL, Johnson KM (1980) Ebola virus: Identification of virion structural proteins. J Gen Virol 49:333–341 l

Lambert DM, Hambor J, Diebold M, Galinski B (1988) Kinetics of synthesis and phosphorylation of respiratory syncytial virus polypeptides. J Gen Virol 69:319–323

Martini GA, Siegert R (1971) Marburg virus disease. Springer, Berlin Heidelberg New York

Mühlberger E, Sanchez A, Randolf A, Will C, Klenk HD, Feldmann H (1992) The nucleotide sequence of the L gene of Marburg virus, a filovirus: Homologies with paramyxoviruses and rhabdoviruses. Virology 187:534–547

Naruse H, Nagai Y, Yoshida T, Hamaguchi M, Matsumoto T, Isomura S, Suzuki S (1981) The polypeptides of mumps virus and their synthesis in chick embryo cells. Virology 112:119–130

Robbins SJ, Fenimore JA, Bussell RH (1980) Structural phosphoproteins associated with measles virus nucleocapsids from persistently infected cells. J Gen Virol 48:445–449

Sänger C, Mühlberger E, Klenk HD, Becker S. Glycosylation of recombinant Marburg virus glycoprotein (in preparation)

Sanchez A, Kiley MP, Holloway BP, McCormick JB, Auperin DD (1989) The nucleoprotein gene of Ebola virus: cloning, sequencing, and in vitro expression. Virology 170:81–91

Sanchez A, Kiley MP, Holloway BP, Auperin DD (1993) Sequence analysis of the Ebola virus genome: organization, genetic elements, and comparison with the genome of Marburg virus. Virus Res 29:215–40

Sanchez A, Kiley MP, Klenk HD, Feldmann H (1992) Sequence analysis of the Marburg virus nucleoprotein gene: comparison to Ebola virus and other non-segmented negative-strand RNA viruses. J Gen Virol 73:347–357

Sanchez A, Trappier SG, Mahy BWJ, Peters CJ, Nichol ST (1996) The virion glycoproteins of Ebola viruses are encoded in two reading frames and are expressed through transcriptional editing. Proc Natl Acad Sci USA 93:3602–3607

Schmidt MFG, Lambrecht B (1985) On the structure of the acyl linkage and the function of fatty acyl chains in the influenza virus haemagglutinin and the glycoproteins of Semliki Forest virus. J. Gen. Virol. 66:2635–2647

Sokol F, Koprowski H (1975) Structure-function relationships and mode of replication of animal rhabdoviruses. Proc Natl Acad Sci USA 72:933–936

Spadafora D, Canter DM, Jackson RL, Perrault J (1996) Constitutive phosphorylation of the vesicular stomatitis virus P protein modulates polymerase complex formation but is not essential for transcription and replication. J Virol 70:4538–4548

Sutter G, Ohlmann M, Erfle V (1995) Non-replicating vaccinia vector efficiently expresses bacteriophage T7 RNA polymerase. FEBS Lett 371:9–12

Takacs AM, Barik S, Das T, Banerjee AK (1992) Phosphorylation of specific serine residues within the acidic domain of the phosphoprotein of vesicular stomatitis virus regulates transcription in vitro. J Virol 66:5842–5848

Tate CG, Blakely RD (1994). The effect of N-linked glycosylation on activity of the Na^+- and Cl^--dependent serotonin transporter expressed using recombinant baculovirus in insect cells. J Biol Chem 269:26303–26310

Towler DA, Gordon JI, Adams SP, Glaser L (1988) The biology and enzymology of eukaryotic protein acylation. Ann Rev Biochem 57:69–99

Veit M, Kretzschmar E, Kuroda K, Garten W, Schmidt MFG, Klenk HD, Rott R (1991) Site-specific mutagenesis identifies three cysteine residues in the cytoplasmic tail as acylation sites of influenza virus hemagglutinin. J Virol 65:2491–2500

Veit M, Schmidt MFG, Rott R (1989) Different palmitoylation of paramyxovirus glycoproteins. Virology 168:173–176

Vidal S, Curran J, Orvell C, Kolakofsky D (1988) Mapping of monoclonal antibodies to the Sendai virus P protein and the location of its phosphates. J Virol 62:2200–2203

Volchkov VE, Becker S, Volchkova VA, Ternovoj VA, Kotov AN, Netesov SV, Klenk HD (1995) GP mRNA of Ebola virus is edited by the Ebola virus polymerase and by T7 and vaccinia virus polymerases. Virology 214:421–430

Volchkov VE, Blinov VM, Netesov SV (1992) The envelope glycoprotein of Ebola virus contains an immunosuppressive-like domain similar to oncogenic retroviruses. FEBS Lett 305:181–184

Will C, Mühlberger E, Linder D, Slenczka W, Klenk HD, Feldmann H (1993) Marburg virus gene four encodes the virion membrane protein, a type I transmembrane glycoprotein. J Virol 67:1203–1210

Yang C, Compans RW (1996) Palmitoylation of the murine leukemia virus envelope glycoprotein transmembrane subunits. Virology 221:87–97

Processing of the Ebola Virus Glycoprotein

V.E. VOLCHKOV

1 Introduction	35
2 Expression Strategy of the Glycoprotein Gene	35
3 Proteolytic Processing and Transport of Glycoprotein	38
4 Release of Ebola Virus Glycoproteins (GPs)	41
References	44

1 Introduction

The glycoprotein (GP) of Ebola virus (EBOV) shows a high degree of polymorphism which is the consequence of an unusual expression strategy involving transcriptional editing of GP specific mRNAs, complex co- and posttranslational modifications and variability at the genomic level. As a result, in addition to mature spike protein and its membrane-bound precursors, a number of soluble proteins are produced that are secreted from infected cells. It is reasonable to assume that the various forms of GP play important roles in virus reproduction and in pathogenicity.

2 Expression Strategy of the Glycoprotein Gene

The GP gene is the fourth (2406 nucleotides; subtype Zaire, strain Mayinga, EMBL Bank accession number U31033) of the linearly arranged genomic RNA. As with all other viral genes, gene four is flanked by consensus sequences at the 5'-(GAT/GGAAGATTAA, mRNA sense) and 3'-(ATTAA/TG/AAAAAA, mRNA sense) terminals (VOLCHKOV et al. 1992, 1995; SANCHEZ et al. 1993, 1996). The deduced amino acid sequence of the GP genes of all known EBOV subtypes shows two large

---

Institut für Virologie, Philipps-Universität Marburg, Robert-Koch-Str.17, 35037 Marburg, Germany

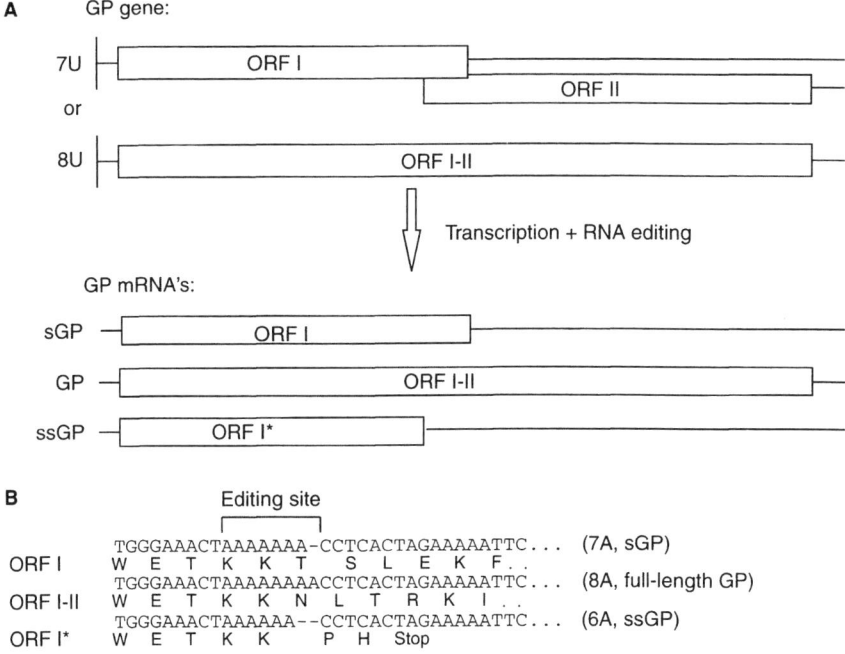

Fig. 1A,B. Strategy of glycoprotein (GP) gene expression of Ebola virus. **A** Wild-type of EBOV (*7U*) and the 8U-variant (*8U*) differ by the numbers of uridine residues at the editing site (VOLCHKOV et al. 1994, 1995; SANCHEZ et al. 1996). The corresponding open reading frame (*ORF*) are shown as *boxes*. *ORF I** represents ORF I fused with only two amino acids from the third reading frame. At least three distinct GP-specific mRNA species exist. The first one encodes for sGP and comprises 80% of total mRNA when transcribed from the 7U genome, but only 10% when transcribed from the 8U genome. This mRNA species has 7A residues at the editing site and is an exact copy of the 7U genome, but requires editing when derived from the 8U genome. The second mRNA species encodes full length GP and comprises only 20% of total mRNA when transcribed from the 7U genome, but 80% when derived from the 8U genome. It is an exact copy of the 8U genome, but requires editing when derived from the 7U genome. The third mRNA species encodes ssGP and has been observed as an editing product of only the 8U genome, or when recombinant GP gene was expressed. ssGP is a glycoprotein of approximately 50 kDa and secreted into the culture medium similar to sGP. **B** The nucleotide and the deduced amino acid sequences of GP at the editing site

overlapping open reading regions (ORFs) (Fig. 1). The first ORF (ORF I) extends from the ATG start codon (EBOV subtype Zaire, position 142, mRNA sense) to a TAA stop codon located approximately in the middle of the gene (position 1232, mRNA sense). The second coding region (ORF II) was found in a −1 shift to ORF I and extends from position 1021 (no start ATG codon) to position 2166. None of these two ORFs are able to encode a protein of the size of GP found in virions (VOLCHKOV et al. 1993, 1995). Analysis of mRNA species isolated from EBOV-infected cells revealed two GP-specific mRNAs, which differed only by a single adenosine residue present in a region of seven consecutive adenosines (at position 1019–1025, mRNA sense). Approximately 20% of GP-specific mRNAs isolated from EBOV-infected cells contain an additional non-template adenosine residue at this specific site (VOLCHKOV et al. 1995, 1997; SANCHEZ et al. 1996) resulting in the expression of full-length GP (676 amino acids). Most of the mRNA (80%), however, is an exact complementary copy of the genomic RNA and encodes for a nonstructural smaller GP (sGP) (VOLCHKOV et al. 1995; SANCHEZ et al. 1996). sGP shares the NH_2-terminal 295 amino acids with GP, but differs in the COOH-terminal by 69 amino acids.

The expression of the EBOV GP gene was examined in detail using a transient system (T7 RNA polymerase/vaccinia virus) and recombinant vaccinia virus. Analysis of mRNA species indicated that transcriptional RNA editing as described for EBOV infection also occurred when GP was expressed in both of the above mentioned recombinant systems. Several DNA-dependent RNA polymerases, such as vaccinia virus polymerase, T7 RNA polymerase (VOLCHKOV et al. 1995) and probably SP6 polymerase (SANCHEZ et al. 1996), recognized the GP editing site and edited the sequence of seven consecutive adenosine residues in synthesized GP mRNA. Different levels of editing activity dependent on different polymerases were confirmed by sequence analysis of mRNA species as well as by SDS-PAGE analysis of GP gene-specific proteins. Whereas T7 polymerase edited more than 5% of GP mRNA transcripts, the efficacy of the vaccinia virus polymerase was less then 1%. In contrast to EBOV infection, deletion of one (six adenosine residues at the editing site, mRNA sense) or insertion of two adenosine residues (nine adenosine residues) was observed indicating that editing by the T7 polymerase may not always lead to (+)1 mRNA transcripts. Thus, accuracy of the editing process is dependent on the respective polymerase. Transcripts with six As and nine As led to termination of the ORF just downstream of the editing site and directed synthesis of an additional protein designated ssGP (VOLCHKOV et al. 1995, 1996). The data indicated that ssGP and sGP lack the membrane anchor and therefore are secreted from cells into the cultural medium. ssGP is the third GP gene-specific product which shares the NH_2-terminal region with GP and sGP, but differs at the COOH-terminal by utilizing a third reading frame (VOLCHKOV et al. 1995, 1996) (Fig. 1).

RNA editing has been described for a variety of paramyxoviruses, such as simian virus 5 (THOMAS et al. 1988), Sendai virus (VIDAL et al. 1990a), measles virus (Cattaneo et al. 1989), mumps virus (PATERSON and LAMB 1990; TAKEUCHI et al. 1990), and parainfluenza virus types 2 and 4 (SOUTHERN et al. 1990; OHGIMOTO et al. 1990; KONDO et al. 1990). With each of these viruses, editing has been shown to occur by the insertion of an additional G residue at the specific sequence 3'-UUU/CUCCC of the P gene. The P gene of paramyxoviruses is reported to be edited exclusively by viral RNA-dependent RNA polymerases (HORIKAMI et al. 1991; PELET et al. 1991; MATSUOKA et al. 1991; VIDAL et al. 1990a). In contrast, the editing site of EBOV GP is also recognized by DNA-dependent RNA polymerases. The mechanism by which editing of the GP gene occurs appears to be similar with RNA- and DNA-dependent RNA polymerases, since both types of enzymes insert the same non-template nucleotide (A, mRNA sense) in exactly the same region. The detected editing site in EBOV genomic RNA (7Us, genomic sense) is different from the editing site found in the P gene of paramyxoviruses, rather resembling the polyadenylation site (transcription stop signal) of EBOV mRNAs (3'-UAAUU-

CUUUUUU, genomic sense). The EBOV editing site is also similar to the transcription stop signal of the vaccinia virus polymerase (3′-UUUUUNU) (Moss 1990). Presumably, temporary pausing of the viral RNA polymerase and both investigated DNA-dependent RNA-polymerases at the editing site of EBOV GP gene enables the transcription complex to slip backward or forward on the viral RNA (vRNA) template before the next nucleotide is incorporated. This mechanism was described for editing of P gene of paramyxoviruses (VIDAL et al. 1990b) and obviously could be used in editing of EBOV GP. In this case the similarity with polyadenylation sites may at least partly explain the relatively broad spectrum of RNA polymerases recognizing the editing site of EBOV GP.

Depending on the propagation history of the virus, the GP gene of the Zaire subtype of EBOV contains either seven (7 U) or eight uridine (8 U) residues at the editing site (genomic sense) (Fig. 1). When an additional U is inserted at this site in the viral genome, translation of the full-length virion GP is facilitated without RNA editing. The co-transcriptional editing in GP mRNA species of this mutant virus has been studied by sequence analysis of cloned viral genomic RNA and mRNAs. Again, evidence has been obtained that a specific insertion of non-templated A residues at the editing site of GP mRNA occurred. However, editing varied from single to several deletions or insertions (up to seven As) of adenosine residues, resembling those observed with T7 or vaccinia virus polymerases. Thus, insertion of the additional U residue in the viral genome at the EBOV GP editing site substantially changed the efficacy and accuracy of editing compared with wild-type virus (EBOV-7 U). The amount of mRNA encoding GP was approximately 80% compared with 20% for EBOV-7 U. Only approximately 10% of mRNA transcripts coded for sGP, whereas about 10% account for different species due to deletions or insertions of adenosine residues (VOLCHKOV et al. 1996).

The functions of nonstructural glycoproteins of EBOV remains to be clarified. Since nonstructural glycoproteins (sGP and probably ssGP) are secreted from infected cells, they may in fact interact with the immune system of the host. sGP can be detected in the blood of acutely infected patients (SANCHEZ et al. 1996) and recently YANG et al. (1998) presented evidence that vector-expressed sGP binds to neutrophils and inhibits their activation. However, the discussed immunomodulator properties of these nonstructural EBOV GPs should be experimentally confirmed in future studies.

3 Proteolytic Processing and Transport of Glycoprotein

GP is a type-I transmembrane protein (VOLCHKOV et al. 1992, 1995, 1998a; SANCHEZ et al. 1993). The NH_2-terminal hydrophobic region of GP appears to be a signal peptide that, similar to MBGV GP (WILL et al. 1993), is cleaved during translocation into the endoplasmic reticulum (ER). The COOH-terminal hydrophobic region (position 651–672) has the characteristics to serve as a transmembrane anchor. The proposed cytoplasmic tail is extremely short and consists of four amino acids with only one charged residue. Recently it has been shown that the two cysteine residues located between the transmembrane anchor and the cytoplasmic tail of MBGV are acylated (FUNKE et al. 1995). This may also be the case with EBOV GP, which contains cysteine residues at the same positions. EBOV GP, similar to MBGV GP (FELDMANN et al. 1991), is present in the viral envelope as a homotrimer (V.E. VOLCHKOV et al., unpublished data).

EBOV GP is highly glycosylated; the N- and O-linked carbohydrates comprise approximately 50% of the total molecular weight which varies depending on the cells used for virus propagation or GP expression. Amino acid sequence comparison of GP of all EBOV subtypes showed high conservation at the NH_2- and COOH-terminal ends, in which also most of the highly conserved cysteine residues are located. The middle region is variable, extremely hydrophilic, and carries the bulk of the glycosylation sites. O-glycans of GP have terminal sialic acid. Differential sensitivities of EBOV GP and MBGV GP to MAA lectin (*Maackia amurensis agglutinin*, recognizing α(2–3)-linked sialic acids) and to SNA lectin (*Sambucus nigra agglutinin*, recognizing α(2–6)-linked sialic acids) may be useful as a diagnostic tool to discriminate both viruses (FELDMANN et al. 1994).

Processing and transport of GP was studied using a recombinant vaccinia virus expression system (VOLCHKOV et al. 1995, 1998a,b). Maturation of the EBOV GP involves a complex sequence of co- and posttranslational processing events. The first precursor form could be detected after pulse-labeling of cells expressing GP as a 110 kDa GP that is converted by endo H treatment into a 75 kDa polypeptide (VOLCHKOV et al. 1995, 1998b). This precursor is full-length GP containing oligomannosidic N-glycans. It represents precursor GP present in the ER and has therefore been designated $preGP_{er}$. The second precursor, preGP, is first observed after a 10–20 min chase period as an endo H-resistant GP with a molecular mass of about 160 kDa. This form represents full-length GP containing complex N-glycans and O-glycans and is present in the Golgi apparatus.

PreGP undergoes posttranslational proteolytic cleavage into the NH_2-terminal 140 kDa fragment GP_1 and the COOH-terminal 26 kDa fragment GP_2 (Fig. 2). GP_1 and GP_2 are disulfide-linked ($GP_{1,2}$). $GP_{1,2}$ is the virion form of GP and it is also present in cells, presumably in the trans Golgi network and on the cell surface. preGP contains the sequence $RTRR_{501}$ which is a cleavage motif of the proprotein convertase furin. The observation that cleavage did not occur when GP was expressed in the furin-defective LoVo cell line, but that it was restored in these cells by vector-expressed furin confirmed the identity of furin as a processing enzyme of GP. The finding that cleavage was inhibited by a sequence-specific peptidyl (RVKR) chloromethylketone or by mutation of the cleavage site supports this concept (VOLCHKOV et al. 1998b).

Furin belongs to the proprotein convertases, a family of subtilisin-like eukaryotic endoproteases that includes also PC1/PC3, PC2, PC4, PACE4, PC5/PC6, and LPC/PC7 (SEIDAH et al. 1996). These enzymes are differentially expressed in cells and tissues, and they display similar but not identical specificity for basic motifs, such as RXK/RR, at the cleavage site of their substrates. Furin is expressed

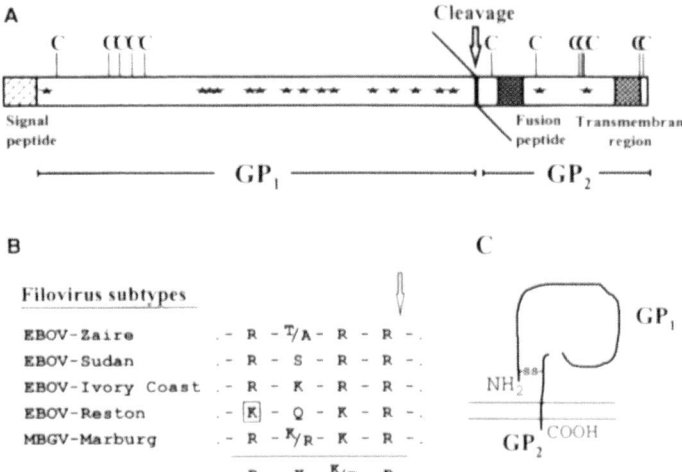

Fig. 2A–C. Filoviral glycoproteins. A Scheme of subtype Zaire EBOV 6P; B structure of cleavage sites, and C proposed structure of the mature glycoprotein (GP) monomer. The NH_2-terminal signal peptide sequence, the COOH-terminal transmembrane domain, and the putative fusion domain are indicated by *gray boxes*. GP is proteolytically cleaved into subunits GP_1 and GP_2. The cleavage site is indicated by an *arrow*. Both subunits are disulfide-linked in the mature molecule $GP_{1,2}$. The altered amino acid in the cleavage site of EBOV subtype Reston is marked by a *box*. *Asterisk*, potential N-glycosylation site; *C*, cysteine residue; GP_1, larger cleavage product of GP; GP_2, smaller cleavage product of GP; *S-S*, disulfide bridged

in most cells and is localized predominantly in the trans Golgi network (MOLLOY et al. 1994; SCHÄFER et al. 1995). Furin appears to be the key enzyme in virus activation (KLENK and GARTEN 1994a), but PC5/PC6 (HORIMOTO et al. 1994) and LPC/PC7 (HALLENBERGER et al. 1997) are also involved in this process. It is also noteworthy that furin, although ubiquitous, is particularly rich in hepatocytes and endothelial cells which are both prime targets of EBOV (GEISBERT et al. 1992). These observations stress the importance of furin as a processing enzyme of GP, but it remains to be seen in future studies if other proprotein convertases can substitute as cleaving enzymes.

Processing by protein convertases is an important control mechanism for the biological activity of viral surface proteins (KLENK and GARTEN 1994a,b). Cleavage often occurs next to a protein domain involved in fusion, and it has long been known that in these cases proteolytic cleavage is necessary for fusion activity. Proteolytic cleavage is the first step in the activation of these fusion proteins and is followed by a conformational change resulting in the exposure of the fusion domain (BULLOUGH et al. 1994; CHAN et al. 1997; WEISSENHORN et al. 1997). The conformational change may be triggered by low pH in endosomes, as is the case with influenza virus (SKEHEL et al. 1982), or by the interaction with a secondary receptor protein at the cell surface, as is the case with HIV (FENG et al. 1996). So far fusion activity of filoviral GP have not been demonstrated experimentally. However, it is interesting to see that GP_2 contains a sequence of 16 uncharged and hydrophobic amino acids at a short distance (22 amino acids) from the cleavage site. This sequence bears some structural similarity to the fusion peptides of retroviruses and has therefore been thought to play a role in EBOV entry (GALLAHER 1996). Marburg virus GP also has a multibasic cleavage site at position 432–435 (RKKR), and preliminary data show that it is cleaved, too (V.E. Volchkov et al., unpublished data).

Finally, it has to be pointed out that proteolytic activation of viral GPs is an important determinant for pathogenicity. Cleavage by furin and other ubiquitous proprotein convertases has been shown to be responsible for systemic infection caused by highly pathogenic strains of avian influenza and Newcastle disease virus (KLENK and ROTT 1988). It is therefore tempting to speculate that cleavage by furin is also an important factor for the pantropism of EBOV and its rapid dissemination through the organism. Furthermore, variations at the cleavage site of GP may account for differences in the pathogenicity of EBOV. The pathogenic strains of subtypes Zaire, Sudan, and Ivory Coast, which have the canonical furin motif RXK/RR at the cleavage site, are highly susceptible to cleavage, whereas strains of subtype Reston, which appear to be apathogenic for humans and only moderately pathogenic for at least some monkey species (FISHER-HOCH et al. 1992), have reduced cleavability because of the suboptimal cleavage site sequence KQKR (VOLCHKOV et al. 1998b). That highly pathogenic variants may suddenly emerge from Reston-like viruses by mutations restricted to the cleavage site is an intriguing hypothesis.

4 Release of Ebola Virus Glycoproteins (GPs)

The release of viral GPs in non-virion form has been observed with several viruses, and the following different mechanisms have been described: (1) Release as a result of cell lysis during viral infection has been demonstrated for soluble VSV proteins that account for about 3% of total viral proteins in an infected culture (KANG and PREVEC 1971; LITTLE and HUANG 1978). (2) Release of viral GPs, as a result of proteolytic cleavage near the membrane anchor by still not known cellular proteases, has been demonstrated for the soluble G protein of VSV (IRVING and GOSH 1982; CHEN et al. 1986) and rabies virus (DIETZSCHOLD et al. 1983; MORIMOTO et al. 1993), soluble hemagglutinin (HA) of measles virus (MALVOISIN and WILD 1994), and soluble G protein of RSV (HENDRICKS et al. 1987, 1988). (3) Shedding of a non-membrane-associated subunit of the processed spike GP has been demonstrated for the HA_1 subunit of influenza A HA (ROBERTS et al. 1993) and the GP120 fragment of the env protein of HIV (VERONESE et al. 1985). (4) Budding of virus-like particles with incorporated viral proteins has been described for the E proteins of a number of flaviviruses (MASON et al. 1991; YAMSHCHIKOV and COMPANS 1993; ALLISON et al. 1995), envelope proteins of lentiviruses (KRÄUSSLICH et al. 1993; YAMSHCHIKOV et al. 1995), and the small envelope protein (E) of coronavirus (VENNEMA et al. 1996). In addition, viral GPs are also released after removal of the transmembrane anchor region by genetic engineering. This has been shown for Rubella virus GP (SETO et al. 1995), influenza virus HA (GETHING and SAMBROOK 1982), and VSV GP (ROSE and BERGMANN 1982).

During EBOV infection, in addition to nonstructural sGP secreted as an antiparallel orientated disulfide-linked homodimer (V.E. Volchkov et al., unpublished data), several forms of GP synthesized from the edited mRNA species are released into the culture medium. Analysis of tissue culture supernatants showed that mature GP was only partly incorporated into virus particles as $GP_{1,2}$ complexes. A significant proportion was released as non-virion molecules consisting of the soluble GP_1 (VOLCHKOV et al. 1998a).

Extracellular appearance of GP_1 was also demonstrated in cells expressing recombinant GP (VOLCHKOV et al. 1998a). GP_1 is released from the complex with GP_2 during the processing pathway as observed with the HA_1 subunit of the influenza HA and with GP120 of HIV. Since GP120 and GP41 of HIV are not linked by disulfide bonds, release of GP120 may occur spontaneously, due to misfolding. In the case of influenza virus an acid-induced conformational change of the cleaved HA in the trans Golgi network and/or in transport vesicles destined for the plasma membrane may make the disulfide linkage between HA_1 and HA_2 susceptible to reduction (ROBERTS et al. 1993). Since $GP_{1,2}$ also contains a disulfide linkage that is formed prior to proteolytic cleavage, such a mechanism is also suggested for the release of GP_1. As an alternative mechanism, disulfide linkage may not always be completed between the GP_1 and GP_2 domains preGP$_{er}$ during early processing of GP. Thus proteolytic processing of these molecules in the Golgi would lead to release of GP_1 molecules. As indicated by the earliest appearance of GP_1 in the culture medium, GP processing and transport take approximately 90–120 min (VOLCHKOV et al. 1998a). Similar maturation kinetics have been observed with several other viral GPs (DOMS et al. 1993). Expression of an anchor-minus mutant of EBOV GP resulted in a complete release of the truncated GP into the culture medium, showing that essential signals for processing and transport are not located in the transmembrane anchor and the cytoplasmic domain.

A minor fraction of non-virion GP ($GP_{1,2}$) was membrane-associated and released as spikes on virosomes, as has been described for flavivirus, lentivirus and coronavirus GPs. The spikes on these vesicles which were seen after vector expression of GP and after virus infection are morphologically indistinguishable from virion spikes. The release of vesicles from plasma membranes of recombinant vaccinia virus-infected cells showed that spike formation is mediated by GP expression alone and independent of the expression of other viral proteins (VOLCHKOV et al. 1998a). Intense proliferation of plasma membranes as observed by GEISBERT and JAHRLING (1995) on EBOV-infected cells might be caused by local accumulation of GP and is responsible for the generation of liposome-like vehicles with surface spikes.

The abundant release of GP_1 and secretion of sGP suggest biological functions of soluble GP which have not yet been determined. During filovirus infection in humans and animals lymphoid depletion and necrosis of lymphoid tissue, including destruction of non-infected macrophages and other antigen-presenting cells (APCs), have been described. Furthermore, a lack of cellular immune response against virus-infected cells has been reported (RYABCHIKOVA et al. 1996). These observations may implicate that macrophages or other APCs are lysed by $CD4^+$-bearing T cells after presenting processed GP_1 and sGP in context with MHC class II antigens. The destruction of these cells might be responsible for the insufficient cellular and humoral immune responses observed in the course of EBOV infection. This concept is supported by the observations that uninfected cells exposing processed soluble viral GPs of HIV (SILICIANO et al. 1988), VSV (BROWNING et al. 1990), and influenza virus (MORRISON et al. 1986, 1988) are efficiently lysed by virus-specific $CD4^+$ T cells (MHC class II-restricted). Class II MHC antigen is mainly expressed on the surface of professional APCs including B lymphocytes and is also found after activation on other cells, such as endothelial cells. The lysis of GP_1- and sGP-exposed endothelial cells by $CD4^+$-bearing T cells may, in addition to destruction by virus replication (SCHNITTLER et al. 1993), further contribute to the bleeding symptoms observed in patients. Released GP_1 and secreted sGP may also interfere with humoral defense mechanisms. Specific antibodies may bind to these molecules and therefore be no longer available for virus elimination. If non-virion GPs are determinants of viral pathogenicity during EBOV infection, this role seems to be more related to shedding of GP_1 than to sGP secretion, as discussed previously. This is supported by the following observations: (1) EBOV-7 U (wild type) and EBOV-8 U (variant), which display high pathogenicity in animal models, such as guinea pigs and monkeys (CHEPURNOV et al. 1995; RYABCHIKOVA et al. 1996), produce comparable amounts of released GP_1, but EBOV-8 U expresses only minute amounts of sGP (VOLCHKOV et al. 1998a). (2) Subtype Reston EBOV produces high levels of sGP (SANCHEZ et al. 1996), but is less pathogenic for humans and some monkeys than the African EBOV subtypes (PETERS et al. 1994; FISHER-HOCH et al. 1992). (3) Marburg hemorrhagic fever is a comparable disease in human and nonhuman primates, but there is no evidence for sGP expression during infections with Marburg virus (FELDMANN and KLENK, 1996). Finally, released GP_1 and sGP may interfere with cellular receptors and, therefore, reduce the efficiency of EBOV replication. This does not seem to hold true for infections in humans and experimental animals, but may play a role in establishing a persistent infection in the yet unknown natural host.

The appearance of virosomes carrying GP spikes also may play a role in the immunopathology of Ebola hemorrhagic fever in experimentally and naturally infected hosts. Full-length GP contains the potential immunosuppressive domain in the COOH-terminal region of the molecule (VOLCHKOV et al. 1992, 1993). This domain consists of 26 amino acids with a high degree of homology to an immunosuppressive motif found in the p15E-related GPs of oncogenic retroviruses. In the retrovirus system this domain has been shown to inhibit the blastogenesis of lymphocytes in response to mitogens, to decrease monocyte chemotaxis and macrophage infiltration, and to decrease proliferation of mononuclear cells (CIANCIOLO et al. 1985; KADOTA et al. 1991). However, whether the presumptive immunosuppressive domain is involved in the pathogenicity of Ebola infection remains to be seen.

References

Allison SL, Stadler K, Mandl CW, Kunz C, Heinz FX (1995) Synthesis and secretion of recombinant tick-borne encephalitis virus protein E in soluble and particulate form. J Virol 69:5816–5820

Browning M, Reiss CS, Huang AS (1990) The soluble viral glycoprotein of vesicular stomatitis virus efficiently sensitizes target cells for lysis by CD4+ T lymphocytes. J Virol 64:3810–3816

Bullough PA, Hughson FM, Skehel JJ, Wiley DC (1994) Structure of influenza haemagglutinin at the pH of membrane fusion. Nature 371:37–43

Cattaneo R, Kaelin K, Baczko K, Billeter MA (1989) Measles virus editing provides an additional cysteine-rich protein. Cell 56:759–764

Chan DC, Fass D, Berger JM, Kim PS (1997) Core structure of gp41 from the HIV envelope glycoprotein. Cell 89:263–273

Chen SSL, Huang AS (1986) Further characterization of the vesicular stomatitis virus temperature-sensitive O45 mutant: intracellular conversation of the glycoprotein to a soluble form. J Virol 59:210–215

Chepurnov AA, Chernukhin IV, Ternovoj VA, Kudoyarova NM, Makhova NM, Azayev MSh, Smolina MP (1995) Attempts at creating a vaccinia against Ebola fever. Prob Virol 40:257–260

Cianciolo GJ, Copeland TJ, Oroszlan S, Snyderman R (1985) Inhibition of lymphocyte proliferation by a synthetic peptide homologous to retroviral envelope protein. Science 230:453–455

Dietzschold B, Wiktor TJ, Wunner WH, Varrichio A (1983) Chemical and immunological analysis of the rabies soluble glycoprotein. Virology 124:330–337

Doms RW; Lamb RA, Rose JK, Helenius A (1993) Folding and assembly of viral membrane proteins. Virology 193:545–562

Feldmann H, Klenk HD (1996) Marburg and Ebola viruses. Adv Virus Res 47:1–52

Feldmann H, Nichol ST, Klenk HD, Peters CJ, Sanchez A (1994) Characterization of filoviruses based on differences in structure and antigenicity of the virion glycoprotein. Virology 199:469–473

Feldmann H, Will C, Schikore M, Slenczka W, Klenk HD (1991) Glycosylation and oligomerization of the spike protein of Marburg virus. Virology 182:353–356

Feng Y, Broder CC, Kennedy PE, Berger EA (1996) HIV-1 entry cofactor: functional cDNA cloning of a seven-transmembrane, G protein-coupled receptor. Science 272:872–877

Fisher-Hoch SP, Brammer TL, Trappier SG, Hutwagner LC, Farrar BB, Ruo SL, Brown BG, Hermann LM; Perez-Ornoz GI, Goldsmith CS, Hanes MA, McCormick JB (1992) Pathogenic potential of filoviruses: role of geographic origin of primate host and virus strain. J Infect Dis 166:753–763

Funke C, Becker S, Dartsch H, Klenk HD, Mühlberger E (1995) Acylation of the Marburg virus glycoprotein. Virology 208:289–297

Gallaher WR (1996) Similar structural models of the transmembrane proteins of Ebola and avian sarcoma viruses. Cell 85:477–478

Geisbert TW, Jahrling PB (1995) Differentiation of filoviruses by electron microscopy. Virus Res 39:129–150

Geisbert TW, Jahrling PB, Hanes MA, Zack PM (1992) Association of Ebola-related Reston virus particles and antigen with tissue lesions of monkeys imported to the United States. J Comp Pathol 106:137–152

Gething MJ, Sambrook J (1982) Construction of influenza haemagglutinin genes that code for intracellular and secreted forms of the protein. Nature 300:598–603

Hallenberger S, Moulard M, Sordel M, Klenk HD, Garten W (1997) The role of eukaryotic subtilisin-like endoproteases for the activation of human immunodeficiency virus glycoproteins in natural host cells. J Virol 71:1036–1045

Hendricks DA, Baradaran K, McIntosh K, Patterson JL (1987) Appearance of a soluble form of the G protein of respiratory syncytial virus in fluids of infected cells. J Gen Virol 68:1705–1714

Hendricks DA, McIntosh K, Patterson JL (1988) Further characterization of the soluble form of the G glycoprotein of respiratory syncytial virus. J Virol 62:2228–2233

Horikami SM, Moyer SA (1991) Synthesis of leader RNA and editing of the P mRNA during transcription by purified measles virus. J Virol 65:5342–5347

Horimoto T, Nakayama K, Smeekens SP, Kawaoka Y (1994) Proprotein-processing endoproteases PC6 and furin both activate hemagglutinin of virulent avian influenza viruses. J Virol 68:6074–6078

Irving RA, Ghosh HP (1982) Shedding of vesicular stomatitis virus soluble glycoprotein by removal of carboxy-terminal peptide. J Virol 42:322–325

Kadota J, Cianciolo GJ, Snyderman R (1991) A synthetic peptide homologous to retroviral transmembrane envelope proteins depress protein kinase C mediated lymphocyte proliferation and directly inactivated protein kinase C: a potential mechanism for immunosuppression. Microbiol Immunol 35:443–459

Kang CY, Prevec L (1971) Proteins of vesicular stomatitis virus. III. Intracellular synthesis and extracellular appearance of virus-specific proteins. Virology 46:678–690

Klenk HD, Garten W (1994a) Host cell proteases controlling virus pathogenicity. Trends in Microbiol 2:39–43

Klenk HD, Garten W (1994b) Activation cleavage of viral spike proteins by host proteases. In: Wimmer E (ed) Cellular receptors for animal viruses. Cold Spring Harbor Laboratory Press, pp 241–280

Klenk HD, Rott R (1988) The molecular biology of influenza virus pathogenicity. Adv Virus Res 34:247–281

Kondo K, Bando H, Tsurudome M, Kawano M, Nishio M, Ito Y (1990) Sequence analysis of the phosphoprotein (P) genes of human parainfluenza type 4 A and 4B viruses and RNA editing at transcript of the P genes: the number of G residues added is imprecise. Virology 178:321–326

Kräusslich HG, Ochsenbauer C, Trauencker AM, Mergener K, Fäcke M, Gelderblom HR, Bosch V (1993) Analysis of protein expression and virus-like particle formation in mammalian cell lines stably expressing HIV-1 gag and env gene products with or without active HIV protease. Virology 192:605–617

Little SP, Huang As (1978) Shedding of the glycoprotein from vesicular stomatitis virus-infected cells. J Virol 27:330-339

Malvoisin E, Wild F (1994) Characterization of a secreted form of measles virus haemagglutinin expressed from a vaccinia virus recombinant. J Gen Virol 75:3603–3609

Mason PW, Pincus S, Fournier MJ, Mason TL, Shope RE, Paoletti E (1991) Japanese encephalitis virus-vaccinia recombinant produce particulate forms of the structural membrane proteins and induce high levels of protection against lethal JEV infection. Virology 180:294–305

Matsuoka Y, Curran J, Pelet T, Kolakofsky D, Ray R, Compans RW (1991) The P gene of human parainfluenza virus type 1 encodes P and C proteins but not a cysteine rich V protein. J Virol 65:3406–3410

Molloy SS, Thomas L, van Slyke JK, Stenberg PE, Thomas G (1994) Intracellular trafficking and activation of furin proprotein convertase: localization to the TGN and recycling from the cell surface. EMBO J 13·18–33

Morimoto K, Iwatani Y, Kawai A (1993) Shedding of Gs protein (a soluble form of the viral glycoprotein) by the rabies virus-infected BHK-21 cells. Virology 195:541–549

Morrsion LA, Lukacher AE, Braciale VL, Fan DP, Braciale TJ (1986) Differences in antigen presentation to MHC class I- and class II-restricted influenza virus-specific cytolytic T lamphocyte clones. J Exp Med 163:903–921

Morrsion LA, Braciale VL, Braciale TJ (1988) Antigen form influences induction and frequency of influenza-specific class I and class II MHC-restricted cytolytic T lymphocytes. J Immunol 141:363–368

Moss B (1990) Regulation of vaccinia virus transcription. Annu Rev Biochem 59:661–688

Ohgimoto S, Bando H, Kawano M, Okamoto K, Kondo K, Tsurudome M, Nishio M, Ito Y (1990) Sequence analysis of P gene of human parainfluenza type 2 virus: P and cysteine-rich proteins are translated by two mRNAs that differ by two nontemplated G residues. Virology 177:116–123

Paterson RG, and Lamb RA (1990) RNA editing by G-nucleotide insertion in mumps virus P-gene mRNA transcripts. J Virol 64:4137–4145

Pelet T, Curran J, Kolakofsky D (1991) The P gene of bovine parainfluenza virus 3 expresses all three reading frames from a single mRNA editing site. EMBO J 10:443–448

Peters CJ, Sanchez A, Feldmann H, Rollin PE, Nichol S, Kziazek TG (1994) Filoviruses as emerging pathogens. Seminars in Virology 5:147–154

Roberts PC, Garten W, Klenk HD (1993) Role of conserved glycosylation sites in maturation and transport of influenza A virus hemagglutinin. J Virol 67:3048–3060

Rose JK, Bergmann JE (1982) Expression from cloned cDNA of cell-surface secreted forms of the glycoprotein of vesicular stomatitis virus in eucaryotic cells. Cell 30:753–762

Ryabchikova E, Kolesnikova L, Smolina M, Tkachev V, Pereboeva L, Baranova S, Grazhdantseva A, Rassadkin Y (1996) Ebola virus infection in guinea pigs: presumable role of granulomatous inflammation in pathogenesis. Arch Virol 141:909–921

Sanchez A, Kiley MP, Holloway BP, Auperin DD (1993) Sequence analysis of the Ebola virus genome: organization, genetic elements, and comparison with the genome of Marburg virus. Virus Res 29:215–240

Sanchez A, Trappier SG, Mahy BWJ, Peters CJ, Nichol ST (1996) The virion glycoprotein of Ebola viruses are encoded in two reading frames and are expressed through transcriptional editing. Proc Natl Acad Sci USA 93:3602–3607

Schäfer W, Stroh A, Berghöfer S, Seiler J, Vey M, Kruse ML, Kern HF, Klenk HD, Garten W (1995) Two independent targeting signals in the cytoplasmic domain determine trans-Golgi network localization and endosomeal trafficking of the proprotein convertase furin. EMBO J 14:2424–2435

Schnittler HJ, Mahner F, Drenckhahn D, Klenk HD, Feldmann H (1993) Replication of Marburg virus in human endothelial cells: a possible mechanism for the development of viral hemorrhagic disease. J Clin Invest 91:1301–1309

Seidah NG, Hamelin J, Mamarbachi M, Dong W, Tadro H, Mbikay M, Chretien M, Day R (1996) cDNA structure, tissue distribution and chromosomal localization of rat PC7: a novel mammalian proprotein convertase closest to yeast kexin-like proteinases. Proc Natl Acad Sci USA 93:3388–3393

Seto NOL, Dawei O, Gillam S (1995) Expression and characterization of secreted forms of Rubella Virus E2 glycoprotein in insect cells. Virology 206:736–741

Skehel JJ, Bayley PM, Brown EB, Martin SR, Waterfield MD, White JM, Wilson IA, Wiley DC (1982) Changes in the conformation of influenza virus haemagglutinin at the pH optimum of virus-mediated membrane fusion. Proc Natl Acad Sci USA 79:968–972

Siciliano RF, Lawton T, Knall C, Karr RW, Berman P, Gregory T, Reinherz EL (1988) Analysis of host-virus interactions in AIDS with anti-gp120 T cell clones: effect of HIV sequence variation and a mechanism for CD4+ cell depletion. Cell 54:561–575

Southern JA, Precious B, Randall RE (1990) Two nontemplated nucleotide additions are required to generate the P mRNA of parainfluenza virus type 2 since the genome encodes protein V. Virology 177:388–390

Takeuchi K, Tanabayashi K, Hishiyama M, Yamada YK, Yamada A, Sugiura A (1990) Detection and characterization of mumps virus V protein. Virology 178:247–253

Thomas SM, Lamb RA, Paterson RG (1988) Two mRNA's that differ by two nontemplated nucleotides encode the amino coterminal proteins P and V of the paramyxovirus SV5. Cell 54:891–902

Vennema H, Godeke GJ, Rossen JWA, Voorhout WF, Horzinek MC, Opstelten DJE, Rottier PJM (1996) Nucleocapsid-independent assembly of coronavirus-like particles by co-expression of viral envelope protein genes. The EMBO J 15:2020–2028

Veronese FD, DeVico AL, Copeland TD, Oroszlan S, Gallo RC, Sarngadharan MG (1985) Characterization of gp41 as the transmembrane protein coded by the HTLV-III/LAV envelope gene. Science 229:1402–1405

Vidal S, Curran J, Kolakofsky D (1990a) Editing of the Sendai virus P/C mRNA by G insertion occurs during mRNA synthesis via a virus encoded activity. J Virol 64:239–246

Vidal S, Curran J, Kolakofsky D (1990b) A stuttering model for paramyxovirus P mRNA editing. The EMBO J 9:2017–2022

Volchkov VE, Becker S, Volchkova VA, Ternovoj VA, Kotov AN, Netesov SV, Klenk HD (1995) GP mRNA of Ebola virus is edited by the Ebola virus polymerase and by T7 and vaccinia virus polymerases. Virology 214:421–430

Volchkov VE, Blinov VM, Kotov AN, Netesov SV (1993) The full length nucleotide sequence of the Ebola virus. In: Abstracts of the IXth International Congress of Virology, Glasgow, Scotland. (Abstract P52-2)

Volchkov VE, Blinov VM, Netesov SV (1992) The envelope glycoprotein of Ebola virus contains an immunosuppressive-like domain similar to oncogenic retroviruses. FEBS Lett 305:181–184

Volchkov VE, Chepurnov AA, Dryga S, Becker S, Blinov V, Kotov AN, Ternovoj VA, Klenk HD, Netesov SV (1994) Molecular characterization of a pathogenicity variant of Ebola virus. Ninth International Conference on Negative Strand Viruses, Estoril, Portugal, (270) 176

Volchkov VE, Slenczka W, Feldmann H, Klenk HD (1996) Ebola virus mutant: effect on RNA editing. EMBO Workshop September 5–8, Maastricht, The Netherlands

Volchkov VE, Volchkova VA, Eckel C, Klenk HD, Bouloy M, LeGuenno B, Feldmann H (1997) Emerging of subtype Zaire Ebola virus in Gabon. Virology 232:139–144

Volchkov VE, Volchkova VA, Slenczka W, Klenk HD, Feldmann H, (1998a) Release of viral glycoproteins during Ebola virus infection. Virology 245:110–119

Volchkov VE, Feldmann H, Volchkova VA, Klenk HD (1998b) Processing of the Ebola virus glycoprotein by the proprotein convertase furin. Proc Natl Acad Sci USA 95:5762–5767

Weissenhorn W, Dessen A, Harrison SC, Skehel JJ, Wiley DC (1997) Atomic structure of the ectodomain from HIV-1 gp41. Nature 387:426–430

Will C, Mühlberger E, Linder D, Slenczka W, Klenk HD, Feldmann H (1993) Marburg virus gene four encodes the virion membrane protein, a type I transmembrane glycoprotein. J Virol 67:1203–1210

Yamshchikov GV, Ritter GD, Vey M, Compans RW (1995) Assembly of Siv virus-like particles containing envelope proteins using a baculovirus expression system. Virology 214:50–58

Yamshchikov V, and Compans RW (1993) Regulation of the late events in flavivirus protein processing and maturation. Virology 192:38–51

The Marburg Virus Outbreak of 1967 and Subsequent Episodes

W.G. SLENCZKA

1	Introduction and History	49
2	Studies on the Etiology of a New Disease	51
3	Structure of the Marburg Virus	52
4	The Virion Proteins	55
5	Virus Classification	56
6	Host Systems and Biological Properties	56
7	Morphological Studies on Viral Replication	57
8	Pathology	58
8.1	Studies on Human Material	58
8.2	Studies in Experimental Animals	60
9	Early Studies on Pathogenesis	61
10	Clinical Features and Patient Management	63
11	Treatment	66
12	Virological Diagnosis of Marburg Virus Disease	67
13	Serological Studies	68
14	Serologic Cross-Reactivity Between Marburg and Ebola Virus	69
15	Investigations into the Epidemiology and Source of Infection	69
16	Safety Considerations	71
References		71

1 Introduction and History

The Marburg virus disease was completely unknown when, in early August 1967, several patients were admitted to the University hospitals of Marburg and Frankfurt in Germany. The patients suffered from a generalized infection of unusual severity. In some cases manifestations of severe hemorrhagic diathesis were observed (MARTINI et al. 1968a; STILLE et al. 1968). The early differential diagnosis

Institut für Virologie, Robert-Koch-Str. 17, 35037 Marburg, Germany included diseases such as typhoid, bacterial dysentery, leptospirosis and even malaria and yellow fever, although none of these patients had ever visited a tropical country. Meningococcal sepsis was also suspected in some patients (MARTINI 1969; STILLE and BÖHLE 1971). All the primary cases were restricted to employees of the Behring company in Marburg and the Paul Ehrlich Institute in Frankfurt. It soon became evident that there was a well-defined epidemiological denominator common to the cases in Marburg and in Frankfurt: contact with captured wild monkeys of the species *Cercopithecus aethiops* (vervet monkeys) which were used for the production of poliomyelitis vaccines in the Behring company in Marburg and for safety testing of vaccines in the Paul Ehrlich Institute in Frankfurt (HENNESSEN et al. 1968). The role of the monkeys became more evident when, in the beginning of September 1967, a similar case occurred at a third location, the institute Torlak, in Belgrade, which was also involved in the production and safety testing of poliomyelitis vaccines (TODOROVITCH et al. 1971; STOIKOVIC et al. 1971). The three institutions had received monkeys from the same monkey trader and from the same shipments (HENNESSEN et al. 1968; HENNESSEN 1969).

Seven persons, five men and two women, died from the disease in the second week of their illness. The case fatality rate amounted to 22%. When it had become evident that none of the human pathogenic agents which were known at that time was causative, the search for a new agent began.

Along with other experimental animals guinea pigs were originally inoculated in Marburg with patient blood, as a rickettsia or a leptospira was suspected as the causative agent. The animals developed fever and soon recovered, but during three consecutive animal passages the pathogenicity increased and finally a lethality of 100% was attained. Human and guinea pig convalescent sera were tagged with fluorescein and were used in direct immunofluorescence stains. A new antigen with intracytoplasmic location was found in the livers and spleens of the inoculated animals and a hitherto unknown virus was revealed by electron microscopy. It was named Marburg virus (SIEGERT et al. 1967a,b).

Since 1967, sporadic cases of Marburg virus disease have been observed in South Africa in 1975 (GEAR et al. 1975) and in Kenya in 1980 (SMITH 1982; CENTERS FOR DISEASE CONTROL 1980) and 1987 (JOHNSON 1996). Three primary and three secondary cases were involved in these occurrences. All the primary cases were fatal, the secondary cases survived. The source of infection was not detected. One case of a fatal laboratory infection has also occurred (NIKIFOROV et al. 1994).

For many centuries monkeys had been introduced into Europe from tropical countries mainly to be kept as pets or to be displayed to the public in zoological gardens or in circuses. During the twentieth century monkeys have been used in rapidly increasing numbers as experimental animals in biomedical studies. Early examples of this usage are the studies of Popper and Landsteiner on the etiology of poliomyelitis, in which it was shown that a filterable virus was causative. Up to the 1980s nearly all the monkeys which were used in biomedicine were captured wild (BEVERIDGE 1971; BONIN 1971). Only very few breeding stations existed, in which monkeys were kept mainly for behavioral studies.

A considerable increase in the use of monkeys has occurred due to the development and introduction of poliomyelitis vaccines in the 1950s. At that time only limited information had been available concerning the risk which monkey pathogens might constitute for human health. To avoid such a risk, poliomyelitis vaccines were routinely tested for the absence of *Mycobacterium tuberculosis*, rabies, herpes B virus and lymphocytic choriomeningitis (LCM) virus. The latter two were known since the 1930s as monkey agents with pathogenic potential for humans. In 1960, when SV40 was discovered, a total of 40 simian viruses was already known, most of which, with the exception of herpes B, rabies and LCM virus, were not regarded as a severe health problem for humans (Hull 1973).

The increasing demand for monkeys has forced African monkey trappers in the 1960s to curtail the length of quarantine and at the same time to catch simians living in remote areas from which no monkeys, up to that time, had been introduced into Europe. Prior to 1967, when Marburg virus was introduced into Europe, monkey traders, public health authorities and scientists had not become aware of the new health problems associated with the importation of monkeys from remote areas (Beveridge 1971; Bonin 1971). Meanwhile the use of captured wild monkeys for biomedical research or vaccine production became a rare exception.

2 Studies on the Etiology of a New Disease

The observation that the disease could be reproduced in guinea pigs was made independently in two laboratories. In Marburg intracytoplasmic inclusion bodies were observed in the organs of infected guinea pigs and their specificity was proven by applying direct immunofluorescence (IF) using human and guinea pig convalescent sera (Siegert et al. 1967a; Slenczka et al. 1968). (This was one of the early examples in which IF was used to identify a new virus.) Enrichment of these inclusions during the initial guinea pig passages was used as the criterion to select material for electron microscopic investigations. In November 1967 (3 months after the outbreak had started) a new virus with a morphology completely unknown for an animal virus at that time was detected and was shown to be the causative agent of the new disease. All the Henle-Koch's postulates were fulfilled to prove its etiological role. It was named Marburg virus, after the place in which it had first appeared and in which it had been isolated and characterized (Siegert et al. 1967a,b). Identical virus particles were detected in the blood and in the organs of inoculated guinea pigs and of human patients during their febrile phase (Siegert et al. 1967a,b). Identification of the causative agent was confirmed in 1968 by several groups (Kunz et al. 1968a; May et al. 1968; Haas et al. 1968; Kissling et al. 1968; Malherbe and Strickland-Cholmley 1968; Zlotnik et al. 1968a).

An English study group found Feulgen-positive intracytoplasmic granules and had "negative virological results." Therefore they claimed that the causative agent was "a rickettsia or a chlamydia" (SMITH et al. 1967). Their conclusion, however, was not confirmed.

3 Structure of the Marburg Virus

Marburg virus is an elongated and enveloped particle of extreme length. In characteristic preparations most particles are curved and appear as hooks or doughnuts, very few particles are linear (SIEGERT et al. 1967a,b; ALMEIDA et al. 1971). At low power magnification undamaged particles have a cloudy surface and do not reveal the nucleocapsid. At high resolution they show a regular arrangement of surface projections reminiscent of trees, each with a thin stem and a spherical treetop. There are always some particles in which the inner structure is revealed. Inside the envelope a matrix is found followed by a structure with transversal striations of 53Å periodicity: the nucleocapsid, which has a helical structure and surrounds an inner axis of 20 nm (PETERS et al. 1971).

The overall dimensions of the Marburg virus particles were measured and compared with standard vesicular stomatitis virus (VSV) particles (180 nm) which were added prior to fixation. Later, Ebola virus, which was isolated in 1976 (PATTYN et al. 1977; JOHNSON et al. 1977) and which is structurally related to Marburg virus, was included in this comparison. Measuring 535 particles, the standard length of Marburg virus was found to be 665 nm and a second maximum was found at 1.318 nm. A total of 506 particles of Ebola virus had a standard length of 805 nm. A second maximum was found at twice, a third at three times and a fourth at four times the standard length (Fig. 1). There were more maxima with even longer particles and the longest rods found in Ebola virus preparations amounted to 20 times the length of a standard virus (D. Peters and W.G. Slenczka, unpublished data).

The diameters of Marburg and Ebola virus were also measured and compared with the corresponding diameters of the VSV particle (SLENCZKA and PETERS 1977) (Table 1). In these experiments Vero cells were infected with the Marburg or with

Table 1. Outer and inner diameters of virus particles

Virus	Projections	Outer ring	Nucleocapsids external	Nucleocapsids internal
Marburg χ	73.5 nm	57.3 nm	22.9 nm	15.3 nm
95% CI	70.0–75.1	56.8–58.6	22.9–24.2	15.3–16.6
Ebola χ	73.5 nm	57.3 nm	20.1 nm	11.4 nm
95% CI	71.8–73.5	56.8–58.6	19.1–20.4	10.2–11.5
VSV χ	60.1 nm	45.1 nm	26.7 nm	20.0 nm
95% CI	60.1–60.1	43.4–46.7	26.7–26.7	16.7–20.0
Random sample	Marburg virus	Ebola virus	VSV	
Particles measured	52–57	55–64	35–40	

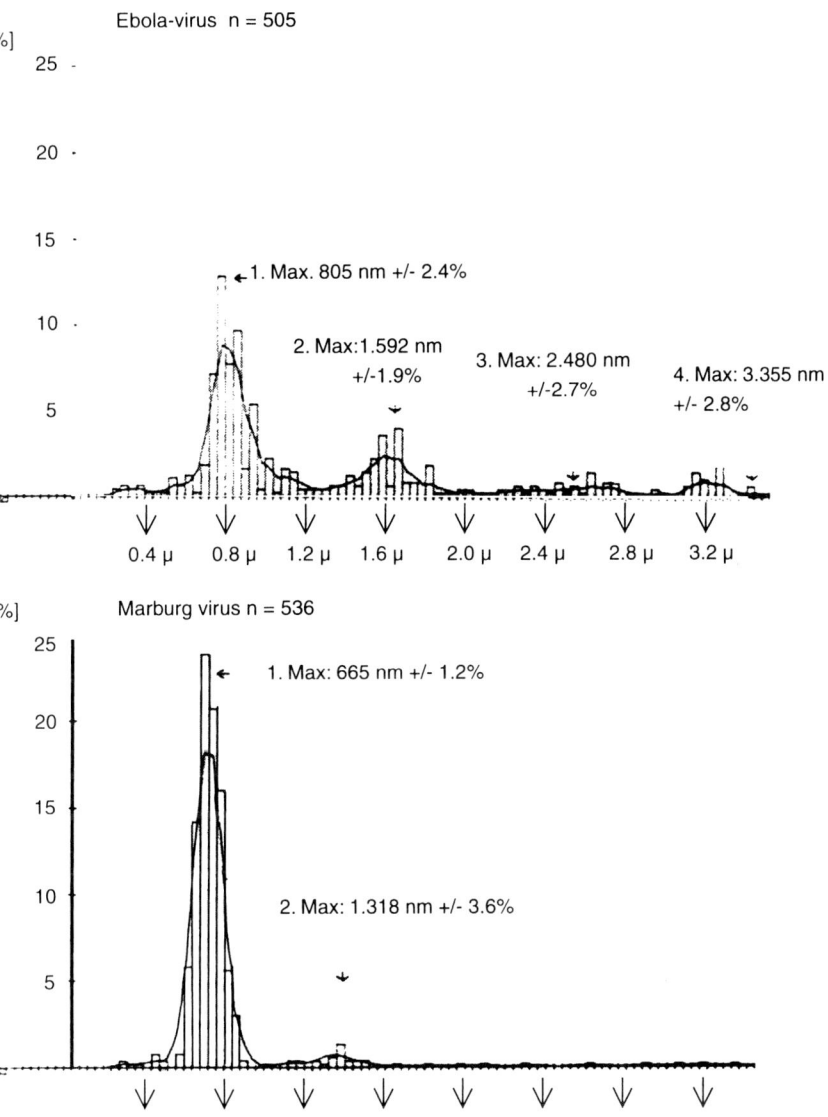

Fig. 1. Using vesicular stomatitis virus (VSV) particles (length 180 nm) as an internal standard the normal length and length distribution of filovirus particles were measured in the electron microscope. Cell cultures infected with Marburg or Ebola virus were superinfected with VSV on day 2 postinfection. On day 3 postinfection, supernatants of infected cells were fixed in 1.25% glutaric aldehyde in cacodylate buffer. Later, 4% formaldehyde was added

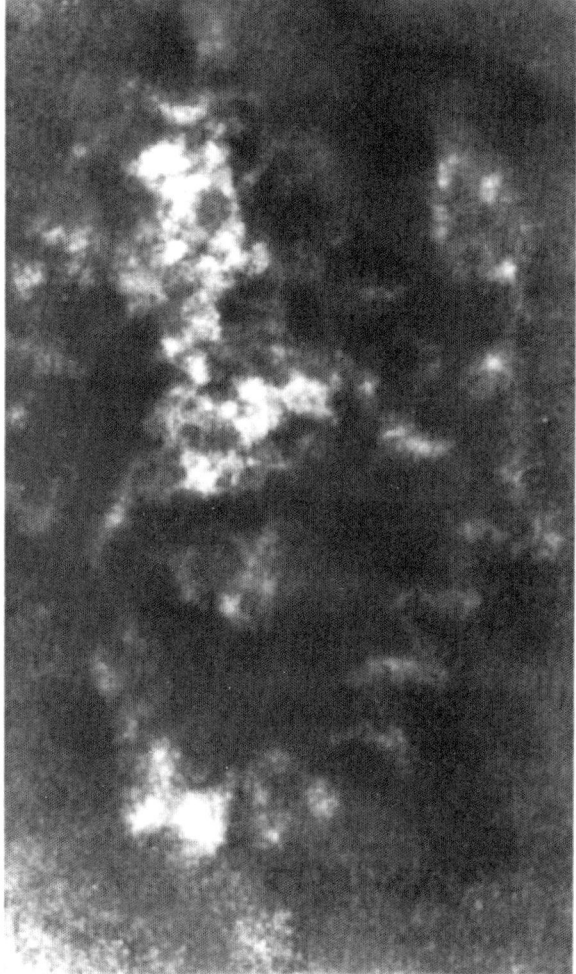

Fig. 2. Marburg virus nucleocapsid. Marburg virus was grown in Vero cells. The supernatant was harvested on day 3 postinfection and was cleared by low speed centrifugation at 2000 rpm for 20 min. Virus was precipitated by centrifugation at 20,000 rpm for 2 h. The sediment was suspended in buffered saline containing 0.5% sodium-deoxycholate. After 10 min 1.25% glutaric aldehyde was added for fixation. Later, 4% formaldehyde was added

the Ebola virus and the same cultures were superinfected with VSV prior to fixation with glutaric aldehyde.

Measuring particles from human or guinea pig blood it was found that the particle diameter was larger than in cell culture virus. This enlargement was due to the presence of antiviral antibodies (SLENCZKA and PETERS 1977). Some particles revealed envelopes with spikes in which no nucleocapsid could be seen and resembling empty envelopes (W.G. Slenczka and D. Peters, unpublished). Incubating virus preparations at 37°C or 4°C resulted in progressive rounding of the virions to spheres (HÜLSER 1969).

Marburg virus is an enveloped virus. The infectivity is destroyed by ether, chloroform and deoxycholate (SIEGERT et al. 1968a; KUNZ et al. 1968a; KISSLING et al. 1968; BOWEN et al. 1969). Upon treatment of purified virus particles in suspension with deoxycholate a spiral structure, presumably the nucleocapsid, was detected (W.G. Slenczka and D. Peters, unpublished data) (Fig. 2). The viral infectivity or integrity was not influenced by treatment of the particles with trypsin, collagenase, thrombin, elastase or hyaluronidase. By treatment with bromelain the infectivity was abolished completely (W.G. Slenczka, unpublished data). β-propiolactone, at a dilution of 1:16000, could be shown to inactivate the viral infectivity completely on overnight incubation. At dilutions of 1:2000 of β-propiolactone inactivation is more rapid and the antigenicity is not impaired (W.G. Slenczka, unpublished data; VAN DER GROEN 1982).

Inhibitor studies with 5-iodo- or bromodeoxyuridine (SHU et al. 1968; KISSLING et al. 1968) and with actinomycin D (SLENCZKA 1969a,b) produced indirect evidence that the viral genome was an RNA. Using an HCl silver methanamine stain for DNA detection (a micro-Feulgen reaction), no DNA could be detected in viral particles or in cytoplasmic inclusions (PETERS et al. 1971). The prediction of an RNA genome was confirmed; the genome was shown to be nonsegmented and single-stranded of negative polarity (REGNERY et al. 1980).

4 The Virion Proteins

Particles of Marburg and Ebola virus were purified by gradient centrifugation in Percoll and analyzed for structural proteins by polyacrylamide gel electrophoresis. Originally six polypeptides were found in Marburg and Ebola virus preparations and were designated according to their size (SLENCZKA and RADSAK 1980) (Table 2).

Later the L-proteins of both viruses were identified with a size of 180,000 or larger and the glycoproteins were therefore designated as VP2 (KILEY et al. 1988; FELDMANN et al. 1992; FELDMANN and KILEY, this volume). An additional

Table 2. Virion proteins of filoviruses

	Marburg virus	Ebola virus
VP2	150,000 (glycosylated)	125,000 (glycosylated)
VP3	110,000	115,000
VP4	32,000	39,000
VP5	30,000	34,000
VP6	27,000	31,000
VP7	23,000	22,000

candidate protein in preparations of Marburg virus, a VP 46.5 kDa, was detected in western blots (HUGHES 1994).

5 Virus Classification

The first descriptions of the Marburg virus morphology had stressed the relationship with members of the rhabdovirus family, mainly with plant rhabdoviruses (SIEGERT et al. 1967; MAY and HERZBERG 1969) and the name "*Rhabdovirus simiae*" was therefore proposed (KUNZ et al. 1968a). However, no serologic cross-reaction with any member of the rhabdovirus family was found (SLENCZKA and PETERS 1978; VAN DER GROEN 1982). Convalescent sera were also tested for antibodies against known arboviruses with negative result (CASALS 1971). This search has included antigens of more than 100 viruses including members of the Alphaviridae, the Flaviviridae and the Bunyaviridae. After the appearance of Ebola virus (JOHNSON KM et al. 1977; PATTYN et al. 1977) a comparison with VSV was made and it was shown that the structure of Marburg and Ebola virus differed markedly from that of VSV (W.G. Slenczka and D. Peters, unpublished) (Fig. 1, Table 1). Significant differences that were found in the genome structure and organization of Marburg and Ebola virus as compared to rhabdoviruses have finally led to the establishment of a new virus family with the Marburg and Ebola viruses as prototypic members. The name Filoviridae (lat. filum a small thread) was coined for this new family (KILEY et al. 1982).

6 Host Systems and Biological Properties

Monkeys of various species, including macaques and *Cercopithecus aethiops* are highly susceptible to infection with Marburg virus by any route of infection (HAAS et al. 1968; HAAS and MAASS 1971; MAASS et al. 1969a; SIMPSON 1969). Survival of an infected monkey has not been reported. The Marburg virus was adapted successfully during several consecutive passages to guinea pigs (SIEGERT et al. 1967a,b) and Syrian hamsters (ZLOTNIK and SIMPSON 1969) resulting in virus strains which produced 100% lethality in these animals. Marburg virus was propagated in mice by intracerebral infection during 20 serial passages without producing signs of illness (HOFFMANN and KUNZ 1970). No evidence of a persistent inapparent infection was found in NMRI mice (SLENCZKA and WOLFF 1971). The virus was inoculated by intrathoracic inoculation into *Aedes egypti* mosquitoes without losing its pathogenicity for guinea pigs during two consecutive passages, which lasted 11 and 21 days, respectively (KUNZ et al. 1968b; HOFMANN and KUNZ 1969). *Anopheles maculipennis* mosquitoes could not be infected under the same conditions.

Several types of cell cultures, primary and persistent, of human, monkey, guinea pig, hamster, mouse and chicken origin can be infected with Marburg virus (HOFMANN and KUNZ 1971), however significant virus titers of 10^6 ID_{50} or more during consecutive passages are achieved mainly in established cell lines from *Cercopithecus* kidney (GMK-AH1 and Vero) (SIEGERT et al. 1968a,b; SLENCZKA 1969a,b) and hamster kidney (BHK21) (ZLOTNIK et al. 1968b). Propagation in established mosquito cell lines, including *Aedes albopictus*, *Anopheles gambiae* (mos 55) and *Anopheles stephensi* (mos 43) was not successful (W.G. Slenczka, unpublished). Virus titration was carried out in guinea pigs (SIEGERT et al. 1968a), in Syrian hamsters (ZLOTNIK and SIMPSON 1969) and by plaque titration in Vero cells (SLENCZKA and PETERS 1977) or in BHK21 cells by fluorescent focus assay (SLENCZKA 1969b). Vero cell cultures persistently infected with the Marburg virus have been subcultured over five consecutive passages (SLENCZKA et al. 1971). A fusion of virus infected cells with the result of syncytia formation was observed occasionally at very low pH levels, but this effect has not yet been reproduced systematically (W.G. Slenczka, unpublished).

To test the possibility that the new virus was a plant rhabdovirus it was attempted to infect several species of plants (tobacco, tomato, potato and sunflowers) with the Marburg virus. Infection of single leaves was tried after mechanical lesion of the cuticula and systemic infection was also attempted. But infectious virus could not be recovered from the plants and no pathology resulted from infection (W.G. Slenczka and G. Wolff, unpublished data).

Erythrocytes from several species (chicken, duck, goose, monkey, humans, guinea pig, rat and mouse) were tested for hemagglutination by the Marburg virus at pH values ranging from 5.75 to 7.4 without positive result (W.G. Slenczka, unpublished data). Human or guinea pig thrombocytes were not agglutinated by the virus.

7 Morphological Studies on Viral Replication

Studies on viral replication were carried out using infected Vero cells. The virus yield was titrated in a plaque assay, and the cells were inspected by immunofluorescence and by electron microscopy (W.G. Slenczka, unpublished data; PETERS et al. 1971). After inoculation at a multiplicity of infection (MOI) of 0.01, the first evidence of viral infection was seen in a few cells after 28 h: tiny inclusions (about 1 μm) which were found by IF to be situated in the cytoplasm in the proximity of the nucleus. In the following hours inclusions appeared in more and more cells, their size increased to more than 10 μm in diameter, and they migrated to the periphery of the cells. In some cells fusion of inclusions seemed to occur, in other cells a diffuse or fine granular staining of the cytoplasm was observed. The cell nucleus did not contain viral antigen. Cytoplasmic inclusions could also be visualized by phase contrast microscopy or by staining with rhodamine or with methylene-fuchsin stain (SLENCZKA 1969a,b).

The electron microscopic appearance of the inclusions was characterized by aggregations of tubular material with a diameter of 25 nm, believed to be the nucleocapsids. The tubules were cut longitudinally or in cross-sections. There were at least two types of tubules found in the same inclusions: some were electron-dense, others were only poorly contrasted. The electron-dense tubules were found in the periphery of the inclusion bodies and they were transported to the cytoplasmic membrane where they were enveloped. In other inclusion bodies the electron-dense and poorly contrasted tubules were equally distributed. In some types of inclusion bodies the tubular structures were found in a concentric arrangement (PETERS et al. 1971). Immunofluorescent staining of infected cells with intact cytoplasmic membrane at 48 h post-infection (p.i.) revealed some cells with membrane bound antigen. Infectious virus was found in cultures after 48 h and increased to 10^5–10^6 pfu/ml on day 3 and subsequently to 10^6–10^7 pfu (SLENCZKA 1969a). Foci of infected cells grew from single cells on day two to 50 cells and to confluence on day 4–5 post infection. Cytopathic effects (rounding of cells), when they occurred, started on day 3 and comprised more than 50% of the cells on day 5. The rounded cells exhibited a much more intensive membrane fluorescence than did infected cells with low cytopathology (SLENCZKA 1969a).

Upon virus propagation in undiluted or low dilution passages an auto-interference phenomenon was seen: every cell was infected in the first cycle, the cytoplasmic inclusions did not grow and no membrane fluorescence could be detected. In these cells, no cytopathic effect (CPE) and no infectious virus appeared; no extracellular virus was found using the electron microscope (W.G. Slenczka, unpublished data).

8 Pathology

8.1 Studies on Human Material

The pathomorphology of Marburg virus disease has been described in great detail (GEDIGK et al. 1968, 1971; JACOB and SOLCHER 1968; RIPPEY et al. 1984). These descriptions are based on six patients who died between days 7 and 16 after onset of the disease.

Macroscopic findings: Signs of marked hemorrhagic diathesis were found in all postmortems mainly on the skin and mucous membranes but only occasionally in parenchymatous organs. Stomach and intestines were filled with blood and black stool, although ulcers or erosions as a possible bleeding source were not found. All of the deceased had a dark blue discoloration in the region of the external genitals.

There was no icterus, the gallbladders were enlarged and filled. Livers were not atrophic and appeared normal in aspect and consistency. Only two out of five spleens were slightly enlarged. There was a generalized but moderate swelling of the lymph nodes. A pale swelling of the kidneys was found in all cases. Two patients had bronchopneumonia. Irrespective of the hemorrhages, the macroscopic findings were not indicative of any specific type of disease (GEDIGK et al. 1968).

The histological picture was characterized by focal necroses, a plasmacytoidal and monocytoidal transformation of the lymphatic tissues, basophilic bodies (1–2 µm in diameter) of unexplained nature, tubular kidney insufficiency, hemorrhagic diathesis and diffuse encephalitis with perivascular lymphocytic infiltration (GEDIGK et al. 1971).

Necroses were found more or less marked in all parenchymatous organs except for the lungs, skeletal muscles and the skeleton; they were most prominent in the liver and lymphatic organs. There were no signs of inflammatory reaction and the parenchymatous cells were more affected than the mesenchymal structures (BECHTELSHEIMER et al. 1970). Rapid replacement of necrotic cells and efficient regeneration of the parenchyma were seen in those deceased who had survived until the end of the second week and also in liver biopsies taken from convalescents (BECHTELSHEIMER et al. 1971).

Follicular necrosis in the lymph nodes and cellular depletion of the red pulp in the spleen and medullar parts of the lymph nodes were most prominent in the early phases of the disease. In later stages plasmacytoidal and monocytoidal infiltration prevailed and the deposition of a finely granular eosinophilic material was noted. The exact nature of this material remains to be determined (GEDIGK et al. 1968). Electron microscopic findings point to the possibility that not only fibrin and thrombocytes are constituents of this material but virus particles as well (PETERS et al. 1971).

The nature of the 1–2 µm in diameter basophilic particles constitutes another unresolved problem (GEDIGK et al. 1971). The conclusion, that these particles might be identical with viral inclusion bodies, is certainly not justified, since cytoplasmic viral inclusions as a rule stain eosinophilic rather than basophilic. Eosinophilic cytoplasmic inclusions were also described as being situated in the periphery of the cells and surrounded by a clear halo (BECHTELSHEIMER et al. 1971). These might be identical with viral inclusions, whereas the basophilic bodies seemed to be the result of extensive nucleoclasia in infected cells (GEDIGK et al. 1968). Nucleoclasia was never found in infected cell cultures, even when pronounced CPE was present. SMITH and coworkers detected basophilic bodies giving a positive Feulgen-reaction in the organs of infected guinea pigs and therefore believed they were dealing with rickettsiae or with chlamydiae (SMITH et al. 1967).

In the liver the extent of parenchymal cell destruction correlates well with the corresponding levels of SGOT. No biliary necrosis was found and consequently alkaline phosphatase and bilirubin were not elevated in the serum. Liver cell destruction might be preceded by Kupffer cell destruction (BECHTELSHEIMER et al. 1970). In the kidneys parenchymal necrosis and signs of tubular insufficiency were found in all cases (GEDIGK et al. 1971).

Surprisingly, damage of the myocardium was not seen. Only an interstitial edema was found, which might be causative of changes in the electrocardiogram

(GEDIGK et al. 1971). The absence of a considerable lung pathology in four out of five cases is another surprising result considering the fact that most of the infections were believed to have been acquired via the nasopharyngeal or oropharyngeal route (GEDIGK et al. 1968). In guinea pigs infected by the aerogenic route, it has recently been described that the macrophages are the only targets of Marburg virus infection in the lungs (RYABCHIKOVA et al. 1996).

Two of the fatal cases had shown signs of a subacute comatose encephalitis, the brains of the deceased revealed the histological picture of a panencephalitis with glial nodules and with a slight perivascular lymphocytic infiltration. A similar but less pronounced process was found in combination with focal hemorrhages in a patient who had died without losing consciousness and without neurologic signs (BECHTELSHEIMER et al. 1968; JACOB and SOLCHER 1968; JACOB 1971).

In three cases, the final cause of death was believed to be encephalitis with swelling of the brain and the consequences of hepatic and renal failure. One patient also showed a hemorrhagic and necrotic bronchopneumonia and the last patient had preexisting lung emphysema and bronchiectases with right ventricular hypertrophy. Cardiac insufficiency may have been a major cause of death in this case. The relative contribution of hemorrhagic diathesis in the course of fatal outcome could not be evaluated at the postmortems (GEDIGK et al. 1971).

8.2 Studies in Experimental Animals

Guinea pigs, hamsters and monkeys (mainly rhesus and *Cercopithecus aethiops*) have been used as animal models to study the course of Marburg virus disease. These studies have largely reproduced and confirmed the pathologic changes which had been found in humans and they have added valuable information on the development of pathologic changes in the course of disease as correlated with clinical observations. However, experimental infections were, as a rule, done with high dose inocula, which were administered by the abdominal, subcutaneous or even intracerebral route. In a critical review of these experiments it seems nearly impossible to decide which observations are due to the species of experimental animal, the route of inoculation, the virus dose, the passage history of the virus strain or the stage of disease. The same criticism is also pertinent to some of the reports in which morphological changes of experimental Ebola virus disease and Marburg virus disease were compared.

In experiments with captured wild monkeys pre-existing infections such as tuberculosis, among others, have influenced the outcome and the pathomorphological picture (OEHLERT 1971).

There are only a few studies in which attempts were made to analyze the influence of the degree of viral adaptation, the virus dose and the route of infection on the course of disease and on the development of pathological changes (KORB et al. 1971a,b; LUB et al. 1995; RYABCHIKOVA et al. 1996). In agreement with findings in human postmortems no necroses were found in the lungs, even after aerogenic infection of guinea pigs (LUB et al. 1995). The central role of virus infected monocytes/macrophages not only in reproduction and dissemination of the virus but also in causing hemorrhage and blood vessel permeability is emphasized (FELDMANN et al. 1996; RYABCHIKOVA et al. 1996). Viral infection of endothelial cells, which was shown to occur in cultured endothelial cells, may be another important cause of diapedesis and hemorrhage (SCHNITTLER et al. 1993).

9 Early Studies on Pathogenesis

Unfortunately, no reliable data have become available on the state of health of the monkeys that had been the source of human infections with the Marburg virus. An excess mortality of the monkeys was only observed in the institution Torlak in Belgrade, where a total of 95 out of 288 monkeys (33%) resulting from three shipments died within 4–6 weeks after arrival but were not examined for the etiology (STOIKOVIC et al. 1971). Excess mortality was not recorded in Frankfurt and Marburg; however the "normal" mortality data were not published. At the postmortems done on the sacrificed monkeys, no signs of the disease were found (HENNESSEN 1971). It has been proposed that the excess mortality in Belgrade might have been due to a defect in acclimatization (HENNESSEN 1971). Seroconversions against the Marburg virus were observed in the Belgrade monkey population using the antigen of infected guinea pigs in the complement fixation (CF) test (STOIKOVIC et al. 1971). This finding however, could not be confirmed on reinvestigation (SLENCZKA et al. 1971). Since no monkey has ever survived an experimental infection with Marburg virus, regardless of the virus dose or route of inoculation, it seems to be a miracle that no excess mortality was observed either in Marburg or in Frankfurt. A possible explanation for this discrepancy may be the fact that in Marburg and in Frankfurt, the monkeys were used within a few days after arrival. As no signs of overt disease such as fever, enlargement of liver and spleen or hemorrhagic diathesis were found in the sacrificed monkeys (HENNESSEN 1971), the possibility remains that some of the monkeys might – in contrast to general supposition – have survived acute Marburg virus disease prior to exportation from Uganda and that the virus was still active at the time of sacrifice. An agent, pathogenic for guinea pigs, was isolated from the organs of some of the incriminated monkeys but unfortunately was not identified (ROBIN et al. 1971). It is known that human convalescents may transmit the virus even 120 days after the onset of the disease (MARTINI et al. 1968b; SLENCZKA et al. 1968). Results on antibody findings in the monkeys were not conclusive (STOIKOVIC et al. 1971; SLENCZKA et al. 1971).

There is another unresolved discrepancy in the question of whether aerogenic transmission of Marburg virus might occur in a monkey population. It is a well documented fact that a virus infection by the aerogenic route may be achieved in experimental conditions (LUB et al. 1995; RYABCHIKOVA et al. 1996), but does it also occur in monkeys that are cage neighbors without the possibility of physical contact? This was not observed in one study, but was observed in another study (HAAS et al. 1969; SIMPSON 1969). The fact that none of the animal caretakers were infected in Marburg, Frankfurt or Belgrade (HENNESSEN 1971) although monkeys excrete a high dose of virus via the urine (SIMPSON 1969), argues against a significant role for aerogenic transmission under natural conditions. Regarding the fact that many patients stayed at home during the first week of their disease it is remarkable that during this phase of disease and prior to hospitalization, no case of a family infection was reported even though most of the patients were married and lived with their families. The second patient in Belgrade was a family contact; however, she was reported to have drawn blood from her infected husband for a leukocyte count (TODOROVITCH et al. 1971). The second patient in Johannesburg may have contracted the disease prior to hospitalization of the index patient. Her companion had been ill for 4 days before hospital admission and she fell ill 5 days after his admission with myalgia (GEAR et al. 1975).

In experiments with guinea pigs it was observed that the noninfected cage mate of an infected animal would invariably be infected. As a rule the cage mate would survive twice as long as the infected animal. While the first animal was inoculated with a high virus dosage by the parenteral route, infection of the cage mate must have occurred with a lower virus dose and via a less efficient route of infection. It is not known at what stage of disease the virus from the first animal was transmitted to the cage mate (W.G. Slenczka, unpublished data).

Observations on the Marburg virus infection in guinea pigs proved that the course of disease closely resembles the course of disease in humans. Damage of hepatic cells is indicated early by the increased levels of serum transaminases with a quotient of SGOT/SGPT of more than 7/1 (SIEGERT et al. 1968a).

The severe coagulopathy observed in human patients in the course of Marburg virus disease was believed to be due to disseminated intravascular coagulation (DIC) and in three out of five postmortems platelet and fibrin thrombi were found. In guinea pigs development of a hemorrhagic diathesis was monitored with the thrombo-elastogram. In addition thrombocytes were counted and the activity of coagulation factors was determined (EGBRING et al. 1971). While significant reductions of clotting factors and platelets were already detected on the first day of disease (fever), a prolongation of the reaction time (r-time) and of the clot formation time (k-time) in the thrombo-elastograms was not measurable before day 3. There was also an indication of increased fibrinogenolysis and of secondary fibrinolysis. These findings are compatible with the hypothesis that the severe bleeding disorder in Marburg virus disease is due to DIC (EGBRING et al. 1971).

A primary dysfunction of the platelets could not be demonstrated. By mixing platelet-rich plasma or heparinized blood with Marburg virus no direct influence of the virus on bleeding time or on clot retraction was detectable. There was also no indication of autoantibodies against platelets which might have been produced in the course of disease (R. Egbring and W.G. Slenczka, unpublished data). Virus-infected megakaryocytes were not detected by electron microscopy of the bone marrow from infected guinea pigs and no virus was seen inside the platelets (W.G. Slenczka and D. Peters, unpublished data). In the organs of guinea pigs but not in the organs of humans, the homogeneous PAS-positive material which was found in necroses could be stained by von Kossa's stain, which is believed to detect calcium (KORB et al. 1971a). It was observed that in guinea pig blood, which would not show coagulation, some coagulation capacity could be restored by addition of calcium. The fatal course of the disease in guinea pigs could, however, not be influenced by administration of calcium solution (W.G. Slenczka, unpublished data).

Theoretically, it should be possible to prevent the development of DIC by early treatment with heparin. However heparin treatment did not show any effect on the course of the disease or the development of coagulopathy in guinea pigs (W.G. Slenczka, unpublished data). An attempt to save the life of a Marburg virus patient in Johannesburg by heparin treatment was not successful (GEAR et al. 1975).

In order to assess the possible effect of cytotoxic T cells on the outcome of the disease, infected animals were treated with low dose cyclophosphamide, which did not show any effect (W.G. Slenczka, unpublished data).

Another unresolved problem is the question of whether protective immunity can be demonstrated in humans or in animals who survived the disease. In an early experiment it was shown that guinea pigs can be protected from a lethal dose of Marburg virus when human convalescent serum is administered simultaneously (SIEGERT et al. 1968a). Whereas a neutralizing antibody against the Marburg virus has not been conclusively demonstrable by in vitro assays, the protection of guinea pigs by passive administration of human antiserum might be a consequence of antibody-dependent cellular cytotoxicity. However it has not been possible to protect monkeys by immunization using killed virus vaccines (MALHERBE and STRICKLAND-CHOLMLEY 1971). There are some unpublished observations on guinea pigs who survived an infection with wild-type virus from nonadapted strains but they died after challenge with a lethal virus dose, although they had developed Marburg virus antibodies as a result of the first infection (W.G. Slenczka, unpublished data). This problem has not yet been studied systematically.

10 Clinical Features and Patient Management

Clinical observations on Marburg virus disease in humans are based on not more than five episodes which were confirmed by virus isolation and which involved only 36 patients (Table 3).

The reports on clinical observations of the patients with Marburg virus disease present a relatively uniform picture although the patients were of varying age groups and constitution and although the data were collected from the reports of four clinical departments (MARTINI et al. 1968a,b; STILLE et al. 1968, 1971; TODOROVITCH et al. 1971; GEAR et al. 1975). The incubation time, whenever it could be estimated, was on average 8 days with 5 days being the shortest and 9 days being the longest. It is remarkable that the first cases of each local outbreak were

Table 3. Summary of known outbreaks of Marburg virus hemorrhagic fever

Year	Location	Lethality	Secondary cases per total cases	Primary source of infection
1967	Marburg	5/24[a] (21%)	3/24[a] 2× needle stick inoculation; 1× sexual intercourse	Cercopithecus aethiops monkeys and cell cultures
	Frankfurt/Main	2/6 (30%)	2/6 1× needlestick inoculation; 1× cut with a knife at post mortem	Same as above
	Belgrade	0/2	1/2 Nosocomial	Postmortems of infected monkeys
1975	Johannes-burg/SAU	1/3 (30%)	2/3 Nosocomial	Unknown during trip in Zimbabwe
1980	Nairobi	1/2 (50%)	1/2 Nosocomial	Unknown (mount Elgon region)
1987	Nairobi	1/1 (100%)	0/1	Same as above
	Koltsovo	1/2		Laboratory infection
	Total number	12/40 (30%)	9/40 (22.5%)	

[a] Only 23 patients were hospitalized in Marburg. An additional patient was severely ill at the same time but was not hospitalized. He was shown to have had the disease by positive serology in 1982 (W.G. Slenczka, unpublished). His reason for infection was contact with cell cultures for production of the poliomyelitis vaccine.

treated in their homes for up to 10 days although the illness was described as beginning suddenly with extreme malaise, pains in the limbs and headache, and a rapid increase of temperature to >39°C. The fever persisted for at least 1 week. During the initial 3–4 days the clinical picture was essentially unremarkable. Gastrointestinal symptoms (vomiting and diarrhea) conjunctivitis, enanthema and a rash occurred about 3–4 days after the onset of disease. On day 7 the temperature fell to 38°C and persisted in this range for another week. On days 7 and 8, as indicated by SGOT values, liver cell destruction reached its maximum and the lowest counts of platelets and white blood cells were reached. Patients who were admitted to the hospital on day five after the onset of disease had, in some cases, already presented signs of hemorrhagic diathesis, which was severe in about one third of the cases. Mild manifestation with petechiae was found in another 30%. Other patients had low thrombocyte counts but did not develop overt hemorrhagic symptoms. Leukopenia and thrombopenia were severest on day 7. Immature and degenerated neutrophils were seen quite often during the phase of leukocytopenia (HAVEMANN and SCHMIDT 1971). Hepatitis, proteinuria, myocarditis, pancreatitits and encephalitis were seen in the second week of disease and persisted to the third week. Death, when it occurred, followed in the second week of disease, at day 9 on average, with a range of from day 7 to day 16 after onset of disease. Orchitis, a typical late symptom, appeared in the third week or even as a relapse 5 weeks after onset. Another typical sequela of disease was hepatitis, as evidenced by increase of SGOT. In one case uveitis appeared 2 months after the acute disease and virus was isolated from the anterior eye chamber (KUMING 1977). It seems to be a common experience in cases of Marburg virus disease that relapses may occur during a period of 3 months after the disease. Early relapse accompanied by fever and hepatitis was seen in a patient 1 month after the onset of disease. Viral antigen was found in a biopsy specimen of the liver at this time (SLENCZKA et al. 1968). Some patients had orchitis during the convalescence phase. One patient infected his wife by sexual intercourse 3 months after onset of his disease and viral antigen was found in his semen (SLENCZKA et al. 1968). The occurrence of relapses later than 3–4 months after the acute disease has never been reported. It seems that probably no protective immunity exists during the first 3 months after the disease. The question is whether protective immunity will develop after this period. Some challenge experiments in convalescent guinea pigs were probably carried out too early after convalescence. Human convalescents must be warned that they will continue to be a risk to other people for at least 3 months after the disease: they should have protected sexual intercourse only, and they are not allowed to give blood for transfusions.

Out of nine secondary cases of Marburg virus diseases two were due to inadvertent pricks with a used needle, one person cut himself with a postmortem knife. Four other people who had unprotected contact with patients and with their blood were thought to have caught the infection in this way. Two cases of family infections occurred, one was transmitted in the first week of disease (GEAR et al. 1975), another was transmitted via semen 3 months after the husband had the disease (MARTINI et al. 1968b). Viral antigen was found in his seminal fluid at this time (SLENCZKA et al. 1968). The second case in Belgrade was also a family infection but was believed to be nosocomial (TODOROVITCH et al. 1971). All the secondary cases had a mild course of disease: no fatality was recorded. The severest disease seems to have occurred in the patient who had cut himself at the postmortem. The cut was 15 cm long (patient, personal communication) and would certainly have transmitted a larger quantity of virus as compared with a needle prick. The wife who was infected by her husband also had a serious course of disease (MARTINI 1971). It remains unclear whether viral attenuation or low dose of infection was responsible for the benign courses of disease in the secondary cases.

Patients of all age groups (between 16 and 64 years) were inflicted with Marburg virus disease. A pronounced influence of age on the course of disease is not observable (Table 4), neither is there a pronounced influence of gender on the outcome of disease. Two women (19 and 64 years old) died out of 12 female patients, but four of these were secondary cases with a rather benign course. Eight out of 24 male patients (22 were primary cases) died.

A follow-up study of 18 survivors of the Marburg virus outbreak of 1967 was published 12 years after the outbreak. At this time no residual signs of the disease

Table 4. No influence of age or gender on outcome

Age group (random sample)	< 20 years ($n=9$)	21–30 years ($n=8$)	31–40 years ($n=7$)	41–50 years ($n=7$)	> 51 years ($n=5$)
Disease duration	11.5 days	15.3 days	16 days	25.5 days	17.7 days
Patient fatality	3/9 (33%)	2/8 (25%)	2/7 (29%)	2/7 (29%)	1/5 (20%)

could be detected in these patients. Most of the convalescents had resumed working already during the first 3 months of 1968 (BALZER et al. 1979).

11 Treatment

There is as yet no specific antiviral therapy available against Marburg virus disease (IGNATYEV 1996). Treatment of patients was more or less "ex juvantibus" or supportive in the known outbreaks. Antibiotics such as tetracycline, chloramphenicol, penicillin, cephalotin and streptomycin did not influence the fever or course of the disease. However some antibiotic treatment (cephalotin) was continued in order to avoid secondary infection.

Treatment of hemorrhagic complications was attempted in 1967 with a combination of various measures: fresh blood transfusion, thrombocyte concentrates, fibrinogen, ε-amino-capronic acid and vitamin K. The best results were obtained with PPSB, containing prothrombin, proconvertin, Stuart factor and antihemophilic globulin B. This preparation had some effect on severe hemorrhagic complications at least for a few hours (MARTINI et al. 1968a; MARTINI 1971). Two of the three patients who were seen in 1975 in Johannesburg (secondary cases) received heparin treatment early in the course of the disease in order to prevent DIC. In the second case fibrin degradation products accumulated and it was believed that DIC had developed in spite of this treatment (GEAR et al. 1975).

Anti-yellow fever immune-plasma was administered without effect to four patients in Frankfurt, when it was believed that they might be suffering from yellow fever (STILLE et al. 1968). Two cases of Marburg virus disease in Johannesburg were treated with anti-Lassa fever immune serum; however, no improvement was recorded (GEAR et al. 1975). Convalescent phase serum was taken 2 weeks after the onset of disease from patient F in Frankfurt and was used to treat four other patients. A single dose of 200 ml of this serum was given to two Frankfurt patients with secondary infection on day 3 of their disease and it was believed to have been beneficial (STILLE et al. 1968, 1971). The same treatment was also given to the two patients in Belgrade (TODOROVITCH et al. 1971). It must be kept in mind that three of the treated patients were secondary cases and that the course of the disease was milder in all the secondary cases. Beneficial effects of treatment with convalescent serum are also reported from the Ebola outbreak in Kikwit (MUPAPA et al. 1998).

The most urgent problem in supportive treatment is electrolyte balancing, and potassium in particular had to be substituted. Human serum albumin (30 g/day) was given to several patients with severe hypoproteinemia. Diuresis was no problem but in some patients mannitol was administered. In one patient anuria occurred and therefore peritoneal dialysis was performed (MARTINI 1971). Hemodialysis was also performed on one of the two patients in Koltsovo and was believed to have been beneficial; the patient survived (NIKIFOROV et al. 1994).

Several patients received digitalis or strophanthin. In two patients prednisone was administered in the final stage but without success. Antipyretic treatment with Novalgin (phenyl dimethyl pyrazolone) was used to very good effect with many patients: the fever went down and their condition improved (MARTINI 1971).

Chloroquine was administered to a patient in Johannesburg on day 7 of his disease because a malaria was assumed, although no plasmodia had been found. This patient died 2 days later (GEAR et al. 1975). In one patient uveitis appeared 2 months after the acute disease and Marburg virus was isolated from the anterior eye chamber. She was treated with steroid and with atropine drops but the condition persisted for several weeks (KUMING 1977).

12 Virological Diagnosis of Marburg Virus Disease

Marburg virus was isolated from clinical material by intraperitoneal inoculation of guinea pigs. Body temperature was measured every day and blood was drawn from animals that had a rise in temperature. The blood was inoculated into other animals and passaged till severe disease resulted (SIEGERT et al. 1967a,b). Direct demonstration of Marburg virus in the blood of patients and of infected guinea pigs is easily achieved by electron microscopy. Using a special centrifugation technique, the virus particles can be sedimented directly onto the grid from formalinized plasma (MÜLLER 1969). Virus infection of the guinea pigs was confirmed by IF staining of acetone-fixed smears which were prepared from spleens and livers (SIEGERT et al. 1967b; SLENCZKA et al. 1968). Blood specimens were drawn from 14 patients during their first week of disease and were 100% positive for infectious virus. Specimens were drawn from four other patients between days 13 and 15 of their disease and of these only two were positive. Out of four urine specimens taken during the first week of disease, one was positive; out of six throat garglings, only two were positive for infectious virus. It must be mentioned that all the specimens had been stored at −65°C for several weeks before they were inoculated into guinea pigs (SIEGERT et al. 1967b, 1968b). Infectious virus was found in semen and in the anterior eye chamber more than 100 days after the onset of disease (SIEGERT et al. 1968b; KUMING 1977).

IF tests have been successfully used for the demonstration of viral antigen in the organs of infected guinea pigs (SLENCZKA et al. 1968; SIEGERT et al. 1971) and in human postmortem specimens (WULFF et al. 1978), but there is only limited experience with detection of viral antigen in clinical specimens taken during the early phase of disease. However, viral antigen was detected with ease in specimens of liver and seminal fluid during relapses in the convalescent phase (SLENCZKA et al. 1968).

When a complement fixation test was used, no antibody was found in the sera of the first week (SLENCZKA et al. 1970). Using an ELISA test low titers of IgG antibody were already found in sera of days 4 to 7, with a large increase in titer during the second and third weeks (HOFFMANN 1982; SLENCZKA et al. 1984). There is no detailed study on the appearance of IgM antibody over the course of infection.

13 Serological Studies

Antibodies against viral antigens were detected in convalescent sera by direct and by indirect IF techniques using acetone- or methanol-fixed cells as the antigen. Formaldehyde and paraformaldehyde can also be used for fixation (SIEGERT et al. 1967; SLENCZKA et al. 1968). Antigen for the complement fixation test was initially extracted from the livers of infected guinea pigs and was inactivated with β-propiolactone. This material was, however, found to be inappropriate due to the occurrence of liver-specific autoantibodies, which were found in convalescent sera and also in some control sera (SLENCZKA et al. 1970, 1971). Much better results were achieved when antigens were extracted from infected Vero cells (HOFMANN and KUNZ 1969; SLENCZKA et al. 1970). Infected Vero cells were also used as a source of antigen preparation for more advanced serological techniques such as ELISA (SLENCZKA et al. 1984; HOFFMANN 1982) and western blotting (HUGHES et al. 1986). In general, antigens used for these techniques are inactivated with sodium lauryl sulfate. However, in our own experience, virus inactivation with β-propiolactone diluted 1 to 2000 and with 1% Triton X-100 will yield more reliable results in the ELISA test than those obtained with SDS-inactivated material (W.G. Slenczka, unpublished data).

Specificity in serologic reactions is usually controlled by a comparison with a control antigen extracted from noninfected cells. When sera are tested from persons without a known history of the disease and especially when sera of African origin are tested, nonspecific reactions may create severe problems. In these cases preabsorption of sera with control antigen may be useful. To prove serologic specificity preabsorption of specimens positive in the ELISA with the homologous or heterologous antigen may be done. When the result is not influenced by preabsorption with an unrelated viral antigen, this test is valid proof of a specific reaction (SLENCZKA et al. 1984).

When indirect IF staining was used for seroepidemiologic studies, non-specific staining of the inclusion bodies occasionally occurred. Therefore this technique is not approved for serosurveys. It is certainly more specific when membrane fluorescence of infected cells is also observed (W.G. Slenczka, unpublished data). However some convalescents have never reacted with the Marburg virus glycoprotein. Therefore this reaction is certainly not very sensitive. In our own experience we have found inclusion body staining of Ebola-infected cells, not of Marburg virus-infected cells, when we stained with a 1:100 dilution of a serum derived from a girl who had been bitten by a monkey. Some days later an acute Epstein Barr virus infection was diagnosed in this patient and the staining of Ebola virus antigen was believed to be due to heterotypic antibodies (W.G. Slenczka, unpublished data).

Neutralizing antibody against Marburg virus was found in some sera when a fluorescent focus assay was used to detect infected cells; however the result was not positive in all the convalescent sera (SLENCZKA 1969b). Attempts to reproduce these results in a plaque reduction assay were not successful. Even in the presence of guinea pig complement or in the presence of anti-human immunoglobulins no significant neutralizing activity could be detected (W.G. Slenczka, unpublished data). (Neutralizing antibody against Ebola virus can be detected in a plaque reduction test; W.G. Slenczka, unpublished data). Neutralizing antibody against Marburg virus was detected in monkey sera from Uganda using Vero cells and reading CPE as the endpoint (HENDERSON et al. 1971).

14 Serologic Cross-Reactivity Between Marburg and Ebola Virus

In order to characterize viral antigenicity, suspensions of virus particles were subjected to ether treatment and a sediment of this material was used to immunize guinea pigs. The resultant sera were tested by IF and stained only inclusion bodies, not the membrane of virus-infected cells. Antisera raised against nucleocapsid material did not reveal serological cross-reactivity between Marburg and Ebola virus (SLENCZKA and PETERS 1977). In these experiments the homologous antibody titers did not exceed 32.

When 100 human sera from 18 convalescents of Marburg virus disease were compared in their reactivity with Marburg and Ebola virus antigens in an ELISA test, it was found that they all reacted with the Ebola antigen although none of these persons had ever experienced this virus (HOFFMANN 1982). The reactivity was exceeded by the homologous virus but was significantly above the cutoff value, which was produced by testing these sera against control antigen and by testing 120 normal sera with the virus antigens. This cross-reactivity was not only found in early convalescent sera but was also detectable in sera taken some years after the disease (HOFFMANN 1982). The molecular target of this cross-reactivity is still unexplained. Recently it was found that antibody against recombinant VP 30 may be cross-reactive (OGAN et al. 1997).

15 Investigations into the Epidemiology and Source of Infection

The *Cercopithecus aethiops* monkeys from Uganda were very soon recognized as the source of Marburg virus infections because there was no other common denominator in Marburg, Frankfurt and Belgrade. All of the three institutions had already used monkeys from the same trading company for several years. Monkeys from two shipments, received on July 21st and 28th, were the source of infection in all the cases which occurred in Marburg and in three cases in Frankfurt. The fourth case in Frankfurt was attributed to a shipment on the 10th of August. However in Frankfurt and also in Belgrade, animals from different shipments were caged in the same rooms and, therefore, the spread of infection from shipment to shipment might have been possible through aerogenic transmission (HENNESSEN et al. 1968). However, aerogenic transmission was only observed in one out of two controlled studies (HAAS et al. 1969; SIMPSON 1969). Published observations on the state of health of the incriminated monkeys are discussed in the chapter on viral pathogenesis.

Monkeys from the same trader which were captured in the same regions have also been exported to laboratories in Japan, Italy, Sweden, Switzerland and USA. Nothing happened at these places (HENNESSEN et al. 1968). Later it became known that an island in Lake Victoria was used by the monkey trappers as a quarantine station. Animals that appeared ill were transported there and whenever more monkeys were needed than were available, some healthy looking animals were caught from the quarantine station to fill up the shipment. This practice might very well be the reason why some monkeys which had previously been ill were included in some of the shipments.

Another epidemiological peculiarity was due to the political situation in the Middle East. During the Six Day War, the monkey shipments were not transported directly to Frankfurt but to London and then from there to Germany. The length of stay of the monkeys in London was between 9 and 46 h for different shipments. In the animal house at the airport in London care was taken, not to put *Cercopithecus aethiops* monkeys, which were destined for biomedical use, in the same rooms with rhesus monkeys in order to avoid transmission of herpes B virus. However the African monkeys were kept in contact with two langur monkeys from Ceylon and with finches (Fringilidae) of South American origin. One of the langur monkeys died in London from an unexplained disease. An examination result of the second langur monkey has never been published (HENNESSEN et al. 1968; HENNESSEN 1969).

For seroepidemiological studies, serum specimens were collected from several species of old world monkeys. The serological tests used in these studies included a CF test (KALTER et al. 1969; STRICKLAND-CHOLMLEY and MALHERBE 1971; STOIKOVIC et al. 1971), neutralization test (KAFUKO et al. 1970; HENDERSON et al. 1971) indirect IF test (JOHNSON et al. 1981, 1982) ELISA (HOFFMANN 1982) and ELISA and western blotting (BECKER et al. 1992). Whereas some seropositive animals were identified in most of these studies, there is no conclusive explanation for the fact that, regardless of the dose and route of inoculation, no monkey has ever survived an infection with Marburg virus under experimental conditions. Serological data which resulted from CF tests with antigens prepared from infected guinea pigs could not be confirmed when a more specific test was used (SLENCZKA et al. 1971).

A considerable number of serological surveys were conducted on human serum specimens from several populations in equatorial Africa in order to determine populations at risk for filovirus infections. Indirect IF tests were mainly used in the early studies (JOHNSON et al. 1983; JOHNSON et al. 1993, among others). Some studies were published in which ELISA or ELISA plus western blotting was used (SLENCZKA et al. 1984; RIETSCHEL 1987, HUGHES et al. 1986, GONZALEZ et al. 1998). The pygmee population in Central Africa (JOHNSON ED et al. 1993; GONZALEZ et al. 1998) and people living in an artificial rain forest (Firestone Rubber Plantations) region in Liberia (HUGHES et al. 1986) were shown to include a high percentage (more than 30%) of seropositives for filovirus antibodies. However in a study conducted in East Africa antibody carriers were most frequently found in a semi-desert area (JOHNSON BK et al. 1983).

There is at present no convincing conclusion which can be drawn from these seroepidemiological studies. An improved surveillance of clinical cases of hemorrhagic fevers would be needed in many countries of equatorial Africa combined with facilities for the identification of etiologic agents.

Regarding the fact that contact with infected monkeys is the only known source of primary filovirus infection in humans, dogs might be of value as sentinel animals, because they are fed with the offals from human food. Therefore it would be recommendable to test collections of dog sera for the presence of filovirus antibody.

16 Safety Considerations

As a consequence of the Marburg virus outbreak, safety aspects for the handling of subhuman primates have been discussed (BONIN 1971; BEVERIDGE 1971). It was demonstrated that cell cultures prepared from the kidneys of experimental monkeys were infected with the Marburg virus (MAASS et al. 1969a,b; MALHERBE and STRICKLAND-CHOLMLEY 1971). Poliomyelitis vaccines which were produced in August 1967 might have been infected with the Marburg virus. They were discarded without having been tested for the presence of Marburg virus (W. Hennessen, personal communication). Technical equipment and legal regulations for the safe handling of dangerous pathogens have meanwhile been developed.

References

Almeida JD, Waterson AP, Simpson DIH (1971) Morphology and morphogenesis of the Marburg agent. In: Martini GA, Siegert R (eds) Marburg virus disease. Springer, Berlin Heidelberg New York, pp 84–97

Balzer G, Slenczka W, Stöppler L, Schmidt-Wilke HA, Hermann E, Siegert R, Martini GA (1979) Marburg Virus Krankheit: Verlaufsbeobachtungen über 12 Jahre (1967–1979). Verhandlungen der Deutschen Gesellschaft für Innere Medizin. 85. Band. JF Bergmann, Munich, pp 1203–1206

Bechtelsheimer H, Jacob H, Solcher H (1968) Zur Neuropathologie der durch grüne Meerkatzen (*Cercopithecus aethiops*) übertragenen Infektionskrankheiten in Marburg. Dtsch Med.Wschr 93:602–604; German Med Monthly XIV:10–12 (1969)

Kunz C, Hofmann H, Aspöck H (1968b) Die Vermehrung des Marburg Virus in Aedes aegypti. Zbl Bakt I Orig 208:347–349

Lub MY, Sergeev AN, Pyankova OG, Pyankov OV, Petrischenko VA, Kotlyarov LA (1995) Clinical and virological characterization of the disease in guinea pigs aerogenically infected with Marburg virus. Probl Virol 3:119–121

Maass G, Haas R, Oehlert W (1969a) Experimental infections of monkeys with the causative agent of the Frankfurt Marburg syndrome. Lab Anim 4:155–165

Maass G, Müller J, Seemayer N, Haas R (1969b) Production of kidney tissue cultures from African green monkeys, experimentally infected with the causative agent of Frankfurt-Marburg-syndrome. Amer J Epidem 89:681–690

Malherbe H, Strickland-Cholmley M (1968) Human disease from monkeys (Marburg virus) Lancet ii:1434

Malherbe H, Strickland-Cholmley M (1971) Studies on the Marburg virus. In: Martini GA, Siegert R (eds) Marburg virus disease. Springer, Berlin Heidelberg New York, pp 188–194

Martini GA (1969) Marburg agent disease in man. Trans Royal Soc Trop Med Hyg 63:295–302

Martini GA (1971) Marburg virus disease. Clinical syndrome. In: Martini GA, Siegert R (eds). Marburg virus disease. Springer, Berlin Heidelberg New York, pp 1–9

Martini GA, Knauff HG, Schmidt HA, Mayer G, Baltzer G (1968a) Über eine bisher unbekannte, von Affen eingeschleppte Infektionskrankheit: Marburg-Virus-Krankheit. Dtsch Med Wschr 93:559–571; German Med Monthly XIII:457–470

Martini GA, Schmidt HA (1968b) Spermatogene Übertragung des Marburg Virus (Erreger der "Marburger Affenkrankheit"). Klin Wschr 46:398–400

May G, Herzberg K (1969) Vergleich eines Affenseuche Erregers mit einem Virus der Vesicularstomatitis-Gruppe. Zbl Bakt I Abt Orig 211 133–143

May G, Knothe H, Hülsser D, Herzberg K (1968) Elektronenmikroskopische Befunde bei einer Affenseuche. (Cercopithecus aethiops). Zbl Bakt I Abt Orig 207:145–148

Müller G (1969) Elektronenmikroskopische Partikelzählung in der Virologie. I. Zentrifugierröhrchen zur direkten Sedimentation von Viren auf Netzträger. Arch Ges Virusforsch 27:339–351

Nikiforov VV, Turovski YI, Kalinin PP, Akinfeeva LA, Katkova LR, Barmin VC, Ryabchikova I (1994) A case of laboratory infection with Marburg hemorrhagic fever. J Microbiol Epidemiol Immunol 3:104–106

Oehlert W (1971) The morphological picture in livers, spleens and lymphnodes of monkeys and guinea pigs after infection with the "vervet agent". In: Martini GA, Siegert R (eds) Marburg virus disease. Springer, Berlin Heidelberg New York, pp 144–156

Ogan M, Feldmann H, Volchkov V, Slenczka W (1997) Expression rekombinanter Antigen von Filoviren. Chemoth J 6:46

Pattyn S, Jacob W, van der Groen G, Piot P, Courteille G (1977). Isolation of Marburg virus like virus from a case of haemorrhagic fever in Zaire. Lancet 1(8011):573–574

Peters D, Müller G, Slenczka W (1971) Morphology, development and classification of the Marburg virus. In: Martini GA, Siegert R (eds) Marburg virus disease. Springer, Berlin Heidelberg New York, pp 68–83

Regnery RL, Johnson KM, Kiley MP (1980) Virion nucleic acid of Ebola virus. J Virol 36:465–469

Rietschel M (1987) Seroepidemiologische Untersuchungen auf Marburg Virus und Ebola Virus Antikörper an Humanseren aus Afrika. Inaugural dissertation. Marburg

Rippey JJ, Schepers NJ, Gear JH (1984) The pathology of Marburg virus disease. S Afr Med J 66:50–54

Robin Y, Brès P. Camain C (1971) Passage of Marburg virus in guinea pigs. In: Martini GA, Siegert R (eds) Marburg virus disease. Springer, Berlin Heidelberg New York, pp 117–122

Ryabchikova E, Strelets L, Kolesnikova L, Pyankov O, Sergeev A (1996) Respiratory Marburg virus infection in guinea pigs. Arch Virol 141:2177–2190

Schnittler HJ, Mahner F, Drenkhahn D, Klenk HD, Feldmann H (1993) Replication of Marburg virus in human endothelial cells: a possible mechanism for the development of viral hemorrhagic disease. J Clin Invest 91:1301–1309

Shu HL, Siegert R, Slenczka W (1968) Zur Pathogenese und Epidemiologie der Marburg-Virus-Infektion. Dtsch Med Wschr 93:2163–2165; Germ Med Monthly XIV:7–10

Siegert R, Shu HL, Slenczka W (1968a) Isolierung und Identifizierung des Marburg Virus. Dtsch Med Wschr 93, 604–612; German Med Monthly XIII:514–518

Siegert R, Shu HL, Slenczka W (1968b) Nachweis des Marburg Virus beim Patienten. Dtsch Med Wschr 93:616–618; German Med Monthly XIII:521–524

Siegert R, Shu HL, Slenczka W, Peters D, Müller G (1967a) Detection of the so-called green monkey agent. Proc IV Congreso Latinamericano de Microbiologia, Lima, Peru, Nov. 26–Dec. 2

Siegert R, Shu HL, Slenczka W, Peters D, Müller G (1967b) Zur Ätiologie einer unbekannten, von Affen ausgegangenen menschlichen Infektionskrankheit. Dtsch Med Wschr 92:2341–2343; Germ Med Monthly XIII:1–3

Simpson DIH (1969) Marburg virus disease: experimental infection of monkeys. Lab Anim H 4:149–154

Slenczka W, Wolff G (1971) Biological properties of the Marburg virus. In: Martini GA, Siegert R (eds) Marburg virus disease. Springer, Berlin Heidelberg New York, pp 105–108

Slenczka W (1969a) Zum Verhalten des Marburg Virus in Vero-Zellen. Zbl Bakt I Abt Ref 215:545–546

Slenczka W (1969b) Growth of Marburg virus in Vero cells. Lab Anim H 4:143–147

Slenczka W, Peters D (1977) Ebola-Virus. Ein neuer Vertreter der Marburg-Virus-Gruppe. Tropenmed Parasit 28:260

Slenczka W, Piepenburg G, Siegert R (1968) Antigen-Nachweis des Marburg Virus in den Organen infizierter Meerschweinchen durch Immunfluoreszenz. Dtsch Med Wschr 83:612–616; German Med Monthly XIII:524–529

Slenczka W, Radsak K (1980) Polypeptides of Marburg- and Ebola-viruses. Fourth Workshop of the Section Virology of the DGHM. Zbl Bakt Hyg I Abt Orig 248:21 [Abstr]

Slenczka W, Rietschel M, Hofmann C, Sixl W (1984) Seroepidemiologische Untersuchungen über das Vorkommen von Antikörpern gegen Marburg und Ebola-Virus in Afrika. Mitt Oesterr Ges Tropenmed Parasitol 6:53–60

Slenczka W, Siegert R, Wolff G (1970) Nachweis komplementbindender Antikörper des Marburg Virus bei 22 Patienten mit einem Zellkultur-Antigen. Arch Ges Virusforsch 31:71–80

Slenczka W, Wolff G, Siegert R (1971) A critical study of monkey sera for the presence of antibody against the Marburg virus. Amercian J Epidem 93:496–505

Smith CEG, Simpson DIH, Bowen ETW, Zlotnik I (1967) Fatal human disease from vervet monkeys. Lancet ii:1119–1121

Smith DH (1982) Marburg virus disease in Kenya. Lancet 1:816–820

Stille W, Böhle E (1971) Clinical course and prognosis of Marburg virus (green monkey) disease. In: Martini GA, Siegert R (eds) Marburg virus disease. Springer, Berlin Heidelberg New York, pp 10–18

Stille W, Böhle E, Helm E, van Rey W, Siede W (1968) Über eine durch Cercopithecus aethiops übertragene Infektionskrankheit (Grüne Meerkatzen-Krankheit). Dtsch Med Wschr 93:572–582

Stoikovic LJ, Bordjoski M, Glicic A (1971) Two cases of cercopithecus monkeys-associated haemorrhagic fever (some data on etiology, epidemiology and epizootology). In: Martini GA, Siegert R (eds) Marburg virus disease. Springer, Berlin Heidelberg New York, pp 24–33

Strickland-Cholmley M, Malherbe H (1971) Examination of South African Primates for the presence of Marburg virus. In: Martini GA, Siegert R (eds) Marburg virus disease. Springer, Berlin Heidelberg New York, pp 195–202

Todorovitch K, Mocitch M, Klasnja R (1971) Clinical picture of two patients infected by the Marburg vervet virus. In: Martini GA, Siegert R (eds) Marburg virus disease. Springer, Berlin Heidelberg New York, pp 19–23

van der Groen G (1982) Use of betapropiolactone-inactivated Ebola Marburg and Lassa intracellular antigens in immunofluorescent antibody assay. Ann Soc Belg Med Trop 62:49–54

Wulff H, Slenczka W, Gear JHS (1978) Early detection of antigen and estimation of virus yield in specimens from patients with Marburg virus disease. Bull WHO 56:633–639

Zlotnik I, Simpson DIH (1969) The pathology of experimental monkey disease in hamsters. Brit J Exp Path 50:393–399

Zlotnik I, Simpson DIH, Howard DMR (1968a) Structure of the vervet monkey disease agent. Lancet ii:26–s28

Zlotnik I, Simpson DIH, Bright WF, Bowen ETW, Batter Hatton DEE (1968b) Growth of vervet monkey disease agent in BHK cell cultures. Brit J Exp Path 49:311–314

Ebola Virus Outbreaks in the Ivory Coast and Liberia, 1994–1995

B. LE GUENNO[1], P. FORMENTY[2] and C. BOESCH[3]

1	Introduction	77
2	Background	78
3	Identification of the Ivory Coast Virus	78
4	Human Infection with Ebola Ivory Coast Virus	79
4.1	Clinical Aspects	79
4.2	Laboratory Diagnosis	80
4.3	Comparison with Other Hemorrhagic Fevers	80
4.4	Transmission Among Humans	81
4.5	The Liberian Case	81
5	Natural History of Ebola Ivory Coast Virus	82
5.1	Pathogenesis in Chimpanzees	82
5.2	Natural Transmission Among Chimpanzees	82
5.3	The Colobus Hypothesis	83
6	Perspective for the Reservoir Search: The WHO Taï Project	83
References		84

1 Introduction

In the Taï National Park, Ivory Coast, the behavior of a community of free-living chimpanzees has been studied since 1979 by a group of ethologists led by C. BOESCH and H. BOESCH (1989). In early November 1994, this team found several corpses of apes and noted the absence of further individuals. An epidemiological survey was conducted by the Laboratoire National d'Appui au Développement de l'Agriculture (LANADA) and the Centre Suisse de Recherches Scientifiques (CSRS) to explain the cause of the deaths. The Pasteur Institute in Paris was contacted for virological support. Their joint efforts led to the identification of a new strain of Ebola virus (LE GUENNO et al. 1995), the clinical and biological description of the

---

[1]WHO Collaborating Center for Arboviruses and Haemorrhagic Fevers, Institut Pasteur, 25 Rue du Dr Roux 75724, Paris Cedex 15, France
[2]Organisation Mondiale de la Santé, 01 BP 2494, Abidjan 01, Ivory Coast
[3]Centre Suisse de Recherches Scientifiques, BP 1303, Abidjan, Ivory Coast

infection in humans, and, for the first time, the collection of objective data about the natural cycle of this virus. In December 1995 another human case was diagnosed 100km south of Taï.

2 Background

The Taï National Park (436,000ha) represents the last and largest area of what remains of the tropical rain forest belt in West Africa. It is located near the Liberian border in the southwest Ivory Coast. This park, classified as a Biosphere Reserve, is one of the UNESCO World Heritage Sites. It is unique in its biological diversity and the presence of many endemic species. Canopy trees average 30m in height and are dominated by emergent trees of 40–60 meters. Average rainfall is 1900mm per year and the average temperature is 25°C. The climate in this area is characterized by a constant humidity and two periods of light rainfall : from December to February, with less than 70mm per month, and again in August, with 140mm.

Hundreds of chimpanzees live in this forest and one of these groups has been followed by C. Boesch. This troop, which numbered 80 animals in 1987, now has 33 individuals. Two episodes of severe mortality were noted, in November 1992 (eight deaths) and then in November 1994 (12 deaths). Several dead chimpanzees with obvious signs of hemorrhage were found, but decomposition was too advanced to collect useful biological samples. One newly dead chimpanzee was discovered on November 16, 1994 and autopsied in the field by three scientists. Formalin-fixed tissues from this animal were sent to France. One of these three scientists, a 34-year-old female who autopsied the chimpanzee, developed a "dengue-like" syndrome on November 24th. She was hospitalized in Abidjan on the 26th for acute fever resistant to anti-malaria treatment.

3 Identification of the Ivory Coast Virus

One of the hypotheses to explain the deaths among the chimpanzees was that they had been infected by a human virus from the team. To test this hypothesis, sera from the ethologists and blood samples from the two older males and one female of the chimpanzee troop were sent to the Pasteur Institute on December 14th for serologic tests. The serum of the sick ethologist, drawn on November 27th during the febrile phase, was among them. All the serologic tests were negative against the antigens for the main African hemorrhagic fever viruses and arboviruses. Despite the less than optimal storage and transport conditions of the febrile patient's serum, we attempted viral isolation by inoculation of Vero E6 (monkey kidney) and AP61 (*Aedes pseudoscutellaris* mosquito) cells. Since no obvious cytopathic effect was noted after 12 days, a blind passage was done. In the Vero E6 subculture, some cells became refractive and detached from the monolayer after 5 days. Immunofluorescent assay (IFA) performed on these cells with the immune ascitic fluids available in the lab were negative, but we had no Ebola reagent then. However, hematoxylin-eosin staining showed large eosinophilic cytoplasmic inclusions in many cells. A patient's late serum from December 16, 1995 gave bright fluorescence on large cytoplasmic inclusions by IFA. Electron microscopy performed on the Vero cells from the same subculture revealed the characteristic morphology of a filovirus. Polyclonal antibodies from rabbit immunized with Ebola Zaire provided by the CDC verified characterization of an Ebola strain. We prepared an ELISA (enzyme-linked immunosorbent assay) antigen by a borate/triton X-100 extraction of the membrane proteins of the infected cells. The IgG titers of the patient's sera both by IFA and ELISA revealed large antigenic differences between the new strain and the three known Ebola viruses. The most reactive of the other antigens was Zaire. However, the IgM antibodies from the patient reacted only with the strain isolated from the patient in an ELISA immunocapture assay. This difference was confirmed by using monoclonal antibodies, then by sequencing. The Zaire strain is the closest with respect to sequence. In our lab we found 8.3% divergence with the nucleotide sequence of the polymerase gene of Zaire 76. SANCHEZ et al. (1996) recently reported 40% divergence in the envelope glycoprotein.

4 Human Infection with Ebola Ivory Coast Virus

4.1 Clinical Aspects

On November 24th (day 1), the patient started shivering with fever. She treated herself for malaria. On day 3, she was admitted to a clinic in Abidjan with persistent chills, headache and myalgia. Physical examination of the abdomen, heart, lung, throat and tongue was normal. On day 4, she was treated with quinine IV for suspected resistant malaria. Her temperature remained around 40°C and quinine was discontinued due to progressive deafness.

On day 5 the patient developed diarrhea, then nausea, vomiting and anorexia. A rash appeared on her left shoulder then became generalized. She also suffered from central nervous disorders such as temporary loss of memory, anxiety, confusion and irritability. Urinary output failed from day 5 to day 7. Blood cultures and smears remained negative for parasites and bacteria. She was rehydrated intravenously. On day 6 there was no improvement in the patient's condition and she was repatriated to Switzerland the next day in an ambulance jet. On admission to the University Hospital of Basel, she was tired but awake. Physical examination revealed tender spleen and liver on palpation; an ultrasound scan of the abdomen was normal. She was treated with ciprofloxacin IV and doxycycline IV for suspected gram-negative sepsis, leptospirosis or rickettsial disease. On day 9 she became afebrile for 2 days, but on day 11 fever recurred, diarrhea changed to constipation and she started to eat normally.

On day 15, she was discharged from the hospital having lost 6 kg (10% of her initial weight). She fully recovered only after 6 weeks. A month after onset, the patient's hair became dry, lost its elasticity and began falling out in large quantities. Hair loss lasted about 3 months.

4.2 Laboratory Diagnosis

Persistent thrombocytopenia and an early lymphopenia followed by neutrophilic hyperleukocytosis were observed. The most notable biochemical finding was ASAT activity threefold higher than ALAT activity; γ-glutamyl-transferase was normal and alkaline phosphatases were slightly elevated. Creatinine kinase was normal but lactate dehydrogenase was 20 times above normal on day 7. Amylase levels increased from day 11 to day 14. After day 15, all values progressively stabilized. Hematuria was observed on day 5 and proteinuria on day 10.

Serologic tests were requested for dengue, hantavirus, hepatitis B, C and E, leptospirosis, rickettsia, brucellosis, Epstein-Barr virus and cytomegalovirus and were negative. Due to the lack of bleeding, diagnosis of hemorrhagic fevers had not been required.

4.3 Comparison with Other Hemorrhagic Fevers

The clinical course, the length of the disease and the signs during convalescence are consistent with the nonlethal Ebola infections described among Yambuku's survivors (SUREAU 1989), an English laboratory infection (EMOND et al. 1977) and the most recent Kikwit and Gabon outbreaks. Loss of memory, central nervous system disorders and loss of hair during the convalescent phase were reported in Ebola and in the other hemorrhagic fevers, Marburg, Lassa and Argentinean HF. The loss of weight is a constant feature of Ebola infection due to asthenia, anorexia and probably also to a direct action of the virus on tumor necrosis factor (TNF) production.

Raised levels of transaminases but normal levels of GGT and bilirubin indicate that hepatic damage was not major. These findings suggest extra-hepatic targets of infection. Higher ASAT than ALAT activity may suggest a myocardial involvement during the first week of illness. This correlates with the increased LDH and the initial above normal values of creatinine kinase. Damage to the myocardium was reported in natural Marburg and Ebola infection and in inoculated monkeys (GEAR et al. 1975; FISHER-HOCH et al. 1983). Slightly elevated creatinine levels suggest dehydration rather than major impairment of renal function.

The mild form of the disease may be due to the strain and/or to the mode of contamination and/or to the biological response of the patient. Regarding the mortality among chimpanzees, Ebola Ivory Coast is certainly a dangerous virus.

4.4 Transmission Among Humans

The patient was probably infected during autopsy of the Ebola-positive chimpanzee. All three researchers wore gloves but not masks or gowns. Two of them wore latex examination gloves but the patient wore household gloves that were in poor condition. No wounds or punctures were noted. It is therefore highly probable that she became contaminated as a result of direct contact with chimpanzee blood either on the hand or by projection of droplets onto the face. Contact persons were defined as those having had direct face-to-face contact with the patient either 2 days before onset or during illness. Although containment measures were not strict and there were no specific precautions taken during laboratory tests, no secondary cases appeared among the contacts. Sera were sampled from 18 contact persons in the Ivory Coast and 52 in Switzerland. No antibodies against Ebola were found in any of the 70 sera. For transmission, Ebola virus requires physical contact with the patient or contact with blood or secretions, as described in 1976 (WHO 1978) and confirmed during the Kikwit outbreak. Universal precautions and barrier nursing are effective in preventing infection.

4.5 The Liberian Case

Due to the civil war, refugees have left Liberia to settle in the Ivory Coast. The quality of hygiene is poor and cholera circulates on both sides of the border. In a cholera camp organized by *Médecins sans Frontières* (MSF), a patient suspected of having yellow fever was hospitalized by the end of November 1995. He complained of fever, nausea and diarrhea on November 15th and came from the town of Pleebo. Once in the MSF camp, he presented with hematuria. A blood sample drawn on the 29th was sent to the Pasteur Institute and found positive for IgM and IgG against Ebola, but viral isolation and polymerase chain reaction (PCR) were negative. These results are consistent with the date of sampling, indicating that the patient was in the convalescent state. He was the chief of a troop of 17 warriors, living generally in the bush. A survey in the town of Pleebo and the region on both sides of the border did not reveal any other cases. All the blood samples obtained from contacts in Liberia and health workers in the Ivory Coast were negative for Ebola antibodies.

5 Natural History of Ebola Ivory Coast Virus

Long-term study of the chimpanzees has been centered on behavioral and ecological questions. Their home range is an area about 27km^2 located in the western part of the park. The observational methods used are noninvasive. Observers avoid all physical contact with the chimpanzees, keeping a minimal distance of 3–4meters. Target individuals are followed on a daily basis from dawn to dusk. Behavioral and social data are collected on the selected individuals, as well as their foraging routes. These data have permitted an analysis of mortality on a long-term basis as well as a precise epidemiological investigation of the 1994 outbreak.

5.1 Pathogenesis in Chimpanzees

The corpses of only two individuals could be observed in good conditions during the 1994 outbreak. A 13-year-old female was found in a fetal position on the forest floor on November 1, 1994. The necropsy revealed that the blood within the heart was brown and noncoagulated. No obvious lesions were seen on the viscera. A 45-month-old female was found lying on her side. The rib cage was full of liquid blood and the lungs were dark red. It is from this individual that tissues from different organs were collected. The liver and the spleen showed obvious hemorrhagic signs and necrosis. The liver lesions consisted of many foci of necrosis, and a small number of single, large, amorphous acidophilic inclusions was seen in the cytoplasm of hepatocytes, near the necrotic foci. These results are similar to those reported from human cases autopsied during the 1976 Zaire and Sudan outbreaks and from experimental inoculation of monkeys with Ebola. The Ebola-specific immunohistochemical staining of the liver showed small aggregates mainly in the hepatocytes near the portal ducts and sometimes in the Kupfer cells. Pieces of liver and spleen kept in formalin were not preserved well enough to allow electron microscopy.

5.2 Natural Transmission Among Chimpanzees

A demographic analysis revealed that the community, after a period of stability with 80 individuals, had recently suffered from a constant decrease and was down to 32 individuals. For the period 1991–1994, we observed a clear peak in the annual mortality rate of the community members, with 27% in 1992 and 37% in 1994. In 1991 and 1993, the annual mortality rate was only 4% and 9%, respectively. From 1991 to 1994, the monthly mortality record shows a significant concentration in November which was related to the 1992 and 1994 outbreaks. During November 1992, eight cases were recorded (attack rate 17%) and 12 cases in October–November 1994 (attack rate 28%). During this period the epidemic developed in three waves. Four cases were recovered from October 24th to 30th, followed 3 days later by a second wave with six cases which lasted 12days. Finally, two late cases occurred 10 days after the end of the second episode.

The corpses showed a clustered distribution within a radius of 1.5km inside the home range, but that zone also corresponds to the most frequently used core area of the territory. Both sexes were affected, and attack rates were higher for apes older than 10 years. The nature of the epidemic curve, with three waves, indicated a point source or an intermittent point source contamination. Usual contacts (taking care of a sick animal, touching a corpse or grooming from a case) were not risk factors. It appears then improbable that the epidemic might have spread by simple contact among these primates. The consumption of meat during September–October was the highest risk factor. Moreover, the risk increased with the quantity of meat eaten. Our analysis showed that chimpanzees might have been infected from mammal prey eaten at this period of the year. Chimpanzees regularly hunt monkeys throughout the year, and the annual average is one hunt per week. Hunting season is from August to November with one hunt per day. Western Red Colobus monkeys (*Colobus badius*) are the major prey. A hunting party occurred on October 19 – when a young Red Colobus was eaten, 6days before the first mortality wave. The main consumers were among the victims of this wave. For the third wave, a hunting party was recorded on November 17th and two adult males that had eaten parts of an adult Red Colobus disappeared 7 days later. Since during the end of October we mainly followed females or selected individuals who hunt less frequently than males, some hunts were missed, as was a possible hunting party responsible for the second epidemic wave.

5.3 The Colobus Hypothesis

Colobus could have been the source of Ebola infection for apes. This does not mean that Colobus were Ebola carriers. If so, we would observe epidemics all year round or during the entire hunting season. Colobus might be an intermediate host, itself contaminated by the true reservoir during October and November. A study on Colobus behavior may help to identify the species that is the reservoir. Colobus are strictly vegetarian monkeys (10% fruits, 90% leaves), their home range is about 1km^2 and they live in multi-male groups of 60–100 individuals, spending most of their time in the canopy. They could be contaminated from food, from a specific arthropod of the upper strata, or through contact with rodents, bats or other vertebrate excretions.

6 Perspective for the Reservoir Search: The WHO Taï Project

Our results have permitted us to define a small area where Ebola virus circulates. Other key features are the annual and seasonal cycles of infection. It seems that the reservoir abundance varies quickly or that the conditions for contact between monkeys and the reservoir exist only during a short period of time. A similar seasonal pattern was also observed with other recent outbreaks. The Gabon 1994 outbreak began in November in Mekouka, the first case in Kikwit was an individual infected by end of December 1994, the Liberian patient fell ill in November 1995; the chimpanzee responsible for the Gabon 1996 outbreak in Mayibout was infected by the end of December 1995. Surveys of mortality among chimpanzees and Colobus are ongoing. Platforms have been built at different levels in the forest to study the relationships and population dynamics of species close to these primates. A long-term ecological study of the canopy is the only way to reduce the number of candidate reservoir species. Hundreds of rodents, bats, and arthropods have been caught around Nzara, Yambuku, Tandala, and now Kikwit, without thus far providing any evidence for a specific reservoir. The conclusions obtained from the data collected during the chimpanzee survey demonstrate the value of ethology as a discipline that may help to identify the natural host of Ebola virus.

References

Boesch C, Boesch H (1989) Hunting behavior of wild chimpanzees in the Taï National Park. Am J Phys Anthrop 78:547–573
Emond R, Evans B, Bowen E, Lloyd G (1977) A case of Ebola virus infection. Brit Med J 541:44
Fisher-Hoch SP, Platt GS, Lloyd G, Simpson DIH, Neild GH, Barret AJ (1983) Haematological and biochemical monitoring of Ebola infection in rhesus monkeys: implications for patient management. Lancet i:1055–1058
Gear JSS, Cassel GA, Gear AJ, Trappler B, Clausen L et al. (1975) Outbreak of Marburg virus disease in Johannesburg. Brit Med J 4:489–493
Le Guenno B, Formenty P, Wyers M, Gounon P, Walker F, Boesch C (1995) Isolation and partial characterization of a new strain of Ebola virus. Lancet i:664–666
WHO/International Study Team (1978) Ebola haemorrhagic fever in Zaire; Ebola haemorrhagic fever in Sudan 1976. Bull WHO 56:247–293
Sanchez A, Trappier S, Mahy B, Peters CJ, Nichol S (1996) The virion glycoproteins of Ebola viruses are encoded in two reading frames and expressed through trancriptional editing. Proc Natl Acad Sci USA 93:3602–3607
Sureau P (1989) Firsthand clinical observations of haemorrhagic manifestations in Ebola haemorrhagic fever in Zaire. Rev Inf Dis 11:790–793

Filovirus Diseases

C.J. PETERS and A.S. KHAN

1 Introduction	85
2 Epidemics	86
2.1 Case Fatality Proportion	87
2.2 Case to Infection Ratios	87
3 The Acute Disease	88
3.1 Marburg Virus	88
3.2 Ebola Virus	88
4 Convalescence	89
5 Differential Diagnosis	90
6 Etiologic Diagnosis	90
7 Therapy	91
8 Prevention	92
9 Concluding Remarks	93
References	93

1 Introduction

The filoviruses comprise a unique virus family that occasionally infects primates, usually with manifest disease. However, only limited observations on human disease are available, and they come from different strains of virus and different routes and doses of infectious agents. Even with these caveats, there is sufficient information to outline the general pattern of impact of infection on the primate organism. Most of the cases have occurred in situations that have only permitted the most rudimentary clinical observations, leading to a good deal of uncertainty as to the details of the disease processes and sequelae. There are four subtypes of Ebola virus and two of Marburg virus, but the overall clinical disease is similar among the viral species. We will attempt to synthesize the patterns of clinical impact on hu-

Division of Viral and Rickettsial Diseases, National Center for Infectious Diseases, Centers for Disease Control and Prevention, 1600 Clifton Road, Atlanta, GA 30333, USA

mans but do so with the understanding that there are gaps and generalizations that certainly require much further work to define.

2 Epidemics

The description of human filovirus disease begins with identification of cases. Ascertainment of human infections is best in epidemic situations. Isolated cases with one or two secondary infections provide a window on clinical manifestations but are certainly biased toward the more severe index cases.

Filovirus infection must be confirmed by laboratory testing. Usually, acute cases yield evidence of the presence of virus by isolation, presence of antigen, or amplification of RNA. Unfortunately, antibody detection tests to document past infection have usually been performed with the fluorescent antibody technique; these tests have been misleading and even apparent seroconversions must be regarded with doubt (JOHNSON BK et al. 1986; KENYON et al. 1994; TEEPE et al. 1983). ELISA antibody detection tests are much more reliable and are under evaluation to see if they have unexpected limitations (KSIAZEK et al. 1998a,b).

Because of these considerations, we have concentrated on the major epidemics, or at least the best documented examples of human disease, to describe the spectrum of clinical disease (Table 1). Sporadic human cases have been used to expand our understanding of the diseases but have only been included if they are documented by virus isolation from the index case and/or secondary cases in close contact with the index case (EMOND et al.1977; GEAR et al. 1975; HEYMANN et al. 1980; JOHNSON ED et al. 1996; KIKIFOROV et al. 1994; LE GUENNO et al. 1995; SMITH et al. 1982).

Table 1. Filovirus epidemics of humans

Virus	Country (year)	Case fatality proportion (number of cases)	Reference
Marburg	Germany, Yugoslavia, (1967)	23% (31)	MARTINI 1971
Ebola (Zaire subtype)	Zaire (1976)	88% (318)	WHO 1978b
	Zaire (1995)	81% (315)	KHAN 1998
	Gabon (1994)	64% (44)	AMBLARD 1997
	Gabon (1996)	57% (37)	WHO 1996
	Gabon (1996)	75% (60)	WHO 1997
Ebola (Sudan subtype)	Sudan (1976)	53% (284)	WHO 1978a
	Sudan (1979)	65% (34)	BARON 1983
Ebola (Reston subtype)	USA (1989 1990)	0% (4)	CDC 1990a,b

2.1 Case Fatality Proportion

Case fatality proportions are best-defined in the original 1967 Marburg outbreak, which led to the identification of this family of viruses, in which 23% of the 31 patients died (MARTINI and SIEGERT 1971). This is close to the three (43%) deaths among seven subsequently reported sporadic and secondary contact cases. The Sudan subtype of Ebola virus had a 53% case fatality among 284 patients in 1976 and 65% case fatality among 34 patients in 1979 (BARON et al. 1983; WHO 1978a). The individual with Ebola-Cote d' Ivoire subtype infection documented by virus isolation and antigen detection survived, as did a Liberian patient with positive IgM antibodies but no virus detection (LE GUENNO et al. 1995; WHO 1995).

The highest case fatality has consistently been observed with Zaire subtype-infected patients. In 1976, 88% of 318 patients died, and all patients infected by injections in the hospital setting had a fatal outcome (WHO 1978b). The 1995 Kikwit episode was associated with a 81% case fatality among 315 patients (KHAN et al. 1998). Case fatality proportions in the three Gabon epidemics, caused by the Ebola-Zaire subtype, were reported to be 57%–75% with an overall figure of 67% among 141 patients (AMBLARD et al. 1997; GEORGE-COURBOT et al. 1997a,b; WHO 1997a,b).

Clearly, the filovirus least pathogenic for humans is the Reston subtype of Ebola, which caused four well-documented infections in Reston, Virginia, USA in 1989–1990 (CENTERS FOR DISEASE CONTROL 1990a,b). None were symptomatic. One of the infections followed a scalpel cut during necropsy of a monkey and was studied with serial blood samples. Although virus was isolated from this patient's blood samples (P.E. Rollin, unpublished observations), it was only in very low titers, supporting the idea that the human host is more restrictive for this Ebola subtype than for the other subtypes.

2.2 Case to Infection Ratios

The difficulty in interpreting indirect fluorescent antibody (IFA) tests has hampered the determination of this basic ratio. In the outbreak setting there has been no epidemiological suggestion that subclinical infection is epidemiologically important. The clearest data followed the Ebola (Zaire subtype) epidemics in the country of Zaire in 1976 and 1995; IFA tests in 1976 failed to show any definite evidence of acute infection without serious disease and ELISA IgG tests in 1995 documented only two such patients among households with known infected patients (KHAN et al. 1998; ROWE et al. 1998; WHO 1978b). In Sudan subtype epidemics it has been suggested that IFA antibodies in the apparently normal population represented distant or subclinical infection, but these data are not interpretable (WHO 1978a).

The situation with the Reston subtype of Ebola virus exemplifies the difficulty in interpretation of serological data. Human and monkey populations in the USA and Asia were reported to have IFA titers as high as 1:1000, leading to much speculation (CDC 1990a; HAYES et al. 1992). Application of western blot tests failed to clearly resolve the problem (ELLIOTT et al. 1993). Later development of a sensitive and more specific ELISA for IgG antibodies led to a realization that all available sera from documented Ebola infections with Reston, Zaire, or Sudan subtypes were positive in this test, that detectable antibodies lasted for a longer time than post-infection IFA antibodies, and that the large number of primates (human and nonhuman) with IFA antibodies to Ebola virus were negative in the ELISA (KSIAZEK et al. 1998a,b). Thus, it is not possible to say whether the isolated high IFA titers found in otherwise asymptomatic humans do indeed represent prior infection unless a validated ELISA test or another confirmatory test is available.

3 The Acute Disease

3.1 Marburg Virus

There is a highly characteristic clinical syndrome with sudden onset of fever and headache accompanied by vomiting, and non-bloody diarrhea (GEAR et al. 1975; JOHNSON ED et al. 1996; KIKIFOROV et al. 1994; MARTINI 1971; SMITH et al. 1982; STILLE and BÖHLE 1971; TODOROVITCH et al. 1971). An enanthema and drowsiness are common, with variable evidence of conjunctivitis, myalgia, and lymphadenopathy. The most reliable diagnostic sign is a characteristic maculopapular rash during the end of the first week of illness with late desquamation. A severe hemorrhagic diathesis, leading to hemorrhage from gums, the nose, and puncture sites as well as the gastrointestinal tract, is noted in approximately a third of patients during the period of the lowest thrombocyte count, approximately days 6–12.

Characteristic laboratory abnormalities include leukopenia with a left shift and circulating immunoblasts. All patients have thrombocytopenia and minor abnormalities on routine coagulation testing that are considered to be insufficient to describe the degree of the bleeding diathesis. All patients had an increase in sAST and ALT, ALT greater than AST, with normal bilirubin, alkaline phosphatase, and creatine phosphokinase levels. An elevated amylase has also been reported with significant elevations of blood urea nitrogen and proteinuria characteristic of the most severe cases.

3.2 Ebola Virus

Beginning with a sudden onset of fever, headache, and joint and muscle aches, the disease soon causes initially non-bloody diarrhea (81%), vomiting (59%), pain and dryness of the throat (63%), and abdominal pain (BARON et al. 1983; WHO 1978a,b; KHAN et al. 1998). Chest pain (83%) was a characteristic feature of Ebola-Sudan infected patients but uncommon in Marburg or Ebola-Zaire infected patients. Patients were usually hospitalized by the fifth day of illness with deep set eyes, ghost-like expressionless faces, and extreme lethargy being prominently noted in the 1976 Ebola-Sudan outbreak. A rubeola-like rash is difficult to evaluate in dark-skinned people but occurs in half, by either being noticeable or by the later onset of desquamation, about the fifth day of illness. Hemorrhagic manifestations, usually gastrointestinal and rarely urogenital, are common (71%–78%), being present in half of the recovered patients and in almost all of the fatal cases.

During the 1995 Ebola-Zaire outbreak 103 patients were prospectively studied and were characterized by a similar biphasic clinical syndrome with emphasis on the asthenia as an early sign and the later evolution of hemorrhagic manifestations, hiccups (15%), and, less frequently, neurologic signs (BWAKA et al. 1998). This series reported overall bleeding as 45% and that terminally ill patients had obtundation, anuria, shock, tachypnea, and normothermia.

The damage to the primate host occurs through several mechanisms, including rampant cytopathic effects on parenchymal and endothelial cells, macrophage infection, and release of mediators (FELDMANN et al. 1996; SCHNITTLER et al. 1993; VILLINGER et al. 1998; ZAKI and PETERS 1997).

4 Convalescence

Convalescence has been reported to be prolonged after filoviral infection with resolution of fatigue, anorexia, and cachexia that may take many weeks; tender livers and other hepatic abnormalities have also been reported in the Marburg patients but may have been due to transfusion-associated hepatitides (MARTINI 1971; STILLE and BÖHLE 1971). Myalgia and arthralgia have continued to be reported by Ebola (Zaire subtype) infected patients 2 years after acute illness (ROWE et al. 1998).

Filoviruses may remain sequestered in protected immunologic sites, such as the uveal and seminal tracts of convalescent patients, and cause myelitis, uveitis or orchitis late in convalescence; therefore, these patients necessitate appropriate precautions after hospitalization. Marburg virus was isolated from the semen of one patient 2 months (SMITH et al. 1982) following recovery and from another patient 83 days following disease onset (MARTINI and SCHMIDT 1968) from a spouse implicated in a sexually transmitted case (MARTINI 1971).

Ebola virus has been isolated from semen 61 days but not 74 days after onset of illness in a infected laboratory worker (EMOND et al. 1977), and detected by molecular techniques 101 days after onset in a Zairian patient (ROWE et al. 1998). Marburg virus has also been isolated from the anterior chamber of the eye from a convalescent Marburg patient with uveitis 80 days after the acute disease began (GEAR et al. 1975). Thus, physicians such as ophthalmologists and urologists should recognize that post-infectious sequelae in these patients are often associated with the recrudescence of residual virus.

5 Differential Diagnosis

The differential diagnosis is dependent on the stage of clinical illness, on presentation and its evolution. Needless to say, the differential diagnosis of an acute febrile illness with headache and diarrhea is broader than one that also includes hemorrhagic manifestations. Causes of viral hemorrhagic fevers include dengue, yellow fever, Lassa and South American arenaviral hemorrhagic fevers, Congo-Crimean hemorrhagic fever, Rift Valley fever, and hemorrhagic fever with renal syndrome (ZAKI and PETERS 1997).

The travel and occupational history is critical in narrowing the diagnoses as are reports of clusters of illness or person-to-person transmission. Contact with severely ill humans or with sick or dead nonhuman primates are relevant as is a history of injections in an African hospital. It is imperative that the concern for a filoviral infection not deter an immediate assessment of treatable and more common causes of a febrile illness among travelers such as malaria, typhoid fever, leptospirosis, borelliosis, septicemic plague, tick typhus, and dysentery.

6 Etiologic Diagnosis

All viral hemorrhagic fevers require a specific etiologic diagnosis due to differences in potential for nosocomial transmission and therapeutic options. During the acute phase of illness, virus, viral antigen, or viral RNA should be sought in serum or blood. Most filoviruses are thought to be readily isolated from serum in Vero cell monolayers. Cytopathic effects may be evident or the presence of virus may be sought by electron microscopy of cell culture supernatants, IFA tests on cells, or other techniques (PETERS et al. 1996). Isolation in guinea pigs is possible for most strains but may require several passages until fever, weight loss, or overt disease appear. The Sudan subtype of Ebola virus has been reported to be difficult to isolate and requires passage in cells or guinea pigs (McCORMICK et al. 1983). Only the Zaire subtype of Ebola has been reported to induce lethal disease in suckling mice. Detection of Ebola antigens by an ELISA test is rapid and sufficiently sensitive for most purposes although lower concentrations of viral RNA may be detectable in some circumstances by reverse transcriptase polymerase chain reaction (RT-PCR) (KSIAZEK et al. 1998a,b; ROLLIN et al. 1998; SANCHEZ et al. 1998).

The IFA test becomes positive when the patient improves, around day 8–10, but these tests can be confusing even in acute cases (KENYON et al. 1994). IgM capture ELISA tests are usually positive early in convalescence and can be coupled with rising IgG antibody titers for a more secure diagnosis (KSIAZEK et al. 1998a,b).

7 Therapy

No specific therapy is currently available for filoviral infections. Studies in cell culture and in nonhuman primates and/or guinea pigs have shown no efficacy for interferon-α, convalescent plasma, or ribavirin (JAHRLING et al. 1996, 1998; PETERS et al. 1996). Treatment with convalescent plasma and interferon of an accidentally infected laboratory worker in England was associated with survival, but clinical improvement occurred at the time of expected recovery and blood virus assays were not precise, so it is difficult to relate the therapy to the outcome (EMOND et al. 1977). Similarly, an exposed laboratory worker in Russia received hyperimmune horse serum with recovery, but his lack of active seroconversion casts some doubt as to whether he was actually infected (KUDOYAROVA-ZUBAVICHENE et al. 1998). Towards the end of the 1995 outbreak in Kikwit, Zaire, patients were given whole blood from survivors with apparent improvement in survival (MUPAPA et al. 1998). Unfortunately, no real conclusions can be drawn from these data because there was no control for predictors of survival including day of treatment. The possible role of whole blood vs frozen convalescent plasma deserves consideration in interpreting that experience.

Prior studies have failed to document high-titered neutralizing antibodies in convalescent sera (VAN DER GROEN et al. 1978; L Rodriguez, personal communication) but hyperimmune horse serum to Ebola virus has in vitro neutralizing capacity and has shown some promise in experimental animals treated within hours after inoculation and may be useful in a laboratory setting after a defined exposure (JAHRLING et al. 1996, 1998; KUDOYAROVA-ZUBAVICHENE et al. 1998). Therapies on the horizon include at least one drug with promise against respiratory syncytial virus which has anti-Ebola activity (HUGGINS et al. 1998) and development of high-affinity monoclonal antibodies using recombinatorial techniques (MARUYAMA et al. 1998).

The lower case fatality with Marburg disease may be an inherent feature of this type of infection or reflect the aggressive medical management of these cases, which has included judicious transfusions of whole blood and replacement of coagulation factors. The extensive pathological damage and widespread evidence of virus replication in patients infected with the Ebola (Zaire subtype) virus suggest that case fatality will continue to be high even with supportive therapy unless some effective antiviral measure becomes available (ZAKI and PETERS 1996). Extreme attention must be paid to correction of fluid and electrolyte abnormalities. The observed endothelial damage provides a sufficient cause to infer disseminated intravascular coagulation, but it is not clear that this is an important contributor to the disease process. Heparin had been used with success in two Marburg virus infected patients for suspect disseminated intravascular coagulation; this therapy should only be attempted when adequate monitoring of coagulation parameters are available.

8 Prevention

Use of recombinant DNA techniques to express filovirus gene products, generated in vitro or synthesized de novo, may offer the safest and the most effective means of inducing protective immune response to filovirus infection (CLEGG and SANCHEZ 1997). It is difficult to envision the nature and magnitude of the target population for these vaccines outside of laboratory workers and health care providers in endemic areas without additional information on the reservoir and the people at risk for infection, even without considering their practicality from an economic and public health perspective. However, development of these products is leading the way in defining the immunological response to infection and the specific immune responses that mediate protection. Knowledge of these mechanisms may in themselves help guide development of therapeutic options for acute infections.

As we have transferred medical technology from the developed world to under developed parts of Africa, some critical aspects have been lost through neglect or poverty. This has been a major source of amplification of Ebola virus in humans as well as having grave implications for the spread of HIV and hepatitis C in Africa. The substitution of plastic disposable syringes for glass syringes and use of disposable needles only make sense if they are employed for a single patient and discarded. It is common knowledge that in underdeveloped countries these are widely reused without sterilization and the problem is compounded by the use of multidose vials for drugs. Most plastic syringes warp under even modest heating and disposable needles do not hold their sharpness and cannot be resharpened, so it is not easy to correct the problem. One solution would be to return to the more durable (and more expensive) glass syringes and needles designed for reuse. Probably a more practical approach would be to develop methods of chemical sterilization that would inactivate the common viruses, be compatible with drugs administered, and be safe for use in humans. This should be combined with locally implementable programs to enhance physician awareness of viral hemorrhagic fevers and permit barrier nursing of suspicious patients (LLOYD et al. 1998).

Wider deployment of laboratory diagnostic capability for filoviruses in Africa could be valuable in the future but the current lack is actually not the problem today. Recognition of the filovirus infections has lagged the initial cases considerably in virtually all the known epidemics. It would seem to make more sense to upgrade the diagnostic capabilities for the common diseases (shigellosis, typhoid, malaria) that are usually not specifically diagnosed in the laboratory in order to enable clinicians to concentrate on the recognition of viral hemorrhagic fevers and institution of inexpensive, locally appropriate barrier nursing with a provisional diagnosis. Then, sustainable and accessible diagnostic methodologies for filoviruses and other uncommon but lethal communicable diseases begin to assume more importance.

9 Concluding Remarks

Filoviruses are zoonotic viruses with unknown reservoirs but occasionally are introduced into primate populations in which they cause serious disease with a high case fatality. The clinical manifestations do not permit differentiation from other viral hemorrhagic fevers in an individual patient. Sensitive and specific laboratory tests exist to permit rapid diagnosis. Unfortunately, there is no specific therapy for the diseases but prompt recognition permits protection of medical staff.

References

Amblard J, Obiang P, Prehaud C, Prehaud C, Bouloy M, Le Guenno B (1997) Identification of the Ebola virus in Gabon in 1994. Lancet 349:181–182
Baron RC, McCormick JB, Zubeir OA (1983) Ebola hemorrhagic fever in southern Sudan: hospital dissemination and intrafamilial spread. Bulletin World Health Organ 6:997–1003
Bwaka MA, Bonnet M-J, Calain P, Colebunders R, De Roo A, Guimard Y, Katwiki KR, Kibadi K, Kipasa MA, Kuvula KJ, Mapanda BB, Massamba M, Mupapa KD, Tamfum-Muyembe JJ, Ndaberey E, Peters CJ, Rollin PE, Van den Enden E (1998) Ebola hemorrhagic fever in Kikwit, Democratic Republic of the Congo (former Zaire): clinical observations. J Infect Dis (in press)
Centers for Disease Control (1990a) Update: filovirus infection among persons with occupational exposure to nonhuman primates. Morb Mortal Wkly Rep 39:266–273
Centers for Disease Control (1990b) Update: filovirus infection in animal handlers. Morb Mortal Wkly Rep 39:221
Clegg JCS, Sanchez A (1997) Vaccines against arenaviruses and filoviruses. In: Levine MM, Woodrow GC, Kaper JB, Cobon GS (eds) New generation vaccines. Marcel Dekker, New York, pp 749–765
Elliott LH, Bauer SP, Perez-Oronoz G, Lloyd ES (1993) Improved specificity of testing methods for filovirus antibodies. J Virol Methods 43:85–100
Emond RTD, Evans B, Bowen ETW, Lloyd G (1977) A case of Ebola virus infection. Brit Med J 2:541–544
Feldmann H, Bugany H, Mahner F, Klenk H-D, Drenckhahn D, Schnittler HJ (1996) Filovirus-induced endothelial leakage triggered by infected monocytes/macrophages. J Virol 70:2208–2214
Gear JSS, Cassel GA, Gear AJ, Trappler B, Clansen L, Meyers AM, Kew MC, Bothwell TH, Sher R, Miller GB, Schneider J, Koornhoff HJ, Gomperts ED, Isaäcson M, and Gear JHS (1975) Outbreak of Marburg virus disease in Johannesburg. Brit Med J 4:489–493
Georges-Courbot MC, Lu CY, Lansoud-Soukate J, Leroy E, Baize S (1997a) Isolation and partial molecular characterisation of a strain of Ebola virus during a recent epidemic of viral haemorrhagic fever in Gabon. Lancet 349:181
Georges-Courbot MC, Sanchez A, Lu CY, et al (1997b) Isolation and phylogenetic characterization of Ebola viruses causing different outbreaks in Gabon. Emerg Infect Dis 1:59–62
Hayes CG, Burans JP, Ksiazek TG, Del Rosario RA, Miranda MEG, Manaloto CR, Barrientos AB, Robles CG, Dayrit MM, Peters CJ (1992) Outbreak of fatal illness among captive macaques in the Philippines caused by an Ebola-related filovirus. Am J Trop Med Hyg 46:664–671
Heymann DL, Weisfeld JS, Webb PA, Johnson KM, Cairns T, Berquist H (1980) Ebola hemorrhagic fever: Tandala, Zaire, 1977-78. J Infect Dis 142:373–376
Huggins J, Zhang Z-X, Bray M (1998) Antiviral drug therapy of filovirus infections: S-Adenosylhomocysteine hyrdrolase inhibitors inhibit Ebola virus in vitro and in a lethal mouse model. J Infect Dis (in press)
Jahrling PB, Geisbert J, Swearengen JR, Jaax GP, Lewis T, Huggins JW, Schmidt JJ, LeDuc JW, Peters CJ (1996) Passive immunization of Ebola virus-infected cynomolgus monkeys with immunoglobulin from hyperimmune horses. Arch Virol [Suppl] 11:135–140

Jahrling PB, Geisbert TW, Geisbert JB, Swearengen JR, Bray M, Jaax NK, Huggins JW, LeDuc JW, Peters CJ (1998) Evaluation of immune globulin and recombinant interferon alpha-2b for treatment of experimental Ebola virus infections. J Infect Dis (in press)

Johnson BK, Ochen D, Oogo S, Gitau LG, Wambui C, Gichogo A, Libondo D, Tukei PM, Johnson ED (1986) Seasonal variation in antibodies against Ebola virus in Kenyan fever patients. Lancet i:1160

Johnson ED, Johnson BK, Silverstein D, Tukei P, Geisbert TW, Sanchez AN, Jahrling PB (1996) Characterization of a new Marburg virus isolated from a 1987 fatal case in Kenya. Arch f Virol [Suppl] 11:101–114

Kenyon RH, Niklasson B, Jahrling PB, Geisbert T, Svensson L, Fryden A, Bengtsson M, Foberg U, Peters CJ (1994) Virologic investigation of a case of suspected hemorrhagic fever. Res Virol 145:397–406

Khan AS, Kweteminga TF, Heymann DL, Le Guenno B, Nabeth P, Kerstiens B, Fleerackers Y, Kilmarx PH, Rodier GR, Nkuku O, Rollin PE, Sanchez A, Zaki SR, Swanepoel R, Tomori O, Nichol ST, Peters CJ, Muyembe-Tamfum JJ, Ksiazek TG, for the Commission de Lutte Contrôle des Epidémies à Kikwit (1998) The reemergence of Ebola hemorrhagic fever, Zaire, 1995. J Infect Dis (in press)

Kikiforov VV, Turovskii II, Kalinin PP, Akinfeeva LA, Katkova LR, Barmin VS, Raibchikova EI, Popkova NI, Shestopalov AM, Nazarov VP (1994) A case of a laboratory infection with Marburg fever (in Russian). Zh Mikrobiologie, Epidemiologie, Immunobiologie 3:104–106

Ksiazek TG, West CP, Rollin PE, Jahrling PB, Peters CJ (1998a) Enzyme-linked immunosorbent assays for the detection of antibodies to Ebola viruses. J Infect Dis (in press)

Ksiazek TG, Rollin PE, Swanepoel R, Burt FJ, Rowe A, Feldmann H, Martin M, Williams AJ, Bressler D, Muyembe T, Khan AS, Peters CJ (1998b) Virological and serological studies of Ebola hemorrhagic fever patients, Kikwit 1995. J Infect Dis (in press)

Kudoyarova-Zubavichene NM, Chepurnov AA, Sergeyev NN, Borisevich IV, Mikhailov VV, Mahclay AA, Netesov SV (1998) Preparation and use of hyperimmune serum for therapy of filoviruses. J Infect Dis (in press)

Le Guenno B, Formentry P, Wyers M, Gounon P, Walker F, Boesch C (1995) Isolation and partial characterisation of a new strain of Ebola virus. Lancet 345:1271–1273

Lloyd ES, Zaki SR, Rollin PE, Ksiazek TG, Calain P, Kondé MK, Kwentaminga T, Bwaka MA, Verchueren E, Kabwau J, Ndambe R, Peters CJ (1998) Long-term disease surveillance in Bandundu Region, Democratic Republic of the Congo (former Zaire): a model for early detection and prevention of Ebola hemorrhagic fever. J Infect Dis (in press)

Martini GA (1971) Marburg virus disease. Clinical syndrome. In: Martini GA, Siegert R (eds) Marburg virus disease. Springer, Berlin, Heidelberg, New York, pp 1–9

Martini GA, Schmidt H (1968) Speratogene Übertragung des Marburg Virus. Klin Wschr 46:391

Martini GA, Siegert R (eds) (1971) Marburg virus disease. Springer, Berlin, Heidelberg, New York

McCormick JB, Bauer SP, Elliott LH, Webb PA, Johnson KM (1983) Biological differences between strains of Ebola virus from Zaire and Sudan. J Infect Dis 147:264–267

Mupapa K, Masamba M, Kibadi K, Kuvula K, Bwaka A, Kipasa M, Muyembe T, on behalf of the International Scientific and Technical Committee (1998) Treatment of Ebola hemorrhagic fever with blood transfusions from convalescent. J Infect Dis (in press)

Maruyama T, Parren PWHI, Sanchez A, Rensink I, Rodriguez LL, Khan AS, Peters CJ, Burton DR (1998) Recombinant human monoclonal antibodies to Ebola virus. J Infect Dis (in press)

Peters CJ, Sanchez A, Rollin PE, Ksiazek TG, Murphy FA (1996) Filoviruses. In: Fields BN, Knipe DM, Howley PM (eds) Fields virology, 3rd edn. Lippincott-Raven, Philadelphia, pp 1161–1176

Rollin PE, Williams RJ, Bressler D, Pearson S, Cottingham M, Pucak G, Sanchez A, Trappier S, Peters RL, Geer P, Zaki S, DeMarcus T, Hendricks K, Kelley M, Simpson D, Geisbert TW, Jahrling PB, Peters CJ, Ksiazek TG (1998) Ebola-Reston virus among quarantined nonhuman primates recently imported from the Philippines to the United States. J Infect Dis (in press)

Rowe A, Bertolli JM, Ksiazek TG, Rodriguez LL, Peters CJ, Khan AS (1998) Follow up of convalescent Ebola-Zaire virus infected-patients. J Infect Dis (in press)

Sanchez A, Ksiazek TG, Rollin PE, Miranda MEG, Trappier SG, Khan AS, Peters CJ, Nichol ST (1998) Detection and molecular characterization of Ebola viruses causing disease in human and nonhuman primates. J Infect Dis (in press)

Schnittler HJ, Mahner F, Drenckhahn D, Klenk H-D, Feldmann H (1993) Replication of Marburg virus in human endothelial cells. A possible mechanism for the development of viral hemorrhagic disease. J Clin Invest 91:1301–1309

Smith DH, Johnson BK, Isaäcson M, Swanapoel R, Johnson KM, Kiley M, Bagshawe A, Siongok T, Keruga WK (1982) Marburg virus disease in Kenya. Lancet 1:816–820

Stille W, Böhle E (1971) Clinical course and prognosis of Marburg virus ("Green-Monkey") disease. In: Martini GA, Siegert R (eds) Marburg virus disease. Springer, Berlin, Heidelberg, New York, pp 10–18

Teepe RGC, Johnson BK, Ocheng D, Gichogo A, Langatt A, Ngindu A, Kiley MP, Johnson KM, McCormick JB (1983) A probable case of Ebola virus hemorrhagic fever in Kenya. East Afr Med J 60:718–722

Todorovitch K, Mocitch M, Klašnja R (1971) Clinical picture of two patients infected by the Marburg Vervet virus. In: Martini GA, Siegert R (eds) Marburg virus disease. Springer, Berlin, Heidelberg, New York, pp 19–23

van der Groen G, Webb PA, Johnson KM, Lange J, Lindsey H, Eliott L (1978) Growth of Lassa and Ebola viruses in different cell lines. In: Pattyn SR (ed) Ebola virus haemorrhagic fever. Elsevier/North-Holland, Amsterdam, pp 255–260

Villinger F, Rollin PE, Brar SS, Chikkala NF, Winter J, Sundstrom JB, Zaki SR, Swanepoel R, Ansari AA, Peters CJ (1998) Markedly elevated levels of IFN-γ/α, IL-2, IL-10 and TNF-α associated with fatal Ebola virus infection. J Infect Dis (in press)

World Health Organization (1978a) Ebola haemorrhagic fever in Sudan, 1976. Report of a World Health Organization International Study Team. Bull World Health Organ 56:247–270

World Health Organization (1978b). Ebola haemorrhagic fever in Zaire, 1976. Report of an International Commission. Bull World Health Organ 56:271–293

World Health Organization (1995) Ebola haemorrhagic fever — confirmed case in Côte d'Ivoire and suspect cases in Liberia. Wkly Epidemiol Rec 70:359

World Health Organization (1997a) Outbreak of Ebola haemorrhagic fever in Gabon officially declared over. Wkly Epidemiol Rec 71:125–126

World Health Organization (1997b) Ebola haemorrhagic fever. Wkly Epidemiol Rec 72:7

Zaki SR, Peters CJ (1997) Viral hemorrhagic fevers. In: Connor DH, Chandler FW, Schwartz DA, Manz HJ, Lack EE (eds) The pathology of infectious diseases. Appleton and Lange, Norwalk, CN, pp 347–364

Pathologic Features of Filovirus Infections in Humans

S.R. ZAKI and C.S. GOLDSMITH

1 Introduction	97
2 Pathology and Pathogenesis	99
3 Diagnosis	106
References	115

1 Introduction

Ebola and Marburg viruses are members of the unique negative-strand RNA virus family, the *Filoviridae* (PETERS et al. 1996) and are the cause of a severe and often fatal viral hemorrhagic fever (HF). The term "hemorrhagic fever viruses" is reserved for a special group of viruses transmitted to humans by arthropods and rodents. Viruses associated with the HF syndrome belong to four different families: *Arenaviridae*, *Bunyaviridae*, *Flaviviridae*, and *Filoviridae* (Table 1).

In 1967, an outbreak of HF with a 22% case-fatality rate occurred in Germany and Yugoslavia among laboratory workers and attending medical staff. The disease was linked to exposure to African green monkeys that had been imported from Uganda and were infected with a previously unknown pathogen. The causative agent, Marburg virus, named after the city in Germany where the main outbreak occurred, would be the first recognized member of a new family of viruses, the *Filoviridae*. Since then, only sporadic cases of Marburg virus HF have been recognized in sub-Saharan Africa, as well as a single case of accidental laboratory infection in Russia.

The second member of the family *Filoviridae*, Ebola virus, was first encountered during concurrent outbreaks in Zaire and Sudan in 1976 involving more than 550 cases, with case-fatality rates of 88% in Zaire and 53% in Sudan (WORLD HEALTH ORGANIZATION 1976, 1978). The virus was named after a small river near the site of the 1976 epidemic in Zaire. A smaller outbreak occurred in Sudan in

---

Infectious Disease Pathology Activity, Division of Viral and Rickettsial Diseases, National Center for Infectious Diseases, Centers for Disease Control and Prevention, 1600 Clifton Road, Atlanta, GA 30333, USA

Table 1. Virus families which cause viral hemorrhagic fever

Filoviridae	Marburg
	Ebola
	Zaire
	Sudan
	Reston
	Ivory Coast
Arenaviridae	Junin
	Machupo
	Guanarito
	Sabia
	Lassa
Bunyaviridae	*Nairovirus*
	Crimean-Congo hemorrhagic fever
	Phlebovirus
	Rift Valley fever
	Hantavirus
	Hantaan, Puumala, Seoul, others
	Sin Nombre, Black Creek Canal, Bayou, and others
Flaviviridae	Mosquito-borne
	Yellow fever
	Dengue viruses 1–4
	Tick-borne
	Kyasanur Forest disease
	Omsk hemorrhagic fever

1979 involving 34 cases with 22 fatalities (BARON et al. 1983). No human infections with Ebola virus were recognized for a span of 15 years, until a veterinarian performing a necropsy on a chimpanzee in the Ivory Coast in 1994 contracted the disease (LE GUENNO et al. 1995). Also in 1994 a small outbreak of Ebola virus HF, which was initially thought to be yellow fever, occurred in Gabon (GEORGES-COURBOT et al. 1997). Subsequently, Ebola virus has caused large outbreaks in Zaire in 1995 (CENTERS FOR DISEASE CONTROL AND PREVENTION 1995) and in Gabon in 1996 (WORLD HEALTH ORGANIZATION 1996). By the time both of these recent outbreaks were over, more than 265 fatalities had occurred among the 350 recorded cases.

A few excellent original studies describing the pathologic features of human filovirus infections have been reported (MURPHY 1978; INTERNATIONAL STUDY TEAM 1978; DIETRICH et al. 1978; ELLIS et al. 1978b; RIPPEY et al. 1984). However, for the most part these earlier studies were limited to examinations of specimens from small numbers of patients, largely because of the risk of hemorrhage associated with biopsies and because of biosafety concerns during autopsy (SHIEH et al. 1998). Therefore, during the 1995 Ebola virus HF outbreak in Kikwit, Zaire, an effort was made to collect a variety of tissues at autopsy from a large number of patients (ZAKI et al. 1996a,b; GOLDSMITH et al. 1997). These tissues, as well as archival tissues of the 1976 Ebola virus HF outbreak in Zaire and the 1980 Marburg virus HF cases in Kenya, are the basis for this review on the pathology of filovirus infection in humans. In addition to the light microscopic and ultrastructural characteristics of the diseases, the immunohistochemical distribution of antigens in tissues and pathophysiology of these viral infections will be described.

2 Pathology and Pathogenesis

Laboratory diagnosis and confirmation of filovirus infections can be accomplished by the detection of viral antigens and antibodies by enzyme-linked immunosorbent assay (ELISA), immunofluorescence, and immunohistochemistry (IHC), or by detection of viral RNA by reverse transcription-polymerase chain reaction (RT-PCR) (PETERS et al. 1996; ZAKI and PETERS 1997). Diagnosis can also be accomplished by isolation of the virus from body fluids and tissues by co-culturing with Vero E6 tissue culture cells. When examined with the electron microscope, the infected culture cells are seen to contain characteristic filoviral inclusions composed of arrays of nucleocapsids (Fig. 1A). The helical-shaped nucleocapsids contain a negative sense, single-stranded RNA genome and acquire a lipid envelope by budding through the cell membrane, forming mature filovirus particles (Fig. 1B). The virions are usually in the form of long filaments varying in length up to 14 µm, with diameters of about 80 nm. They sometimes appear as short U-shaped or circular forms (Figs. 1C,D) (MURPHY et al. 1978; ELLIS et al. 1979; GEISBERT and JAHRLING 1995).

Filovirus infections in humans have several pathologic features in common with other viral HFs, such as Rift Valley fever, Crimean-Congo hemorrhagic fever, and Lassa fever (ZAKI and KILMARX 1997; ZAKI and PETERS 1997). They all share variable degrees of hemorrhage and widespread necrosis of internal organs. Necrosis is usually most prominent in liver and lymphoid tissue. Among the hemorrhagic fever viruses, the filoviruses cause the most widespread destructive tissue lesions. Gross pathologic findings at autopsy include widespread petechial hemorrhages and ecchymoses involving skin, mucous membranes, and internal organs. The histopathologic changes are similar in Marburg virus and Ebola virus infections, with necrosis seen in many organs, including liver, spleen, kidney, and gonads (MURPHY 1978; INTERNATIONAL STUDY TEAM 1978; DIETRICH et al. 1978; RIPPEY et al. 1984; ZAKI et al. 1996a,b; GOLDSMITH et al. 1997; ZAKI and KILMARX 1997; ZAKI and PETERS 1997). The necrosis is both ischemic in nature and related to the cytopathic effect of the virus. Viral particles and inclusions are widely distributed in various tissues examined by electron microscopy (ELLIS et al. 1978b; GOLDSMITH et al. 1997; GEISBERT and JAAX 1998). These are seen primarily within endothelial cells, mononuclear phagocytic cells, and fibroblasts, and free within the interstitium. IHC and in situ hybridization (ISH) studies reveal that the distribution of viral antigens and nucleic acids correlates with ultrastructural localization of virus (ZAKI et al. 1996b).

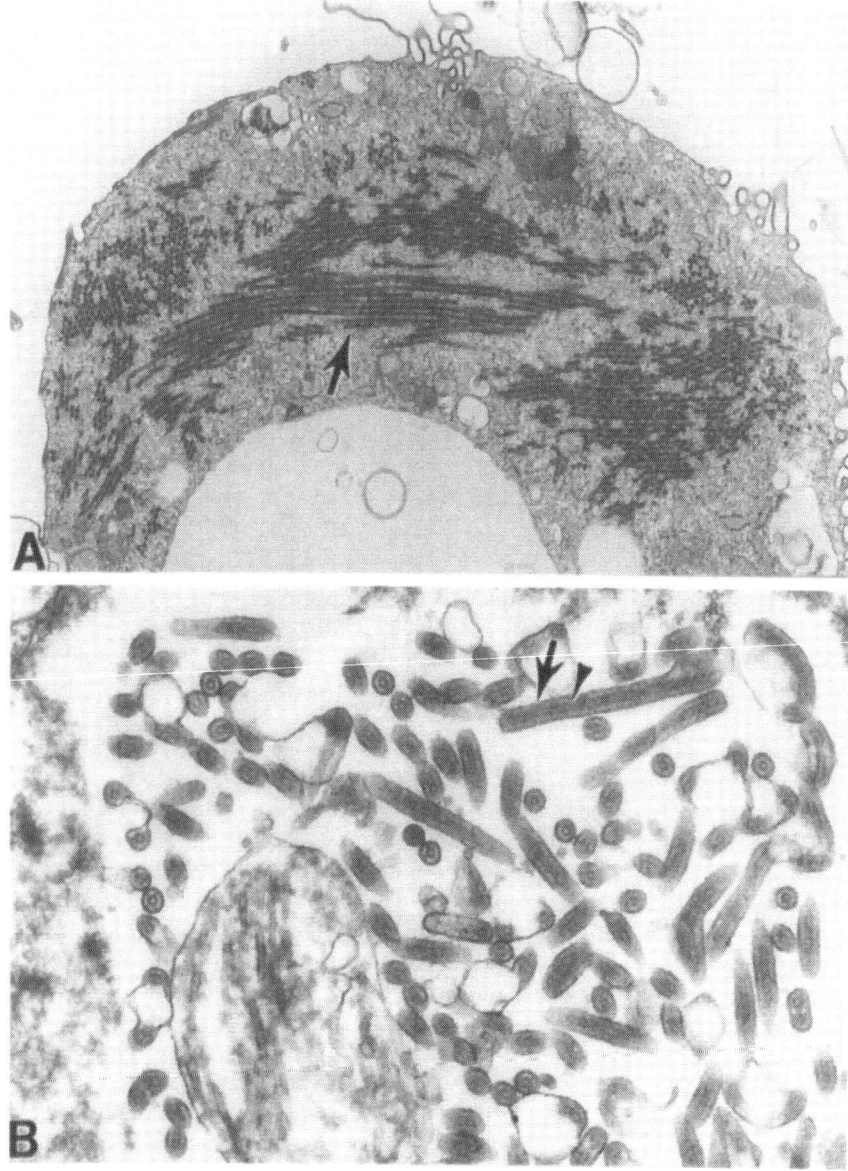

Fig. 1A–D. Ultrastructural characteristics of filoviruses as seen in Vero E6 culture cells. **A** In thin section, Ebola virus inclusions are seen to be composed of cytoplasmic aggregates of nucleocapsids (*arrow*). **B** Thin-section electron micrograph of Ebola virus, showing abundant extracellular filamentous viral particles, approximately 80 nm in diameter. A lipid envelope (*arrow*) surrounds a central nucleocapsid (*arrowhead*). **C** Negative stain and scanning **D** electron photomicrographs showing short and long filaments and an occasional characteristic "6"-shaped form (*arrow*). (A–C, isolate from 1995 outbreak; D, isolate from 1976 outbreak in Zaire; C, courtesy of Charles Humphrey, Infectious Disease Pathology Activity, Centers for Disease Control and Prevention. Magnifications: **A**, ×17,000; **B**, ×50,000; **C**, ×28,000; **D**, ×13,500

Pathologic Features of Filovirus Infections in Humans

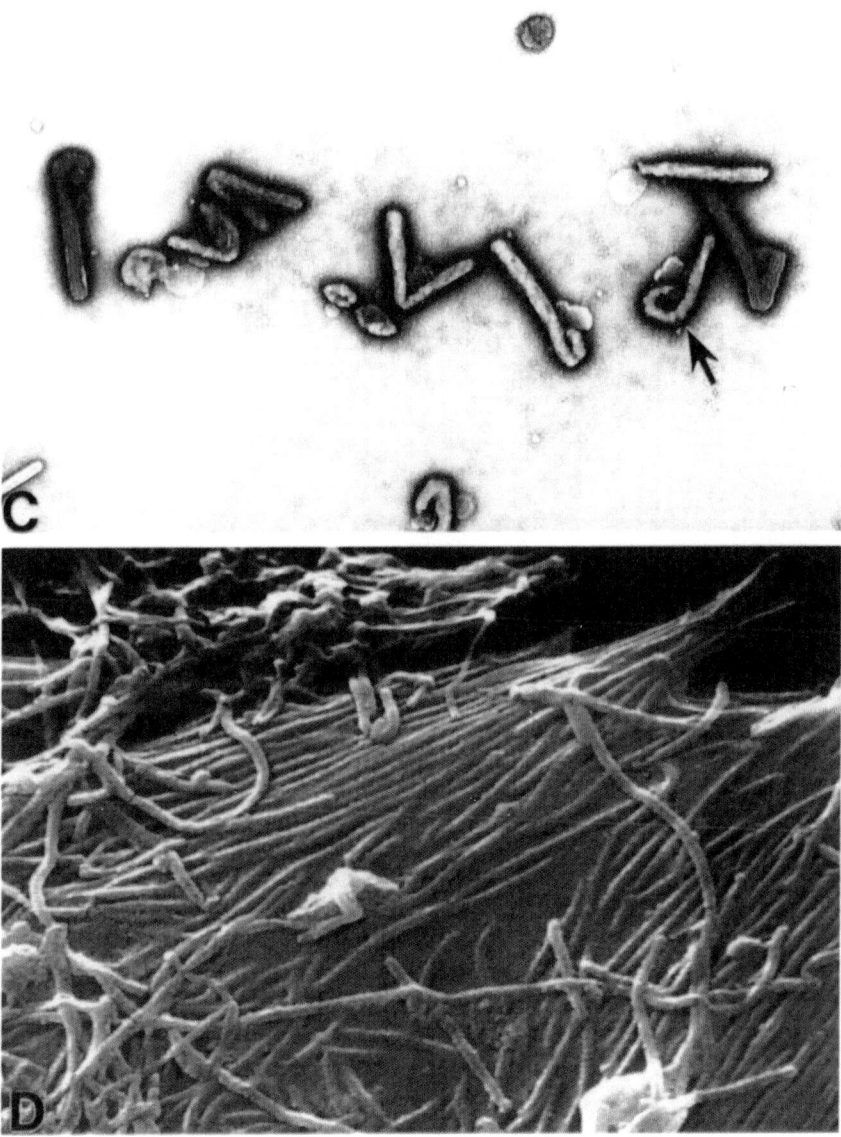

Fig. 1C,D.

The most characteristic histopathologic features are seen in the liver, with widespread hepatocellular necrosis, Councilman bodies, microvesicular fatty change, and Kupffer cell hyperplasia (Figs. 2,3). The portal tracts usually exhibit extensive karyorrhectic debris and a mononuclear inflammatory infiltrate. Filovirus

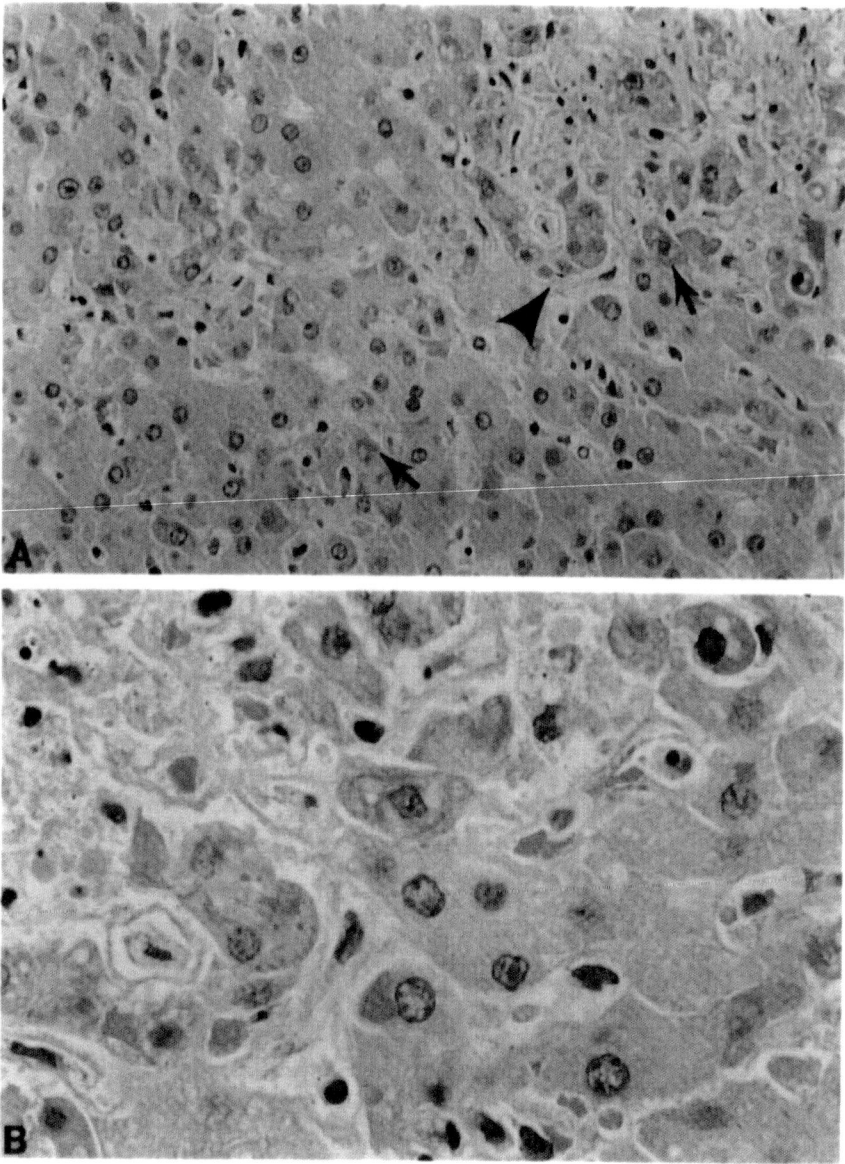

Fig. 2A,B

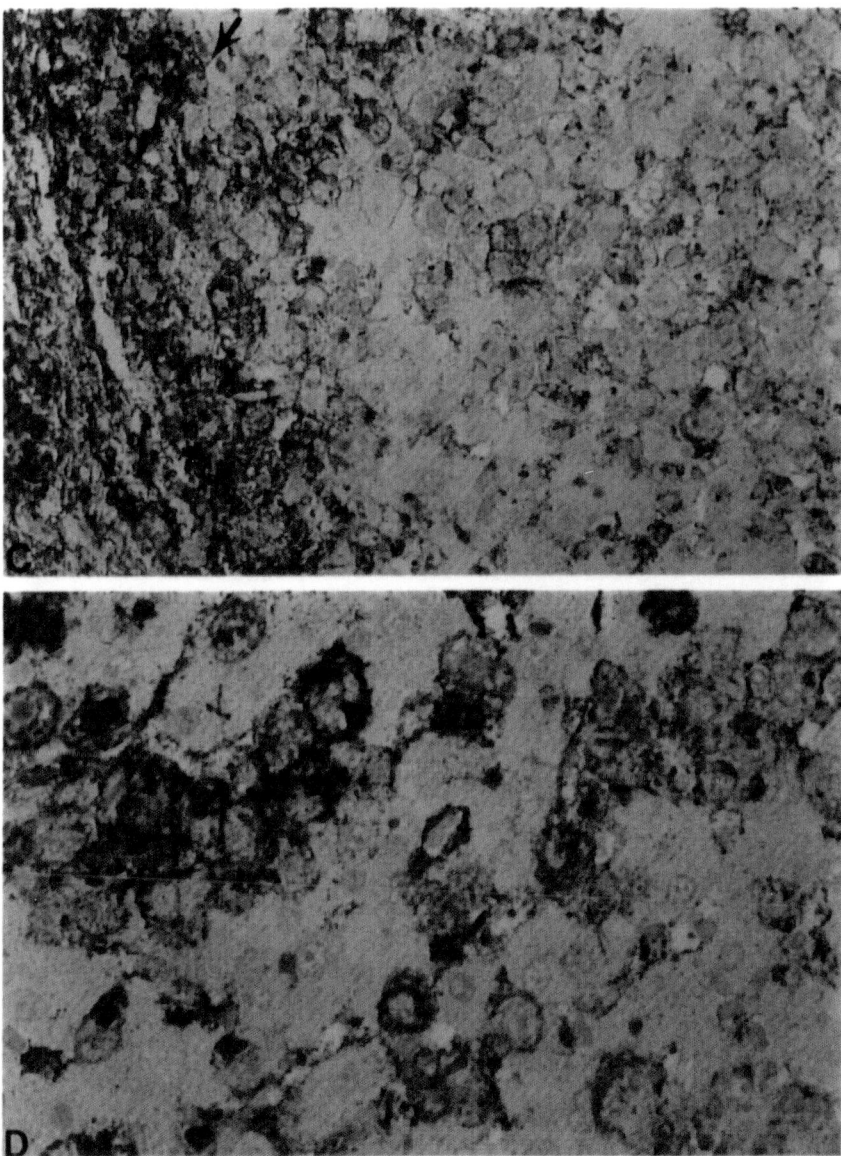

Fig. 2A–D. Pathologic findings in liver in patients with Ebola virus hemorrhagic fever (HF) from the 1995 outbreak. **A** Low-power magnification of liver, showing sinusoidal dilatation and congestion and hepatocellular necrosis. Numerous Ebola viral inclusions (*arrows*) are seen in association with an area of coalescent hepatic necrosis (*arrowhead*) and throughout the tissue. **B** Higher magnification of necrotic area in **A** showing details of the intracytoplasmic eosinophilic filamentous and globular inclusions. **C, D** Ebola antigens are seen throughout tissue section in sinusoids, sinusoidal lining cells, and hepatocytes. Note presence of large amount of extracellular antigen in capsular and subcapsular area (*arrows*). Magnifications: **A**, ×100; **B**, ×250; **C**, ×100; **D**, ×158. **A, B**, hematoxylin and eosin stain; **C, D**, immunoalkaline phosphatase staining, Naphthol fast red substrate with light hematoxylin counterstain

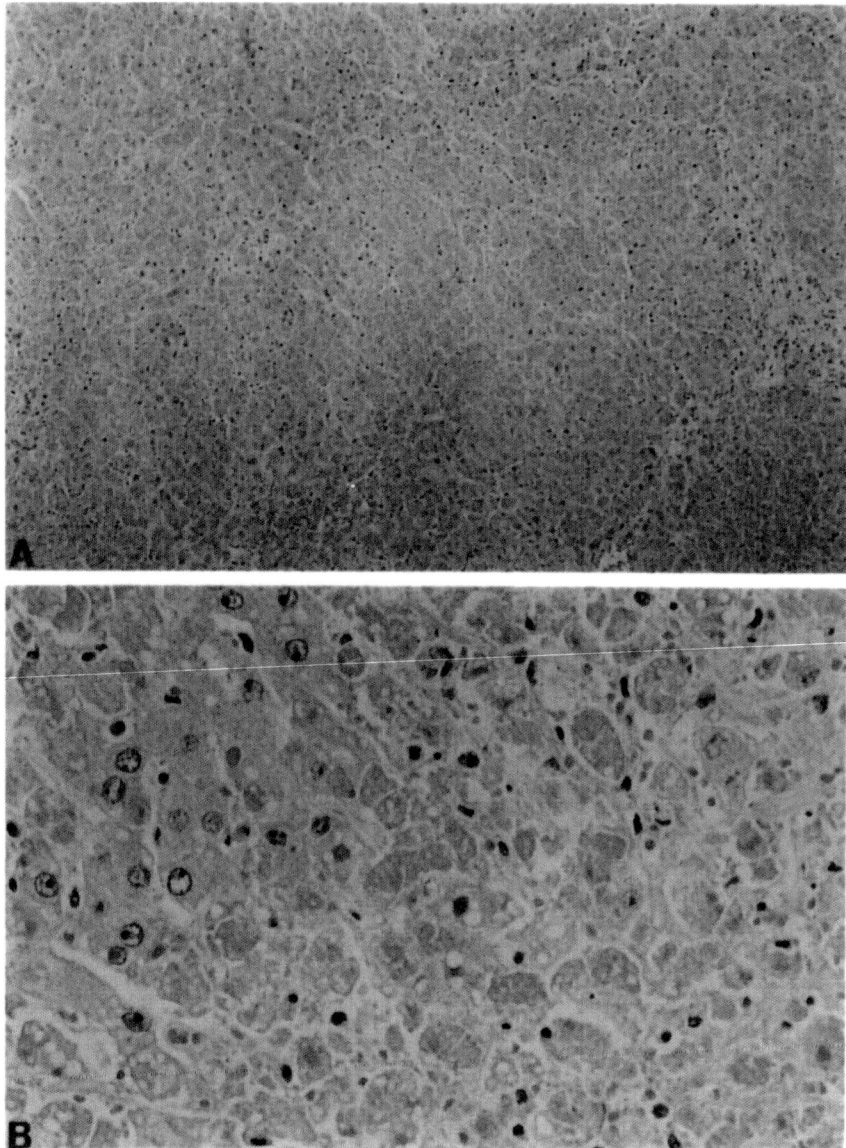

Fig. 3A,B

inclusions are seen within the cytoplasm of hepatocytes and are usually eosinophilic, and filamentous or oval in shape. Typical inclusions are usually easily recognized in the livers of patients with Ebola virus HF, while inclusions in Marburg virus HF are less defined and more difficult to recognize. Inclusions are most numerous in periportal zones and within and surrounding areas of necrosis. In some

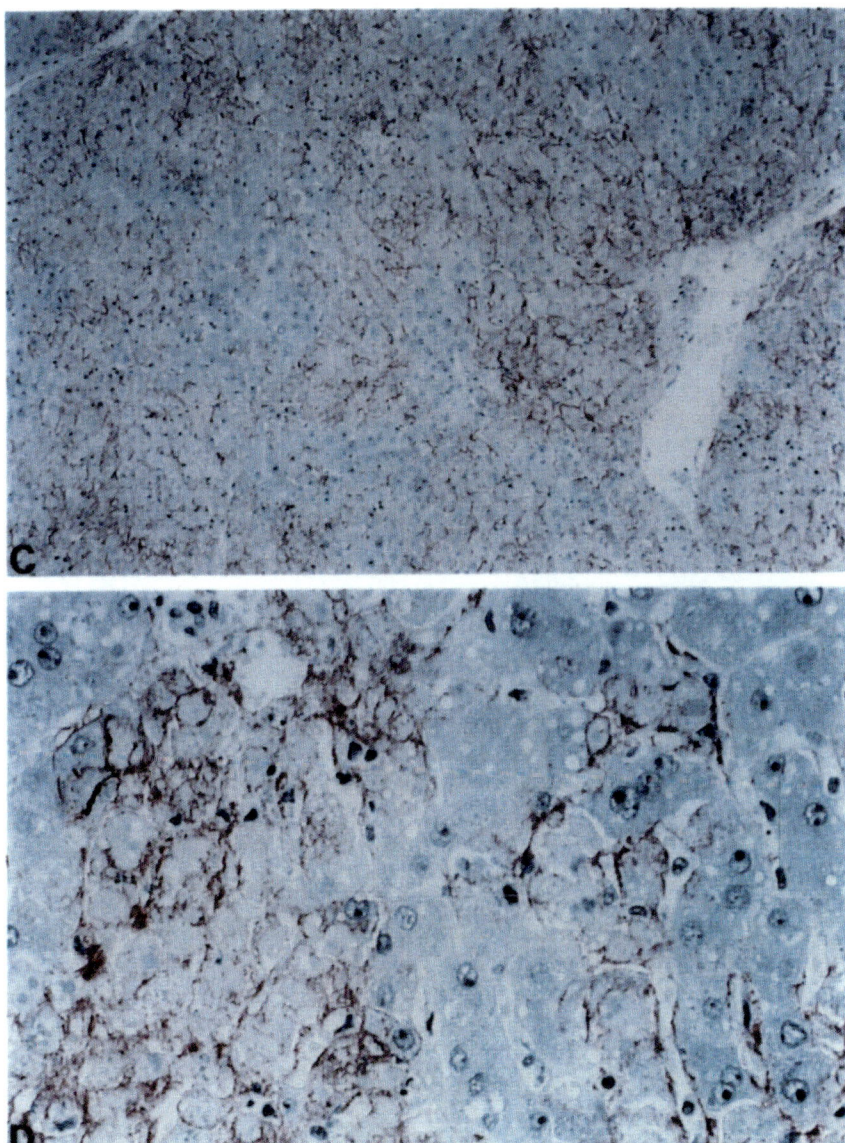

Fig. 3A–D. Pathologic findings in liver in patient with Marburg virus hemorrhagic fever (HF) from 1980. **A** Extensive hepatocellular necrosis with isolated clusters of intact hepatocytes. **B** Higher magnification showing acidophilic necrosis, with Councilman bodies and microvesicular fatty changes. Note absence of discernable filovirus inclusions. **C,D** Abundant Marburg virus antigens are diffusely distributed in hepatic parenchyma, as seen by immunohistochemistry. Antigen is concentrated in areas of hepatocellular necrosis and seen mainly within sinusoidal lining cells. Magnifications: **A**, ×25; **B**, ×100; **C**, ×25; **D**, ×100. **A,B**, hematoxylin and eosin stain; **C,D** immunoalkaline phosphatase staining, naphthol fast red substrate with light hematoxylin counterstain

sections, virtually every hepatocyte is infected as evidenced by the presence of inclusions. Ultrastructurally, they are seen to be composed of aggregates of viral nucleocapsids (Fig. 4). IHC and ISH studies reveal a remarkably high viral load, as evidenced by widespread presence of viral antigens and nucleic acids within hepatic parenchyma and sinusoids (Figs. 2,3).

Spleen and lymph nodes show various degrees of lymphoid depletion with extensive vascular follicular necrosis and necrotic debris (Fig. 5A). Viral antigen-positive cells seen throughout the spleen include cells of the mononuclear phagocytic system, dendritic cells, and fibroblasts. Additionally, large amounts of antigen are present extracellularly, associated with necrotic cells and debris (Fig. 5B). The lungs are usually hemorrhagic and show features of diffuse alveolar damage. Notably absent was evidence of interstitial pneumonitis or inflammatory response (Fig. 6A). Viral inclusions and antigen are most prominent within intra-alveolar macrophages (Fig. 6B). Viral particles can be seen free within the interstitium and within the alveoli (Figs. 6C,D). The kidneys frequently show evidence of acute tubular necrosis. Necrosis is related mostly to shock, although evidence of direct viral infection of tubular cells and cytopathic effect can be observed (Fig. 7). Significant myocardial edema is usually seen but is not associated with any appreciable inflammatory infiltrates. Viral immunostaining is seen primarily within the endocardium, other endothelial cells throughout the myocardium, and in rare myocytes.

The pathologic and immunopathologic findings in human tissues described above provide some insight into the pathogenesis of filovirus infections. Pathogenesis of filovirus infections has also been studied in several animal model systems, including monkeys, guinea pigs, hamsters, and suckling mice. The pathology in these experimentally infected animals is very similar to that observed in humans and provides an opportunity to systematically study these diseases (KISSLING et al. 1970; MURPHY et al. 1971, 1972; BASKERVILLE et al. 1978, 1985; ELLIS et al. 1978a; FISHER-HOCH et al. 1992; GEISBERT et al. 1992; RYABCHIKOVA et al. 1993, 1996; PEREBOEVA et al. 1993; JAAX et al. 1996; SWANEPOEL et al. 1996; DAVIS et al. 1997; XU et al. 1998). Significant injury to the microvasculature and increased endothelial permeability appear to be central to the pathogenesis of the shock syndrome and bleeding seen in filovirus infections. Infection of macrophages and other cells of the mononuclear phagocytic system is also thought to play a critical role in the pathogenesis of filovirus infections through the secretion of physiologically active substances, including cytokines and other inflammatory mediators (PETERS 1996; PETERS et al. 1996, 1997; VILLINGER et al. 1998).

3 Diagnosis

The diagnosis of filovirus HF, suspected by history and clinical manifestations, can also be supported histopathologically. Diagnosis of filovirus infection should be considered in patients who have traveled to sub-Saharan Africa, had direct contact with body fluids of a person or animal with Ebola virus infection, or worked in a

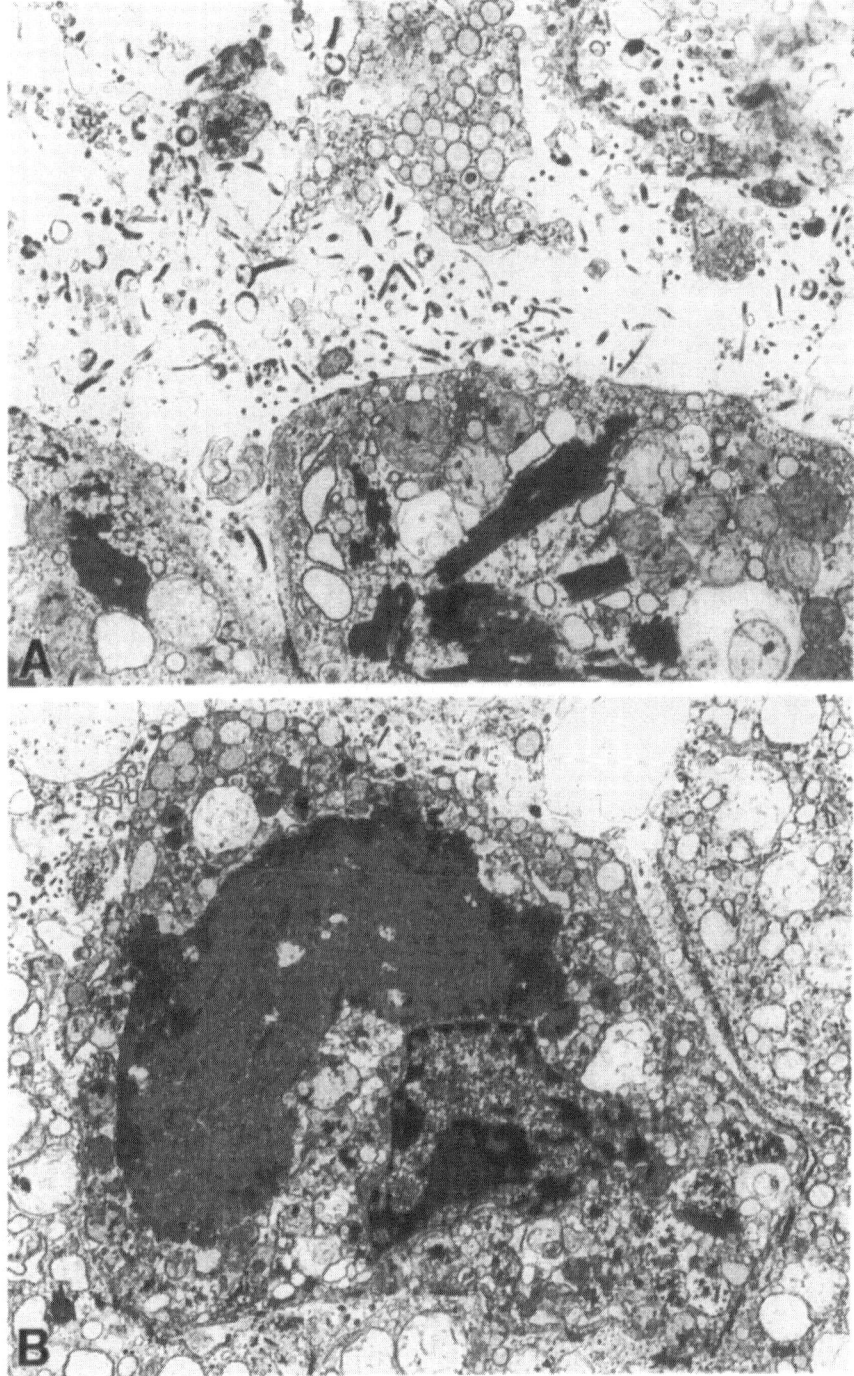

Fig. 4A,B.

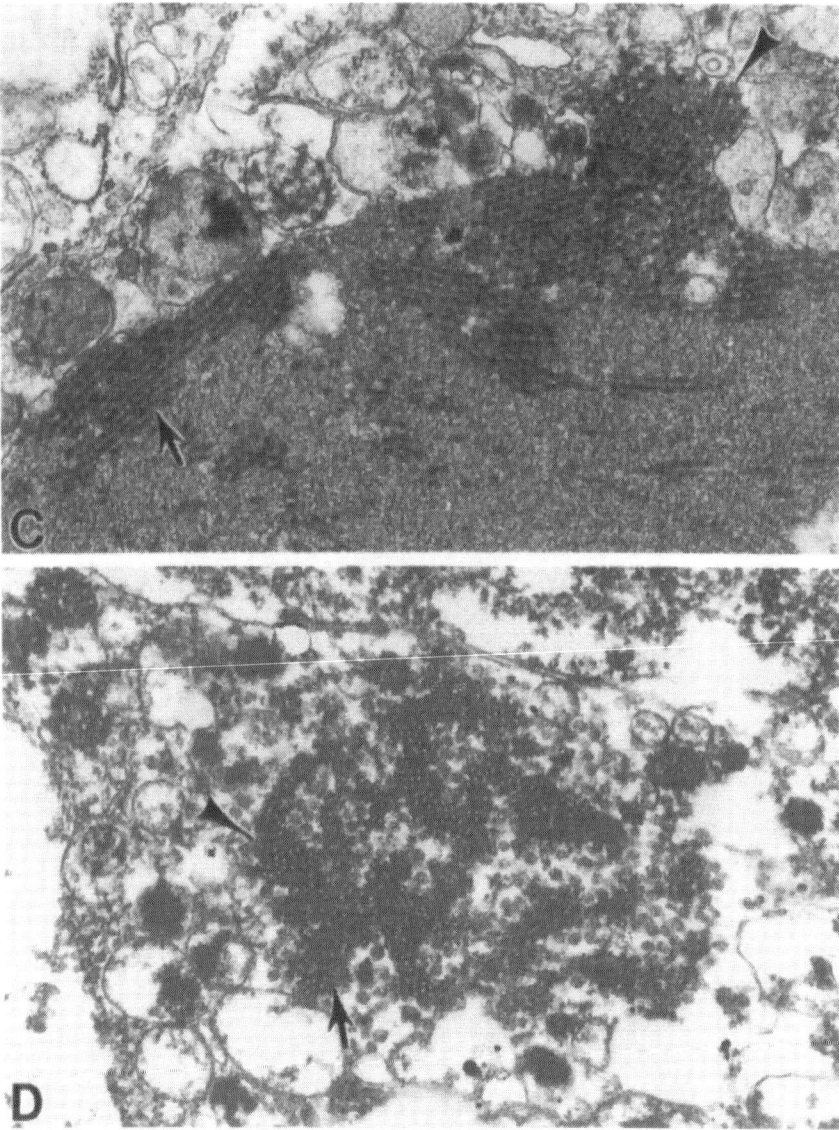

Fig. 4A–D. Hepatic ultrastructural features of filovirus infection. **A** Abundant extracellular Ebola virus particles are seen in hepatic sinusoids. Note variation in size and shape of viral particles associated with necrotic debris. **B** Large Ebola virus inclusion within the cytoplasm of an infected hepatocyte. **C** Higher magnification of **B**; the inclusions consist of lighter staining granular areas admixed with areas of dense, well-formed tubular nucleocapsids as seen in longitudinal (*arrow*) and transverse (*arrowhead*) sections. **D** Marburg virus inclusion showing aggregation of viral nucleocapsids sectioned longitudinally (*arrow*) and transversely (*arrowhead*). Tubular nucleocapsids seen in **C** and **D** are identical in composition and diameter to the core structure of the virus particles (see Fig. 1B). Magnifications: **A** ×11,000; **B** ×9,000; **C,D** ×28,000

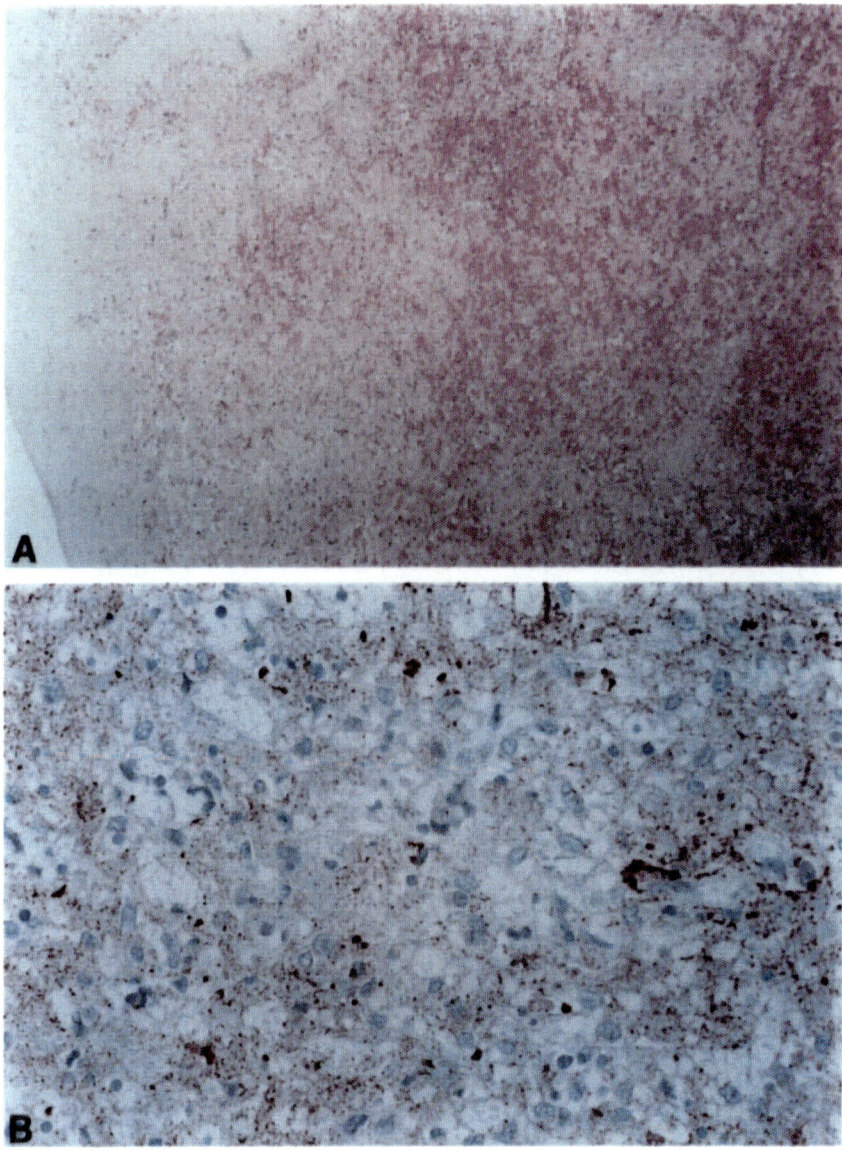

Fig. 5A,B. Pathologic findings in spleen from patient with Ebola virus hemorrhagic fever (HF). **A** Low-power photomicrograph of a spleen showing congestion and virtual absence of lymphoidal elements. **B** Ebola antigens are seen throughout the tissue section, mostly extracelluarly but also within cells of the mononuclear phagocytic system. Magnifications: **A** ×25; **B** ×100. **A** hematoxylin and eosin stain; **B** immunoalkaline phosphatase staining, naphthol fast red substrate with light hematoxylin counterstain

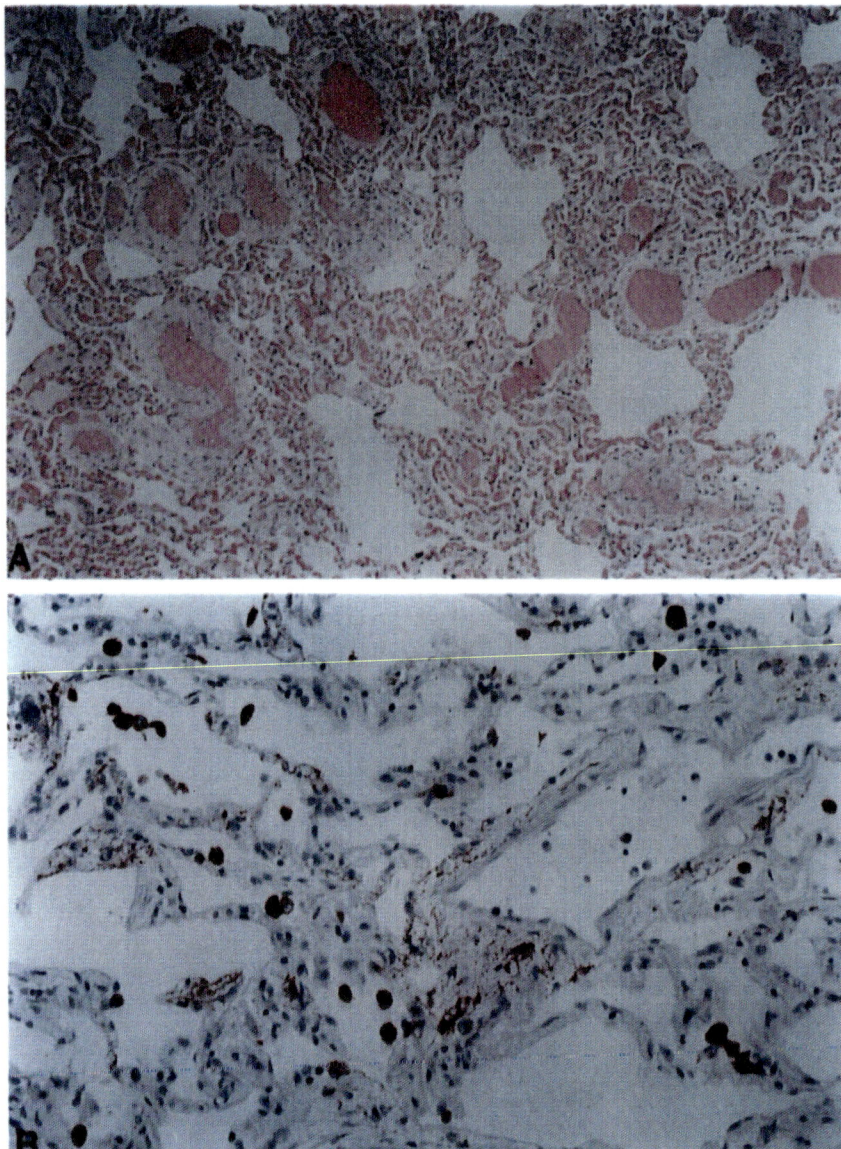

Fig. 6A–D. Lung, Ebola virus hemorrhagic fever (HF). Light microscopic and ultrastructural features. **A** Low-power photomicrograph of lung showing marked vascular congestion and absence of significant inflammatory cellular infiltration. **B** Ebola virus antigens are seen mainly within intra-alveolar macrophages, fibroblasts, and endothelial cells. Extracellular antigens are also seen within interstitium. **C** A thickened edematous interstitium showing collection of Ebola virus particles in association with necrotic debris. **D** High-power magnification showing structural details of a collection of virions within pulmonary interstitium. Magnifications: **A** ×25; **B** ×50; **C** ×10,000; **D** ×52,000. **A** hematoxylin and eosin stain; **B** immunoalkaline phosphatase staining, naphthol fast red substrate with light hematoxylin counterstain

Pathologic Features of Filovirus Infections in Humans   111

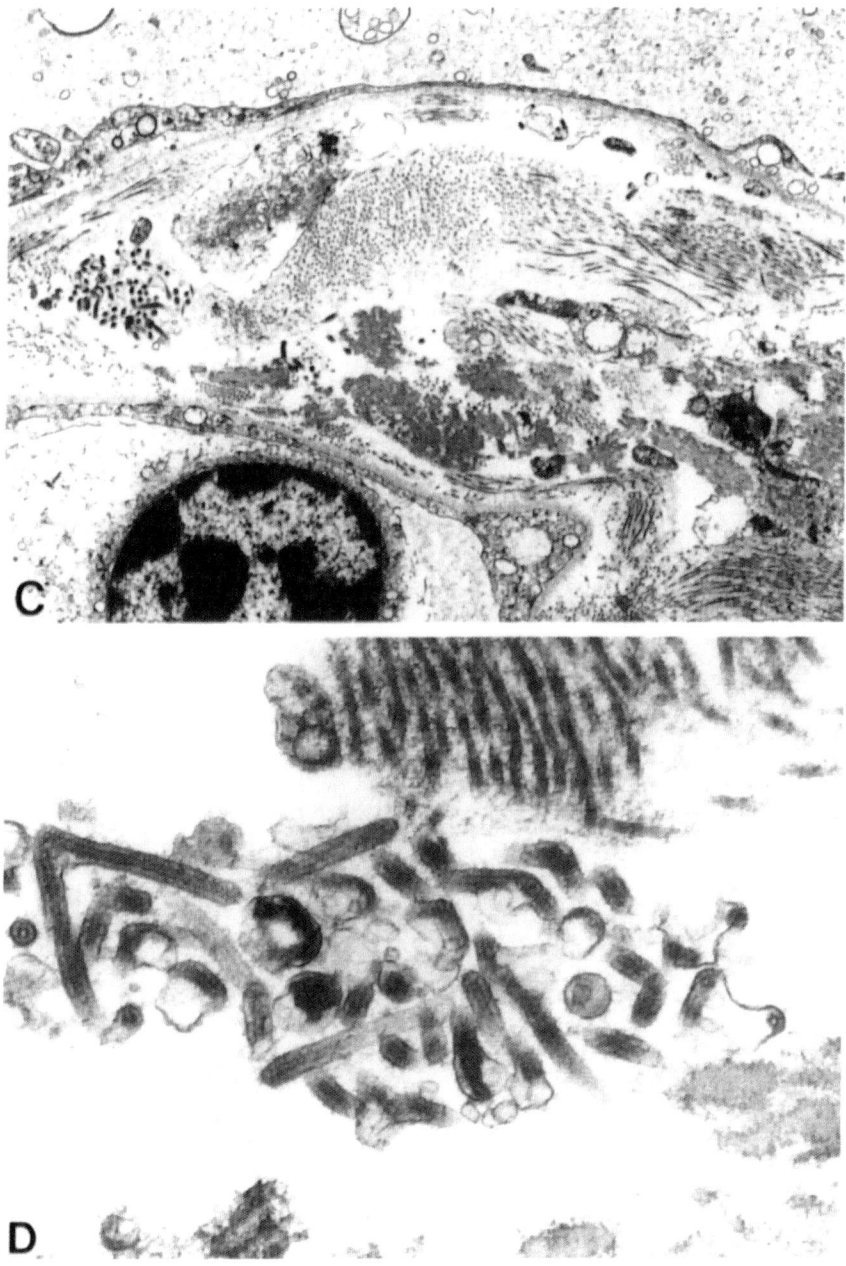

Fig. 6C,D.

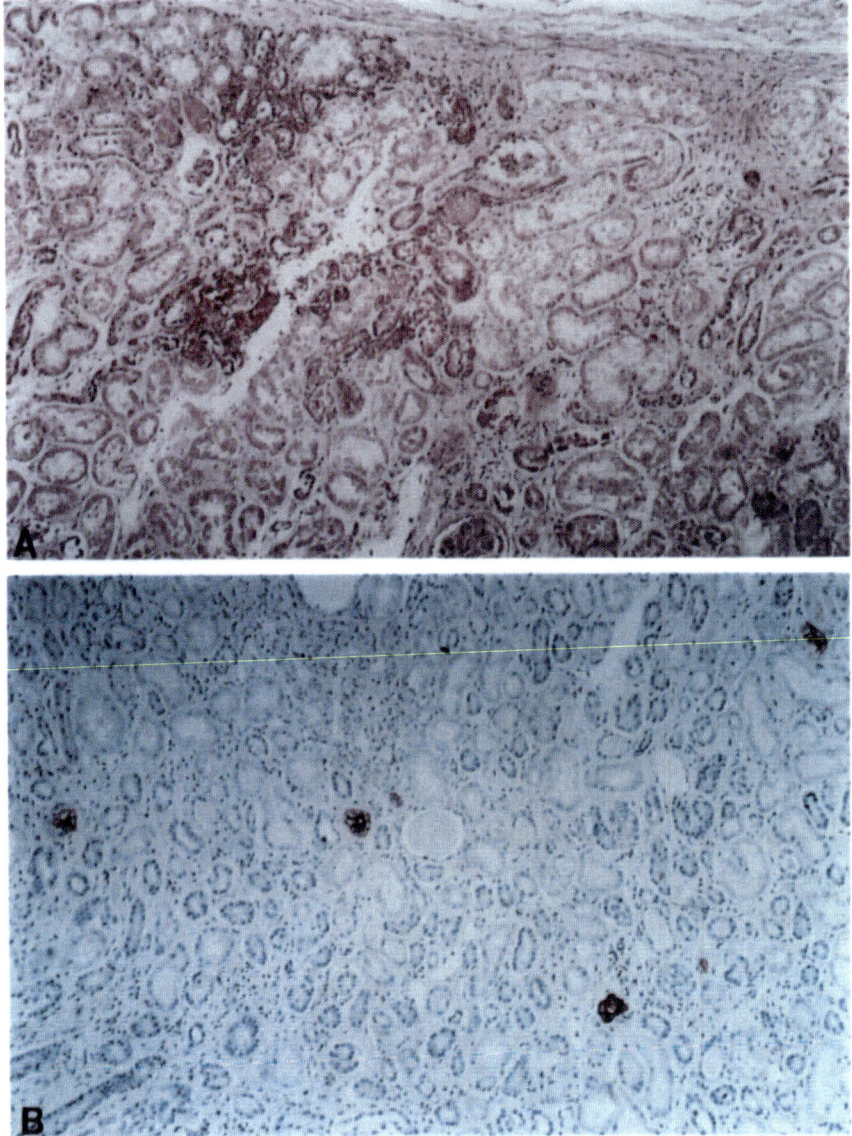

Fig. 7A,B

laboratory or animal facility that handles Ebola virus. The major diseases that must be clinically ruled out are malaria, rickettsial diseases, leptospirosis, shigellosis, typhoid, and other viral HFs. There are characteristic features that may support the histopathologic diagnosis in individual diseases. Histopathologically, characteristic

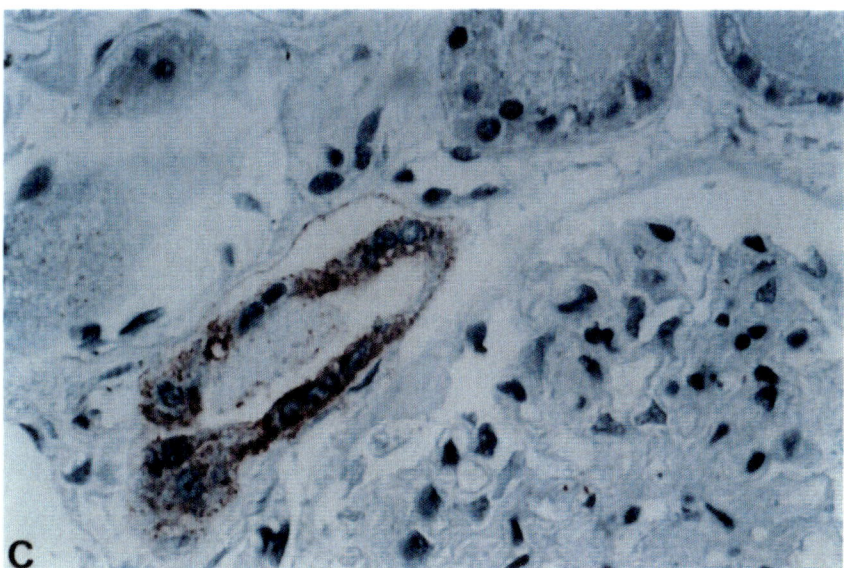

Fig. 7A–C. Renal pathologic features in Marburg virus hemorrhagic fever (HF). **A** Low-power photomicrograph showing vascular congestion, focal hemorrhages, dilatation of tubules, and interstitial edema. Note absence of inflammatory cellular response. **B** Low-power magnification showing focal staining in tubules and endothelial cells in kidney as seen by immunohistochemistry. Tubular staining is seen both within the lining epithelial cells and the intraluminal cast. **C** High-power magnification showing immunostaining within epithelial cells of renal tubule. Magnifications: **A**, **B** ×25; **C** ×158. **A** hematoxylin and eosin stain; **B**, **C** immunoalkaline phosphatase staining, naphthol fast red substrate with light hematoxylin counterstain

hepatic inclusions may suggest a specific diagnosis of filovirus infection. However, because similar pathologic features are seen in filovirus HF and a variety of other viral, rickettsial, and bacterial infections, unequivocal diagnosis can be made only by laboratory tests such as serology, viral culture, molecular methods, electron microscopy, and IHC. The main pathologic differential diagnosis should include viral hepatitis, leptospirosis, malaria, and rickettsial diseases.

Ebola virus antigens can be detected by IHC examination of formalin-fixed tissues by using specific polyclonal and monoclonal antibodies. These methods have a unique role in cases in which archival tissues are the only specimens available for diagnostic testing. During the recent epidemic in Kikwit, Zaire, a novel IHC test was developed at the CDC in which skin biopsy specimens are used for the diagnosis of Ebola virus HF (Fig. 8) (GEORGES-COURBOT et al. 1997; ZAKI et al. 1998; LLOYD et al. 1998). Formalin-fixed biopsy specimens are not infectious and may be sent without special precautions or refrigeration and may be taken in the most basic field conditions. This approach is advantageous over viral cultures and fluorescent antibody tests that are commonly used for surveillance purposes and require special handling and transport of infectious material and a cold chain. This new diagnostic modality and IHC findings open the way to a more practical surveillance mecha-

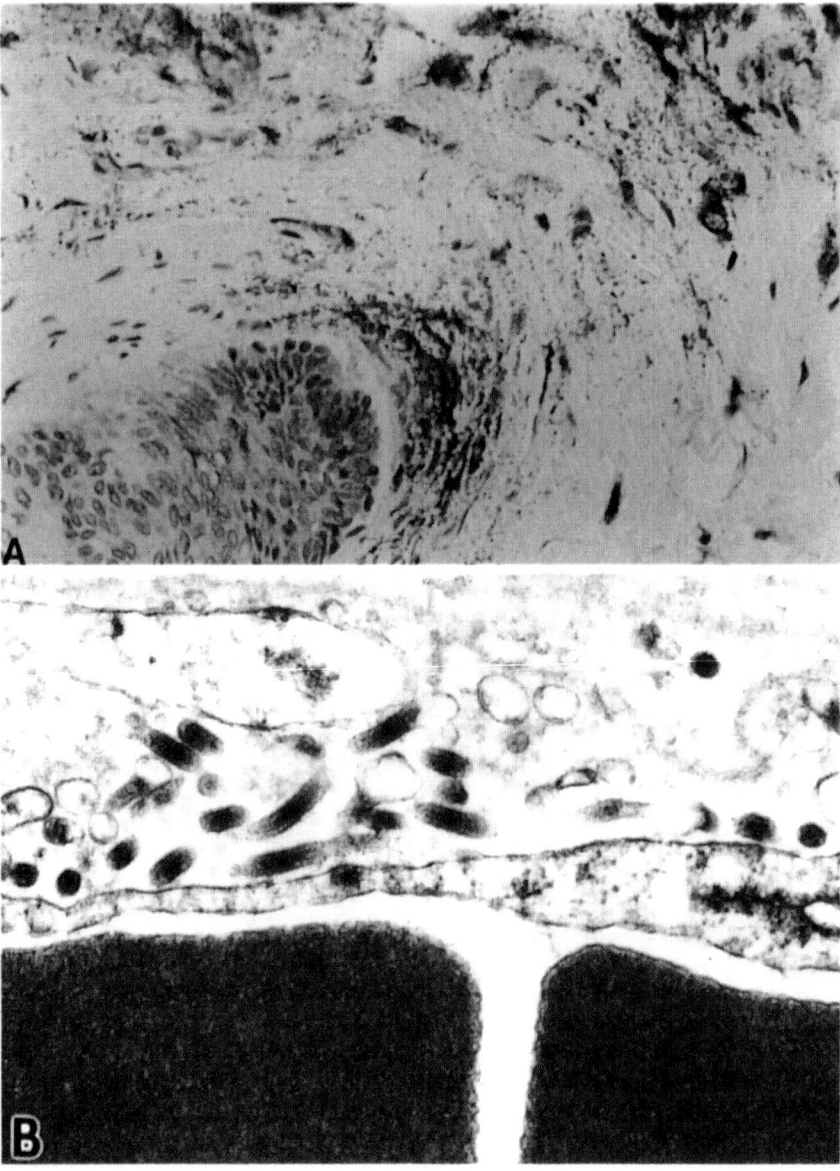

Fig. 8A,B. Skin, Ebola virus hemorrhagic fever (HF). **A** Extensive amounts of Ebola antigen are seen in section of skin. Immunostaining is predominately seen in the extracellular matrix as well as within dermal fibroblasts and endothelial cells. **B** Thin-section electron micrograph showing multiple viral particles in association with dermal blood vessel. Particles are seen in the surrounding connective tissue matrix. Magnifications: **A** ×100; **B** ×45,000. **A** immunoalkaline phosphatase staining, naphthol fast red substrate with light hematoxylin counterstain

nism and provide new insights into a possible epidemiologic role for contact transmission.

References

Baron RC, McCormick JB, Zubeir OA (1983) Ebola virus disease in southern Sudan: hospital dissemination and intrafamilial spread. Bull World Health Organ 61:997–1003
Baskerville A, Bowen ETW, Platt GS, McArdell LB, Simpson DIH (1978) The pathology of experimental Ebola virus infection in monkeys. J Pathol 125:131–138
Baskerville A, Fisher-Hoch SP, Neild GH, Dowsett AB (1985) Ultrastructural pathology of experimental Ebola haemorrhagic fever virus infection. J Pathol 147:199–209
Centers for Disease Control and Prevention (1995) Outbreak of Ebola viral hemorrhagic fever – Zaire, 1995. MMWR Morb Mortal Wkly Rep 44:381–382
Davis KJ, Anderson AO, Geisbert TW, Steele KE, Geisbert JR, Vogel P, Connolly BM, Huggins JW, Jahrling PB, Jaax NK (1997) Pathology of experimental Ebola virus infection in African green monkeys: involvement of fibroblastic reticular cells. Arch Pathol Lab Med 121:805–819
Dietrich M, Schumacher HH, Peters D, Knobloch J (1978) Human pathology of Ebola (Maridi) virus infection in the Sudan. In: Pattyn SR (ed) Ebola virus haemorrhagic fever. Elsevier/North-Holland Biomedical, Amsterdam, pp 37–41
Ellis DS, Bowen ETW, Simpson DIH, Stamford S (1978a) Ebola virus: a comparison, at ultrastructural level, of the behaviour of the Sudan and Zaire strains in monkeys. Brit J Exp Pathol 59:584–593
Ellis DS, Simpson DIH, Francis DP, Knobloch J, Bowen ETW, Lolik P, Deng IM (1978b) Ultrastructure of Ebola virus particles in human liver. J Clin Pathol 31:201–208
Ellis DS, Stamford S, Lloyd G, Bowen ETW, Platt GS, Way H, Simpson DIH (1979) Ebola and Marburg viruses: some ultrastructural differences between strains when grown in Vero cells. J Med Virol 4:201–211
Fisher-Hoch SP, Brammer TL, Trappier SG, Hutwagner LC, Farrar BB, Ruo SL, Brown BG, Hermann LM, Perez-Oronoz GI, Goldsmith CS, Hanes MA, McCormick JB (1992) Pathogenic potential of filoviruses: role of geographic origin of primate host and virus strain. J Infec Dis 166:753–763
Geisbert TW, Jahrling PB (1995) Differentiation of filoviruses by electron microscopy. Virus Res 39:129–150
Geisbert TW, Jaax NK (1998) Marburg hemorrhagic fever: report of a case studied by immunohistochemistry and electron microscopy. Ultrastruct Pathol 22:3–17
Geisbert TW, Jahrling PB, Hanes MA, Zack PM (1992) Association of Ebola-related Reston virus particles and antigen with tissue lesions of monkeys imported to the United States. J Comp Pathol 106:137–152
Georges-Courbot M-C, Sanchez A, Lu C-Y, Baize S, Leroy E, Lansout-Soukate J, Tevi-Benissan C, Georges AJ, Trappier SG, Zaki SR, Swanepoel R, Leman PA, Rollin PE, Peters CJ, Nichol ST, Ksiazek TG (1997) Isolation and phylogenetic characterization of Ebola viruses causing different outbreaks in Gabon. Emerging Infec Dis 3:59–62
Goldsmith CS, Rollin PE, Zhang XH, Peters CJ, Zaki SR (1997) Ebola virus hemorrhagic fever, Zaire, 1995: an ultrastructural study. Micros Microanalysis 3:77–78
International Study Team (1978) Ebola haemorrhagic fever in Sudan, 1976. Bull World Health Organ 56:247–270
Jaax NK, Davis KJ, Geisbert TJ, Vogel P, Jaax GP, Topper M, Jahrling PB (1996) Lethal experimental infection of rhesus monkeys with Ebola-Zaire (Mayinga) virus by the oral and conjunctival route of exposure. Arch Pathol Lab Med 120:140–155
Kissling RE, Murphy FA, Henderson BE (1970) Marburg virus. Ann NY Acad Sci 174:932–945
Le Guenno B, Formentry P, Wyers M, Gounon P, Walker F, Boesch C (1995) Isolation and partial characterization of a new strain of Ebola virus. Lancet 345:1271–1274
Lloyd E, Zaki SR, Rollin P, Ksiazek T, Callain P, Konde MK, Tchioko K, Bwaka MA, Verchueren E, Kabwau J, Ndambe R, Peters CJ (1998) Long-term disease surveillance in Bandundu region, Democratic Republic of the Congo (former Zaire): a model for early detection and prevention of Ebola hemorrhagic fever. J Infec Dis (in press)

Murphy FA (1978) Pathology of Ebola virus infection. In: Pattyn SR (ed) Ebola virus haemorrhagic fever. Elsevier/North-Holland Biomedical, pp 43–59

Murphy FA, Simpson DIH, Whitfield SG, Zlotnik I, Carter GB (1971) Marburg virus infection in monkeys. Ultrastructural studies. Lab Invest 24:279–291

Murphy FA, Simpson DIH, Whitfield SG, Zlotnik I, Carter GB (1972) Marburg virus infection in monkeys. Ultrastructural studies. Medecine et Chirurgie Digestives 1:325–332

Murphy FA, van der Groen G, Whitfield SG, Lange JV (1978) Ebola and Marburg virus morphology and taxonomy. In: Pattyn SR (ed) Ebola virus haemorrhagic fever. Elsevier/North-Holland Biomedical, pp 61–82

Pereboeva LA, Tkachev VK, Kolesnikova LV, Krendeleva LI, Ryabchikova EI, Smolina MP (1993) Ultrastructural changes of guinea pig organs in sequential passages of Ebola virus. Voprosy Virusologii 4:179–182

Peters CJ (1996) Pathogenesis of viral hemorrhagic fevers. In: Nathanson N, Ahmed R, Gonzalez-Scarano F, Griffin D, Holmes KV, Murphy FA, Robinson HL (eds) Viral pathogenesis. Lippincott-Raven, Philadelphia, pp 779–799

Peters CJ, Sanchez A, Rollin PE, Ksiazek TG, Murphy FA (1996) Filoviridae: Marburg and Ebola viruses. In: Fields BN, Knipe DM, Howley PM (eds) Field's virology. Lippincott-Raven, New York, pp 1161–1176

Peters CJ, Zaki SR, Rollin PE (1997) Viral hemorrhagic fevers. In: Mandell GL (ed) Atlas of infectious diseases. Churchill Livingstone, Philadelphia, pp 10.1–10.26

Rippey JJ, Schepers NJ, Gear JHS (1984) The pathology of Marburg virus disease. South Afr Med J 66:50–54

Ryabchikova E, Kolesnikova L, Smolina M, Tkachev V, Pereboeva L, Baranova S, Grazhdantseva A, Rassadkin Y (1996) Ebola virus infection in Guinea pigs – presumable role of granulomatous inflammation in pathogenesis. Arch Virol 141:909–921

Ryabchikova EI, Baranova SG, Tkachev VK, Grazhdantseva AA (1993) Morphological changes in Ebola virus infection in guinea pigs. Voprosy Virusologii 4:176–179

Shieh W-J, Demby A, Merdel S, Conteh A, Ferebee T, Goldsmith CS, Bausch D, Farrar B, Lloyd E, Ksiazek T, Rollin P, Peters CJ, Zaki SR (1998) High risk autopsy of fatal Lassa fever cases in Sierra Leone. Lab Invest (in press)

Swanepoel R, Leman PA, Burt FJ, Zachariades NA, Braack LEO, Ksiazek TG, Rollin PE, Zaki SR, Peters CJ (1996) Experimental inoculation of plants and animals with Ebola virus. Emerging Infec Dis 2:321–325

Villinger F, Rollin PE, Brar SS, Chikkala NF, Winter J, Sundstrom JB, Zaki SR, Swanepoel R, Ansari AA, Peters CJ (1998) Markedly elevated levels of IFN-γ/α, IL-2, IL-10 and TNF-α associated with fatal Ebola virus infection. J Infec Dis (in press)

World Health Organization (1976) Ebola hemorrhagic fever in Sudan, 1976. Bull World Health Organ 56:247–270

World Health Organization (1978) Ebola haemorrhagic fever in Zaire, 1976. Report of an international commission. Bull World Health Organ 56:271–293

World Health Organization (1996) Outbreak of Ebola hemorrhagic fever in Gabon officially declared over. Wkly Epidemiol Rec 71:125–126

Xu L, Sanchez A, Yang ZY, Zaki SR, Nabel GJ (1998) Immunization for Ebola virus infection. Nat Med 4:37–42

Zaki SR, Greer PW, Goldsmith CS, Coffield LM, Rollin PE, Callain P, and others (1996a) Ebola virus hemorrhagic fever: pathologic, immunopathologic and ultrastructural studies. Proceedings of the International Colloquium on Ebola Virus Research. Sept 4–7, Antwerp, Belgium

Zaki SR, Greer PW, Goldsmith CS, Coffield LM, Rollin PE, Callain P, Khan AS, Ksiazek TG, Peters CJ (1996b) Ebola virus hemorrhagic fever: pathologic, immunopathologic, and ultrastructural study. Lab Invest 74:133 A

Zaki SR, Greer PW, Shieh W-J, Goldsmith CS, Ferebee T, Rollin PE, Khan AS, Callain P, Lloyd ES, Ksiazek TG, Peters CJ (1998) A novel immunohistochemical assay for detection of Ebola virus in skin: implications for diagnosis, spread, and surveillance of Ebola hemorrhagic fever. J Infec Dis (submitted)

Zaki SR and Kilmarx PH (1997) Ebola virus hemorrhagic fever. In: Horsbugh C Jr, Nelson AM (eds) Pathology of emerging infections. ASM Press, Washington, DC, pp 299–312

Zaki SR and Peters CJ (1997) Viral hemorrhagic fevers. In: Connor DH, Chandler FW, Schwartz DA, Manz HJ, Lack EE (eds) Diagnostic pathology of infectious diseases. Appleton and Lange, Stamford, CT, pp 347–364

Experimental Filovirus Infections

S.P. Fisher-Hoch[1] and J.B. McCormick[2]

1 General Observations	117
2 Nonhuman Primates as a Model for Filovirus Disease	118
3 Handling of Nonhuman Primates Infected with Filoviruses	119
4 Species of Nonhuman Primates Used in Filovirus Studies	119
5 Marburg Virus Infection in Nonhuman Primates	120
6 Ebola Virus Infection in Nonhuman Primates	121
7 Studies of the Pathogenesis of Filovirus Infection	122
8 The Immune Response to Filoviruses in Nonhuman Primates	125
9 Comparison of Filovirus Strains in Laboratory Animals	126
10 Transmission of Filoviruses Among Nonhuman Primates	128
11 Persistence of Filoviruses	132
12 Protection Studies in Nonhuman Primates	136
13 Smaller Laboratory Animals in Filovirus Studies	137
14 Biosafety in Experimental Filovirus Infections	139
15 Challenges for the Future	139
References	141

1 General Observations

Filoviruses are zoonoses, but at the moment of writing the natural reservoir or reservoirs are undefined. Bats have long been suspect on epidemiologic grounds, and recent reports by Swanepoel and others suggest that bats may play a central role in the filovirus ecosystem (SWANEPOEL et al. 1996). Natural infections in primates including humans range from asymptomatic to severe, almost invariably

---

[1] Directrice Laboratoire Jean Mérieux SL4, Fondation Marcel Mérieux, 17 rue Bourgelat, 69002, Lyon, France
[2] Unité d'Epidémiologie, Institut Pasteur, rue du Docteur Roux, 75015, Paris, France

fatal disease, apparently depending on the origin of the infecting virus and the primate species infected. With this variability and uncertainty about the natural history of the virus, it is likely that infections in most commonly used laboratory animals, including rodents, reflect pathophysiologic processes in species which, like humans, are not natural hosts and are therefore not ecologically adapted to the virus.

Much of the data from experimental animals have been obtained using nonhuman primates. The disease processes in all experimental monkey species so far infected closely resemble those in humans, and, with a few reservations, it is valid to extrapolate data from experimental infections in nonhuman primates to human disease. Nonhuman primates are also useful for evaluation of potential therapy of acute disease and safety, immunogenicity and efficacy of vaccine candidates. Smaller, less expensive and more easily handled laboratory species such as guinea pigs and mice have been very useful for filovirus isolation and reagent production. They have also been used for some studies of pathophysiologic and immunologic processes.

2 Nonhuman Primates as a Model for Filovirus Disease

Filoviruses are somewhat unusual in that, historically, infections of captive nonhuman primates can be divided into two overlapping categories: accidental ('natural') and experimental, with the remarkable feature that there have been much larger numbers of animals in the former category. Both have yielded important data, which are usefully considered together. In a few instances 'naturally' infected animals have later been placed on protocol, and data collected in an experimental fashion. Unfortunately, much important data from outbreaks and also from experimental infection remain unpublished and our knowledge is limited.

Despite the fact that any experiment can only be conducted with small numbers of outbred animals, which normally limit statistical analyses, the advantages offered by nonhuman primates in modeling filovirus disease are considerable. Monkeys generally develop severe disease following inoculation and can be reproducibly expected to undergo pathophysiologic and pathologic processes very closely mirroring those observed in humans. Monkeys are generally large enough to allow sufficiently frequent and adequate sampling to monitor hematological and biochemical parameters. Nonhuman primates are also sufficiently closely related to humans that most of the serum proteins and other factors which need to be assayed are biochemically and antigenically similar to those of humans. This means that commercial assays and other reagents prepared with polyclonal or monoclonal antibodies directed against human proteins generally work well with monkey sera and cells. Provided the investigator is careful to use appropriate nonhuman primates as controls and collect sufficient baseline data for whatever assay is planned, the results are usually interpretable and useful. The limited number of animals that can be used in any single study necessitates a protocol which uses pre-inoculation results from each monkey as its own internal control. In practice this can be achieved by careful baseline study of each animal before it goes on biosafety level 4 (BSL4) protocol. For many assays, particularly hematology, there are published veterinary data for common primate species giving normal ranges.

3 Handling of Nonhuman Primates Infected with Filoviruses

Nonhuman primates experimentally infected with filoviruses should be handled by a team of highly trained and experienced individuals and should contain at least one veterinarian or animal handler accustomed to working with nonhuman primates. Cages, even within BSL4, should be placed in laminar flow, double high efficiency particulate air (HEPA) filtered hoods to limit cross-contamination between animals, and to contain the infectious agent as far as possible within an easily defined area. All clinical measurements and specimen collections require the animals to be lightly anesthetized, and to do this safely squeeze-bar cages or other forms of restraint must be used. Anesthesia is needed even for measurement of body temperature using a rectal probe. Daily venipuncture may account for sufficient blood loss to compromise a small animal, and minor trauma in a filovirus-infected animal may induce bleeding. A compromise using 48-hourly measurements under light ketamine/rompol anesthesia has been shown to be acceptable providing there is careful handling of the animal, preferably under the supervision of a veterinary primatologist. Continuous monitoring of temperature can now be performed using implanted probes. In view of the high overall cost of nonhuman primate studies in BSL4 conditions, remote monitoring is highly desirable wherever it becomes practicable.

Nonhuman primates cannot be nursed or supported as would be desirable for a human with filovirus disease, for whom intensive care support is mandatory. This support includes monitoring of fluid balance, hematological and cardiovascular systems and attempts to correct or at least compensate for abnormalities as they occur. Experimental animals also easily become dehydrated and thus volume depleted, and it is not normally possible to correct the fluid balance and hematological abnormalities or prevent or treat shock. The outcome in infected animals is therefore likely to be poor. Unfortunately, most patients are infected and cared for in remote hospitals, and often their care is regrettably often no better than that afforded for the experimental nonhuman primate (ROLLIN 1996).

4 Species of Nonhuman Primates Used in Filovirus Studies

Monkeys are the mainstay of filovirus animal experimentation. Studies in nonhuman primates, however, have been confined to the most plentiful and commonly used laboratory species. These are: the African green monkey, *Cercopithecus aethiops*, wild-caught from Africa, mainly Uganda and Tanzania, in which species Marburg infection was first observed; the rhesus monkey, *Macaca macaca*, originally wild-caught in India, but now facility bred; and cynomolgus species, *Macaca fascicularis*, also known as the crab-eating monkey, from Asia, also mainly wild-caught since this species does not breed easily in captivity. All three of these species develop severe disease when infected with Ebola or Marburg viruses. Baboons may also be infected and are reported to have less severe disease (E. Ryabchikova, personal communication). Chimpanzees infected in the wild experience fatal disease (GEORGES et al. 1996). With increasing and appropriate limitation of, and more stringent transport regulations for wild-caught animals, the cost of these animals is steadily rising. Captive breeding colonies are also expensive to maintain. Smaller primate species might be desirable on the grounds of cost and safety, but experience with filovirus experimentation in candidate species is not available.

5 Marburg Virus Infection in Nonhuman Primates

The very first recorded appearance of a filovirus was in 1967 in Marburg and Frankfurt, Germany, and in Belgrade, Yugoslavia (KISSLING et al. 1968; MARTINI 1971). The outbreak originated from recently imported, presumably accidentally ('naturally') infected, African green monkeys (SIEGERT et al. 1968). Thirty-one animal staff and laboratory workers handling the animals or fresh autopsy tissues fell sick with a fulminating hemorrhagic fever; seven died (MARTINI et al. 1968; MARTINI 1969, 1971). The virus apparently originated in Uganda, where the monkeys had been freshly captured.

The published data following the Marburg outbreaks remain the best documentation of Marburg virus disease in animals. It was found that Marburg virus was equally pathogenic for African green monkeys and for humans. The disease in African green monkeys was acute, with mortality 20%–50% and possibly higher, since the exact number of infected animals was not known. This argued against monkeys being the natural reservoir. The pathophysiology and pathology of the infection in the monkeys closely resembled that observed in the human victims. Transmission was by direct contact with blood, secretions or tissues, and there was no evidence for airborne spread (MARTINI 1971).

Since the original Marburg outbreak was in wild-caught African green monkeys, this species was an obvious choice for modeling human disease. The rhesus monkey, in plentiful supply at that time, was found to be equally susceptible to infection and disease following inoculation with Marburg virus. Not surprisingly the disease in experimentally infected monkeys was identical to that in the 'naturally' infected monkeys; weight loss, fever, hemorrhages, and skin rash, culminating in hypothermia, shock and death. The characteristic petechial rash was best observed on the forehead, chest, axillae and groins of the animals. Edema around the eyes was also accompanied by rash and sometimes conjunctival injection. The incubation period to onset of fever was as short as 2 days. Mortality was 100% in experimentally infected animals, with death between 5 and 8 days after infection, and high titers of virus in most organs, but particularly in liver, spleen and lungs (BOWEN et al. 1978). Autopsy findings resembled those in fatal human cases with lesions in liver, spleen and lymphoid tissue. The uniformly fatal outcome may be related to the high virus inoculum (about 10^4 guinea pig LD_{50}/ml). It may also be related to the parenteral route of inoculation, since it was shown that reducing the inoculum led to slower onset of disease and longer time to death, but no reduction in mortality (HAAS and MAASS 1971). In 1976 the observation that high dose, parenteral inoculation of filoviruses is clearly lethal in humans was recorded during the outbreaks of Ebola virus disease in Zaire where needle inoculation resulted in uniformly fatal disease (WORLD HEALTH ORGANIZATION 1978a).

6 Ebola Virus Infection in Nonhuman Primates

The simultaneous appearance of Ebola viruses in 1976 in Zaire and Sudan drew attention away from Marburg virus studies in animals to its two newly described relatives (WORLD HEALTH ORGANIZATION 1978a,b). The Zaire virus has the highest mortality in humans of all the filoviruses (88%) and, not surprisingly, mortality in experimentally infected nonhuman primates approaches 100%. The Sudan virus (mortality in humans 57%) did not produce uniform fatality in nonhuman primates. (BOWEN et al. 1980) The disease was essentially indistinguishable from that produced by Marburg virus. Ebola viruses produced fever by the third day post inoculation, which peaked on days 7 and 8, persisting in one monkey until day 14 (ELLIS et al. 1978). Monkeys infected with the Zaire virus were moribund by day 6. Diarrhea and weight loss were marked, and a characteristic maculopapular rash was observed. Viremia was detectable within 48h of infection, (the time of the first sample) and animals died between 5 and 8 days following inoculation (BOWEN et al. 1978).

Histopathological observations were similar to Marburg virus infection, with focal acute necrosis in liver, spleen, lymph nodes, lungs and testes. Fibrin deposition in several organs suggested disseminated intravascular coagulation (BASKERVILLE et al. 1978). Tissues from monkeys infected with the Zaire virus showed higher levels of replication of viral particles than samples from Sudan virus-infected monkeys, with widespread involvement of many organs. Sudan virus particles were limited to liver, spleen and lymphoid tissue, and many were 'aberrant' in appearance, suggesting that high proportions of defective viral particles were being produced (ELLIS et al. 1978). However, these conclusions are questionable, since this study consisted of only four monkeys, two inoculated with each virus, and autopsies were performed at very different times during the course of infection, day 6 in the Zaire virus infected animals, and day 12 in the one Sudan virus infected animal autopsied, at a time when it was recovering from the infection.

7 Studies of the Pathogenesis of Filovirus Infection

Pathology of necessity is limited to observation of the static state at a single point in the disease course in the animal or patient, usually after death. Pathophysiology, by contrast, provides continuous monitoring of rapidly changing dynamic processes during the course of the disease. Filovirus infections evolve extremely rapidly, thus pathophysiology experiments with filoviruses are difficult. As a result there are few systematic published data, limited to some older studies conducted using rhesus monkeys as subjects in the 1980s (FISHER-HOCH et al. 1983, 1985). These studies preceded the understanding of cytokines and easily performed assay systems using inactivated specimens. Bioassays therefore predominated. These require large volume specimens and mandatory repetition performed on specimens with high-titer, live virus. The objective of these studies was to gain some understanding of the processes involved in the fulminating disease caused by filoviruses in primates, and, by inference, humans, and to develop rational approaches to therapy based on the findings. The animals were infected with Ebola (Zaire), 10^5 guinea pig LD_{50} (guinea pig passage 3), inoculated intraperitoneally. They were monitored daily using a range of hematological and clinical chemistry estimations designed to describe the evolution of the disease process more precisely. Data were evaluated by a team of virologists, hematologists, intensive care specialists and experts in the pathophysiology of the microcirculation (FISHER-HOCH et al. 1983, 1985). Neutrophilia, depletion of lymphocytes and early failure of platelet aggregation preceded a consumption coagulopathy accompanied by a microangiopathic hemolytic anemia (Fig. 1). Thrombocytopenia was marked, starting about day 4 following inoculation. However, at that time, though the platelet count still fell within the normal range, in vitro aggregation responses to standard challenges of collagen and ADP were impaired or absent (Fig. 2). This demonstrated that significant functional impairment preceded destruction in filovirus infections. As the disease progressed toward death, about day 7 following inoculation, platelet factor 4 (PF4) levels in platelets decreased but increased in serum, suggesting release. Endothelial cell function was also crudely assayed by measuring the ability of biopsies of autopsy specimens of aorta to produce prostacyclin in response to standard stimuli, and found to be impaired. Attempts were made to protect the microcirculation using a range of therapeutic agents which might act as anti-oxidants such as high dose vitamin E, infusions of 6-keto prostaglandin F1, and a thromboxane synthetase inhibitor. Though some differences in clinical chemistry could be detected between treated and untreated animals, all died regardless. The small number of subjects (11) and lack of availability of precise tools to measure accurately responses such as cytokine production, make interpretation of these data uncertain. Liver enzymes, (aspartate transaminase, AST, and alanine transaminase, ALT) were raised, with AST:ALT ratios as high as 11:1, a unique feature of viral hemorrhagic fevers observed also in patients infected with the Marburg virus in 1967 (Fig. 3) (MARTINI 1971; FISHER-HOCH et al. 1995). This ratio, which is the reverse of that observed in infection with hepatitis viruses, appears to be a feature of several hemorrhagic

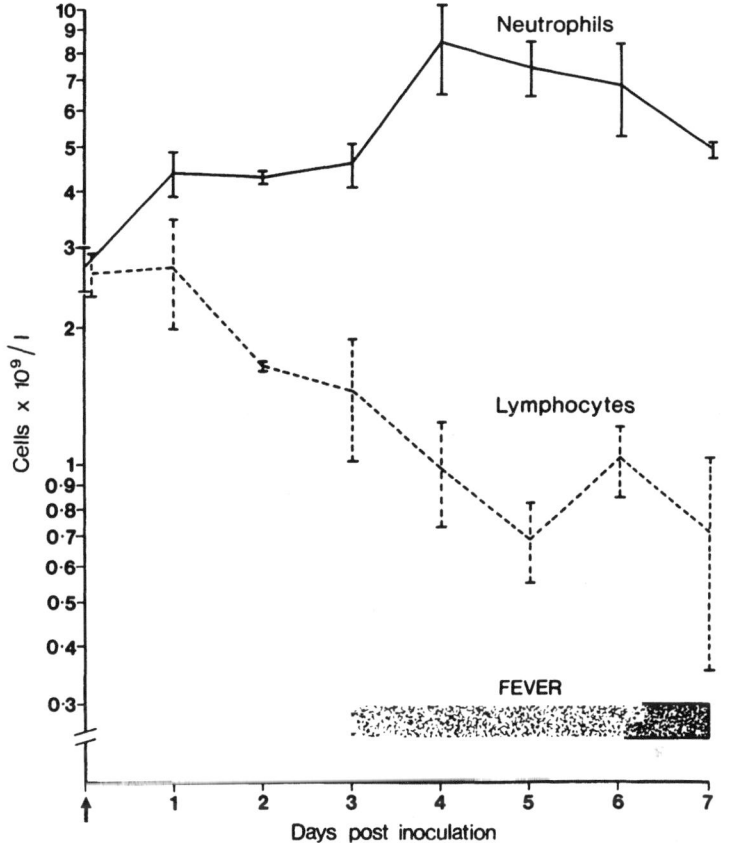

Fig. 1. Evolution of neutrophilia and lymphopenia in rhesus monkeys experimentally infected with Ebola virus

fevers (FISHER-HOCH et al. 1987, 1995). Weight loss was a reliable marker of disease severity in animals living beyond 7–9 days. Only one of the 11 animals survived.

Later experiments confirmed these data. Rhesus monkeys inoculated intraperitoneally with 10^3–10^4 guinea pig infectious units of Ebola (Zaire) virus become febrile the third to fifth day after inoculation, developing a petechial rash on the forehead, face, limbs and chest between the fourth and fifth days (VAN DER GROEN 1987). Severe prostration with diarrhea and bleeding led to rapid death. Though similar in onset, the disease caused by the filovirus from Sudan was characterized by less intensity of viremia and enzyme elevations, and there were some survivors. Petechiae were seen in dying monkeys. Viremias and liver enzymes did not reach the levels seen in Zaire virus infections, but as the illness progressed, severe thrombocytopenia, neutrophilia and lymphopenia developed, accompanied by very high levels of serum AST and LDH. Monkeys were obviously very sick, even the

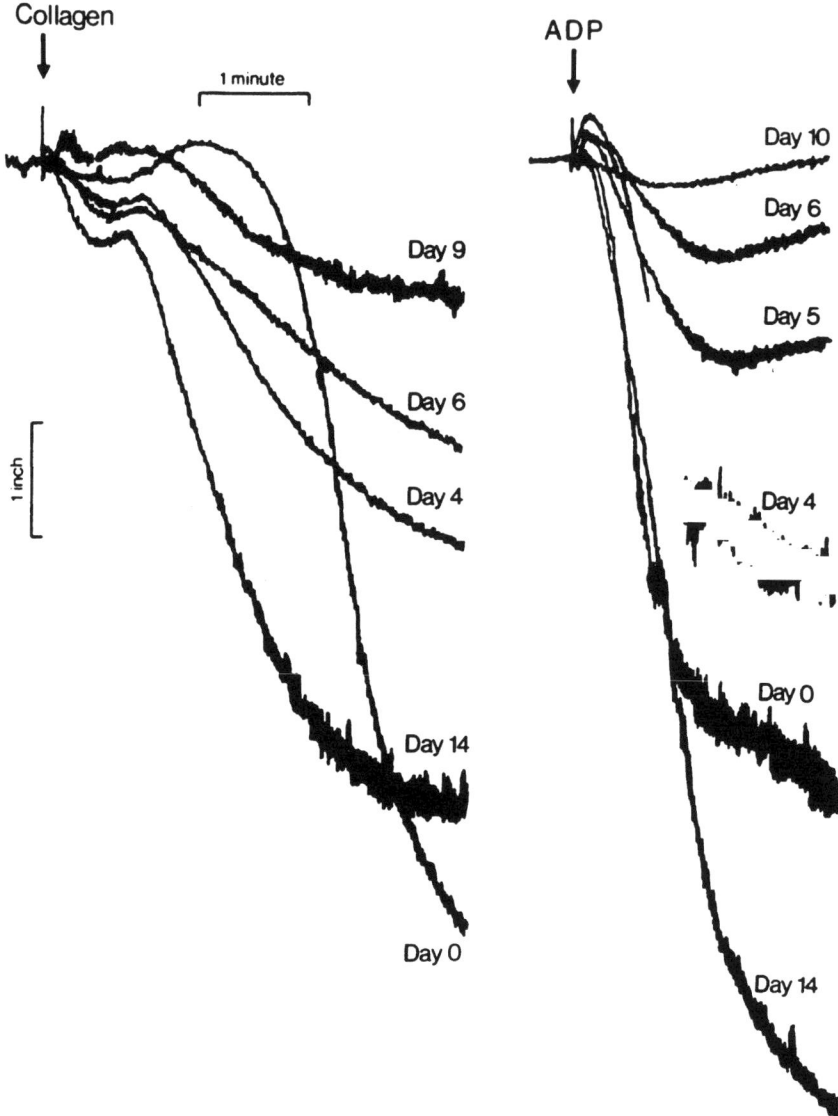

Fig. 2. Evolution of platelet function failure in rhesus monkeys experimentally infected with Ebola virus

eventual survivors, but hematological and biochemical parameters returned essentially to normal by day 20; thereafter recovery was rapid and complete.

The most profound physiologic alteration invariably associated with death is shock manifested by hypotension, effusions and facial edema. Severe, acute intravascular fluid loss often with frank bleeding into the tissue and into the gut is characteristic and results in dehydration, and electrolyte and acid-base imbalance.

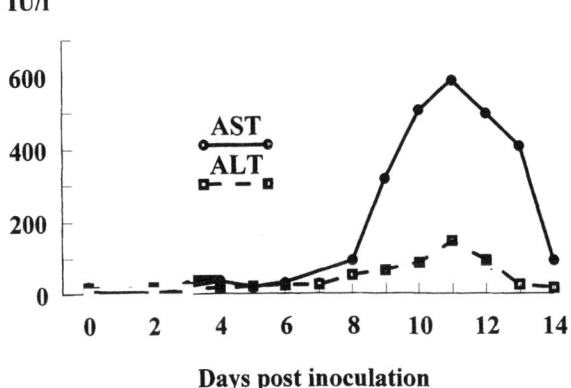

Fig. 3. Changes in serum transaminases during experimental infection with Ebola (Zaire) in nonhuman primates

Thus profound disruption of biochemical integrity of the endothelium is likely. Since there is no evidence for extensive viral destruction of endothelial cells by histology or electron microscopy, it must be assumed that the fluid losses are due to functional changes rather than lytic destruction of the endothelium by replicating virus. Nevertheless, the presence of profuse extracellular antigen suggests overwhelming viral replication, and this may underlie the processes which causing the multisystem collapse seen in Zaire virus infections.

8 The Immune Response to Filoviruses in Nonhuman Primates

Nonhuman primates develop high-titer filovirus-specific antibody by immunofluorescence (≥1024) within 14–21days of infection (FISHER-HOCH et al. 1992b). However, animals infected with the Zaire virus die for the most part before immunofluorescent antibody (IFA) can be detected (usually by day 9 post challenge). Sixty-nine days after challenge titers in animals infected with Sudan virus had decayed to a titer of between 256 to 1024 to Sudan virus antigen, and to a titer of 1024 to Zaire and Reston virus antigens. By 340 days following challenge, titers in animals infected with Asian filoviruses range from 64 to 1024 to the homologous filovirus virus and from 16 to 1024 to heterologous viruses. Sudan virus infected monkeys react to as high a titer to the Reston antigen as to the antigens from Sudan or Zaire viruses, whereas animals infected with Asian viruses react less strongly with the heterologous (African virus) antigens.

Radioimmunoprecipitation analysis of antibody responses showed that though there are cross-reactions between Asian and African filovirus nucleoproteins; no cross-reacting antibody to the glycoprotein is observed (FISHER-HOCH et al. 1992a).

9 Comparison of Filovirus Strains in Laboratory Animals

In 1989 three filoviruses were known; Marburg, Zaire and Sudan viruses. Then, unexpectedly, in November 1989, and again in early 1990, a filovirus closely related to Zaire virus was isolated from cynomolgus monkeys in the United States (JAHRLING et al. 1990). The animals were in quarantine facilities in Reston, Virginia, in Texas and in Pennsylvania (CENTERS FOR DISEASE CONTROL 1990b). The monkeys had recently been imported into the United States from the Philippines. No link with Africa or African animals could be found, and in the absence of such evidence, this is the first Asian filovirus observed.

The pathogenicity of the newly detected Asian filovirus in cynomolgus was unclear because of a high rate of concurrent infection with simian hemorrhagic fever virus (SHFV) (DALGARD et al. 1992). SHFV is a DNA virus which is a known severe, simian pathogen, producing hemorrhagic disease, unrelated to the filoviridae. SHFV does not apparently cause disease in humans. In the first reported SHFV epizootic, 223 of 1050 exposed animals died, with increased handling a risk factor for disease and death. The natural hosts and geographic distribution are also unknown, though the virus is known to infect rodents. The extent to which SHFV alone or synergy between SHFV and the Asian filoviruses was responsible for the illness and death of the original monkeys from which the newly described Asian filoviruses were isolated is altogether uncertain.

The monkeys imported to the United States in 1989 and 1990 were all from a single Philippine export facility (HAYES et al. 1992). Antigen detection ELISA assays using liver homogenates revealed that 85 (52.8%) of 161 of monkeys that died in the Philippine facility over a period of less than 3months were infected with filovirus. Incidence of infection was calculated to be 24.4/100 animals, or 0.6/100 monkey/days of follow-up. Documented case fatality of monkeys infected with filovirus and SHFV at this institution was 82.4%, and survivors developed high titer IFA antibody to Reston virus. Average duration of viremia was 5.6 ± 2.4 days. Diarrhea and respiratory problems were the most frequently recorded manifestations. As in the United States, many of the filovirus infected monkeys were co-infected with SHFV, which means data from these epizootics are again difficult to interpret and conclusions are open to question.

That the Asian filoviruses have a lower pathogenicity in primates is consistent with observations in accidental human infections, which have been uniformly asymptomatic (CENTERS FOR DISEASE CONTROL 1990c,d; MIRANDA et al. 1991). Other biological properties, such as the speed of replication in tissue culture, support the contention that there are clear differences in virulence between African and Asian filovirus. In a Reston-infected monkey, virus particles were observed embedded in the basement membrane of lung alveoli, and replication in the lung may occur with this filovirus, with which respiratory manifestations have been a feature in epizootics (GEISBERT et al. 1992). The nature of the pulmonary involvement needs further study.

These differences in pathogenicity between humans and nonhuman primates are also seen between nonhuman primate species. Consequently a systematic study was designed to address the issue of virus and host species variability in pathogenesis (FISHER-HOCH et al. 1992a). The design was checkerboard, using 32 monkeys of two different species, eight Asian and eight African, infected with four different viruses, two Asian (Reston and Pennsylvania viruses) and two African (Zaire and Sudan viruses). Care was taken to ensure that the Reston and Pennsylvania virus stocks were not contaminated with SHFV. The infected animals were kept in two separate rooms, in adjoining, but separate BSL4 facilities, one for Asian viruses and one for African viruses, and in each room the animals were handled by a separate team in a separate examination facility. In each room the viruses were further separated by species using laminar flow animal isolation hoods to avoid any chance of cross-contamination. The two monkey species were handled in two separate sequential experiments, with thorough decontamination of the facilities between the two exercises. Inocula were prepared from the same virus stocks, which were also back-titrated to ensure uniform challenges.

Four of eight African green and one of eight cynomolgus monkeys infected with African filoviruses survived (Table 1). Overall, survival was lower with African filovirus infection (5/16 animals survived following African filovirus infection and 11/15 following Asian filovirus infection). One animal in the Asian virus-infected group was apparently not infected, and because it was uncertain whether this was due to technical error or resistance to the inoculum, the animal was dropped from the analysis. No animal survived infection by the Ebola virus from Zaire. Regardless of virus strain, African green monkeys showed significantly higher survival (11/15) compared with cynomolgus (5/16).

The median incubation period for development of disease with Asian filoviruses was 7days (range 7–14), and for African filoviruses 3.5days (range 3–7). Highest fevers were seen on day 5 post-challenge in African filovirus-infected animals, and day 7 in Asian filovirus-infected animals. Rises in body temperature ranged from 2.3°C (Ebola) to 3.2°C (Reston). Many animals became hypothermic immediately before death, especially those infected with the Zaire virus which rapidly developed circulatory collapse.

The lowest platelet counts were observed between days 7 and 9 in animals infected with African viruses (median 117,500 cells/ml) and days 10–12 in animals infected with Asian viruses (median 82,000 cells/ml). Cell counts fell earliest in

Table 1. Number of animals challenged with filovirus strains and number of survivors by nonhuman primate species (numbers of deaths/number challenged)

Virus strain	Pennsylvania	Reston	Sudan	Zaire	Total
Primate species					
African green	0/3	0/4	0/4	4/4	4/15
Cynomolgus	3/4	1/4	3/4	4/4	11/16
Total	3/7	1/8	3/8	8/8	15/31

Asian virus-infected animals. In animals which recovered there was a rapid return to normal in platelet numbers by day 20, with rebound thrombocytosis in a few individuals. Giant platelets were frequently seen. Early neutrophilia was marked in animals infected with African filovirus, (highest recorded count 26,885 cells/ml on day 5 in an Ebola-infected animals, 24,924 cells/ml on day 11 in a Sudan-infected animal). High neutrophil counts were also observed in some Asian filovirus-infected animals (highest recorded count day 14 following inoculation in a Reston-infected animal, 20,736 cells/ml). Lowest lymphocyte counts were observed by day 7 post-infection with all viruses, and reactive lymphocytosis from about day 13 in survivors.

In African filovirus infections the highest recorded level of lactate dehydrogenase (LDH) was 16,465IU/l on day 5 following inoculation of an animal infected with Ebola (Zaire) virus, and highest AST, 7,961IU/l, day 5 following inoculation, again of an animal infected with Ebola (Zaire) virus. In Asian filovirus infections peak liver enzyme levels were not reached until about 15days after infection (highest recorded level of LDH 22,164IU/l day 14 following inoculation, and AST 3,219IU/l day 14 following inoculation). Even though the disease was less severe with Asian viruses, peak enzyme levels were nevertheless as high, or higher, in individual monkeys with these infections, than those with African infections. Between days 14 and 19 following inoculation enzyme levels fell rapidly back to normal in all survivors.

In summary, death from African filovirus infections is more rapid, occurring between days 6 and 11 post-challenge, compared with days 11–19 for Asian filovirus infections (Fig. 4). African filoviruses produce a rapid, high-titer viremia, peaking between days 5 and 7 post-challenge, whereas Asian filovirus viremias peak between days 7 and 9 and are cleared more slowly (maximum duration of viremia 7days for Sudan-infected animals and 12days for Asian virus-infected animals, Table 2). Overall, though favorable outcome was more likely in African green than cynomolgus monkeys, African greens did exhibit higher mean viremias than cynomolgus early in disease. Apparently, both monkey species develop viremia at the same time, but cynomolgus monkeys do not control the virus as efficiently as African greens (Fig. 4).

10 Transmission of Filoviruses Among Nonhuman Primates

Published data observed in Belgrade during the Marburg outbreaks are matched only by the reports of observations during ongoing epizootics in a holding facility in the Philippines (MARTINI 1971; HAYES et al. 1992). In the Marburg outbreak, 400–600 animals originating from four shipments reached Europe from Uganda over a 3week period. Frankfurt received only 40–60 animals from two shipments, and Belgrade about 300 animals from three shipments. The remainder went to Marburg. All spent 60–87days in a holding facility in Uganda before being shipped

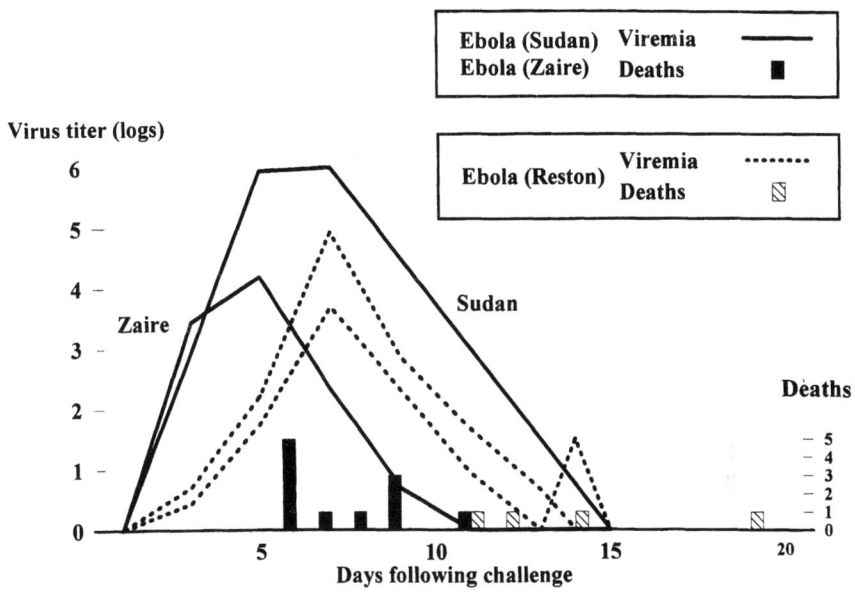

Fig. 4. Comparison of outcome in monkeys challenged with African and Asian filoviruses

Table 2. Mean viremia in nonhuman primates challenged with different filovirus strains

Day	Species	Pennsylvania	Reston	Sudan	Zaire
3	African green	1.333	0.875	4.375	3.875
	Cynomolgus	0.250	0.000	2.500	2.000
5	African green	2.833	2.500	4.875	5.875
	Cynomolgus	1.750	1.000	3.500	6.000
7	African green	4.167	3.625	2.625	6.000
	Cynomolgus	5.500	3.750	2.000	No samples
9	African green	3.000	2.375	1.375	No samples
	Cynomolgus	4.500	2.250	0.000	No samples
11	African green	0.667	0.375	0.000	No samples
	Cynomolgus	2.667	1.500	0.000	No samples
12	African green	No samples	No samples	No samples	No samples
	Cynomolgus	3.000	No samples	No samples	No samples
13	African green	0.667	0.000	0.000	No samples
	Cynomolgus	No samples	No samples	No samples	No samples
14	African green	No samples	No samples	No samples	No samples
	Cynomolgus	0.000	1.500	0.000	No samples
15	African green	0.000	0.000	0.000	No samples
	Cynomolgus	No samples	No samples	No samples	No samples

"No samples" indicates no survivors to sample.

to London, Heathrow, where they spent between 6 and 36 hours in an animal hostel prior to being forwarded to Germany. In Marburg, monkeys were housed in separate rooms with no recirculation of so-called air-conditioned ventilated air. Published data are unclear as to whether ongoing enzootics were observed in Germany nor are there details of animal movement. Data from the Belgrade enzootics are well recorded. Three shipments of monkeys from the same source in Uganda were received (Fig. 5). Unusually high mortality during 6 weeks quarantine was noted in all three shipments; 46/99 animals died from the first, and 20/95 and 30/94 from the second and third shipments, respectively. The Belgrade epizootic was clearly characterized by ongoing transmission by direct contact and daily death of one or more animals. This pattern was closely mirrored in the outbreaks of the Philippine filovirus, the Reston strain of Ebola virus, two decades later.

Epidemiologic studies published after the 1967 Marburg epidemics concluded that two or three infected monkeys would have been sufficient to initiate the epizootic and all three outbreaks of human disease (KISSLING et al. 1970; HENDERSON et al. 1971). It was stated at the time that evidence clearly pointed to transmission between monkeys in quarantine facilities through direct contact with equipment. Direct contact with blood and tissues was documented for all human cases and there was ample evidence against transmission by air. No evidence was ever produced that supported the hypothesis that the monkeys were infected in transit in London from any number of a wide range of mammals and birds that were also

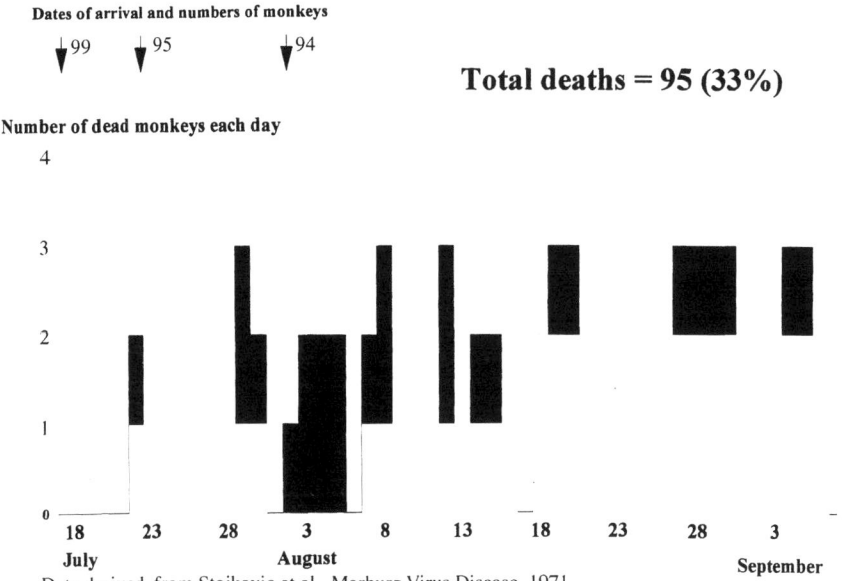

Fig. 5. Epidemic curve of the Marburg epizootic, Belgrade 1967

temporary lodgers at the airport hostel. Furthermore, no evidence could be found of epizootics in Uganda, but later some indirect information emerged at the time of the outbreak that there had been excess deaths in monkey colonies in islands near Lake Kyoga, north of Lake Victoria, to the east of Mount Elgon in Kenya. In Uganda monkeys were captured from this same area and placed in holding facilities there, reportedly in single cages. They were then transported to Entebbe. Here they were held for at least 3 days before shipment. At the time of the outbreak the trade had expanded considerably and holding times had been reduced, and crowding may have ensued. During July and August 1967, 1,772 *C. aethiops* were housed in Entebbe, and 1,290 exported, the majority to Germany and Yugoslavia (HENDERSON et al. 1971).

In early 1990 an outbreak of Reston virus infecting monkeys in a facility in Texas was monitored, and transmission was shown to be via handling and needles used for multiple monkeys (HENDRICKS et al. 1992). Tuberculin was routinely inoculated into newly arrived monkeys using one needle and syringe for seven animals. Monkeys were housed in cages and tested in such a way that every eighth monkey received a fresh tuberculin needle and syringe.

Epidemiologic data from human outbreaks have consistently shown that Ebola virus does not spread by aerosol (BARON et al. 1983; WORLD HEALTH ORGANIZATION 1978a,b). There are also strong epidemiologic data that natural transmission of filovirus between nonhuman primates and to animal handlers is by direct and intimate contact with blood and secretions, particularly inoculation and cuts (EMOND 1977; CENTERS FOR DISEASE CONTROL 1990a-e). Emphasis on avoiding penetrating injuries continues to be the most critical issue in safety. Despite this, transmission of the Zaire strain of Ebola to 'control' monkeys housed in the same room as infected monkeys not in animal isolators has been cited as evidence of aerosol infection (JAAX et al. 1995). In these circumstances it is impossible to exclude fomite, dust, hand or equipment contamination, or spraying of viral particles around the room during sluicing down of feces (DALGARD et al. 1992). It is exceedingly difficult to exclude viral transmission on fomites or personnel in a single room, whatever the virus, particularly in an animal care facility. The value of this report is to emphasize the need to house controls in a separate room. Since control animals are uninfected, they do not have to be in BSL4 facilities. Similarly, experimental infection of monkeys with Ebola virus has been achieved using high dose, forced inhalation (JOHNSON et al. 1995). This is not surprising since forced inhalation of more or less any microorganisms in sufficient quantities suspended in particles of size sufficiently small to reach the terminal alveoli are liable to induce pneumonia. Care needs to be taken to design animal experiments with physiologic relevance. If this essential scientific principle is ignored misleading conclusions are liable to be reached which impede serious scientific understanding of natural modes of transmission, and thus the real issues needed for control.

There are also published data reporting the presence of filovirus particles in lung alveoli (GEISBERT et al. 1992). This observation demonstrates viral pneumonitis, but does not in itself prove aerosol transmission, though this may be possible. Transmission by direct contact rather than aerosol transmission between monkeys is strongly supported by data from Manila where there was epidemiologic evidence against aerosol transmission (HAYES et al. 1992). The most significant risk factor for monkeys being infected in the epizootic in the Philippines holding facility was being an occupant of the same gang cage (sixfold increase of risk $p < 0.001$, OR 5.96, 95% CI 2.87–12.38).

In the absence of epidemiological data in nonhuman primates and strong epidemiological data against aerosol transmission in human outbreaks (WORLD HEALTH ORGANIZATION 1978a; Baron et al. 1983; CENTERS FOR DISEASE CONTROL 1990b,c, 1995; MIRANDA et al. 1991; HAYES et al. 1992), claims of aerosol transmission need to be viewed with circumspection and should not be allowed to distract attention from the importance of avoiding direct blood contact and penetrating injury. Undue attention to the remote possibility of aerosol transmission may lead to inappropriate and unwarranted neglect of patients by inexperienced medical personnel (ROLLIN 1996).

11 Persistence of Filoviruses

Humans and nonhuman primates with documented acute infection with Ebola Zaire, Sudan or Reston filoviruses seroconvert promptly to high-titer IFA to antigens prepared from all three viruses. Low-titer IFA antibody may be associated with filovirus infection in the past, and possibly immunity, but this remains to be demonstrated. Apparently nonspecific reactions are frequently seen with Zaire, Sudan and Reston antigens. This phenomenon is unexplained and is not seen with any other group of viruses. The same sera in a western blot react with most if not all viral proteins homologous with the IFA antigen used, and ELISA tests run in parallel are also positive at low titer. Among several possible hypotheses currently unsupported by data is the existence of unidentified, related viruses, or, alternatively, cross-reactions with host proteins. The result is that serological surveys are extremely difficult to interpret. This phenomenon does not occur with Marburg virus.

Marburg virus has been isolated from semen 7 months after acute infection, and sexual transmission of the virus under these circumstances has occurred (STOJKOVIC et al. 1971). Virus has also been isolated from the anterior chamber of the eye 2 months after acute infection (GEAR et al 1975). However, data concerning persistence are scarce since human infections with Marburg virus have been few, survivors scarce, and most experimentally Marburg virus-infected nonhuman primates have not survived.

Antibody to filoviruses in wild-caught monkeys and other animals has been reported from early studies (JOHNSON et al. 1981, 1982; IVANOFF et al. 1982). Following the importation of filovirus infected monkeys into the United States and Italy in 1989 and 1990, and the observation that up to 10% of nonhuman primates from a variety of sources apparently had antibody to filoviruses, real concern was raised about the potential for persistence of filoviruses in primates (Table 3). Large numbers of wild-caught nonhuman primates are imported into the United States alone (more than 20,000 each year in 1990) (CENTERS FOR DISEASE CONTROL 1990e). Many of the protocols on which they are placed involve immunosuppression, which might be thought to provide opportunities for reactivation of persistent or latent virus.

To address these concerns, 27 nonhuman primates were studied for up to 600 days following recovery from documented infection with Reston, Pennsylvania or Sudan filovirus viruses (FISHER-HOCH et al. 1992b). Sixteen had been experimentally infected and had survived, six had survived an outbreak in a quarantine facility and five were wild-caught African green monkeys from Tanzania with high-titer antibody, or seroconversion or fourfold rise in IFA titer to filovirus antigens (Table 4). Efforts to detect filoviruses were made using repeated culture on Vero E6 monolayers, by cocultivation techniques and by a reverse transcriptase polymerase chain reaction (RT-PCR) specific for Reston and Pennsylvania filovirus RNA. Serial specimens were obtained, including kidneys obtained at elective unilateral nephrectomy with liver and spleen biopsies (Table 5).

Two hundred and forty nine serum samples, 53 liver biopsies, 40 kidney and 40 spleen biopsies were cultivated for filoviruses, and examined for filovirus RNA by RT-PCR. From day 5 to day 14 following challenge, 37/121 serum specimens from Reston virus-infected animals contained virus. Filovirus could not be isolated from tissues beyond 19 days following challenge and from serum beyond 14 days (Table 2). Similarly viral genome could not be detected by RT-PCR in serum after day 14. No virus could be detected by RT-PCR or virus isolation in 55 serum specimens, 20 liver specimens, 17 spleen or 17 kidney specimens taken between days 19 and day 600 following infection.

The conclusions from these studies were that transmission in nature is not likely to be from persistently infected monkey to monkey or persistently infected

Table 3. Antibody to filovirus antigens measured by immunoflouresence

Monkey species	Antigen specificity	Number studied	Immunofluorescence assay titer			Percent positive
			16–32	64–128	>256	
Cynomolgus	Zaire	2,799	145	97	51	9.5%
	Sudan	2,958	58	44	33	4.4%
	Reston	2,724	165	116	79	11.7%
African green	Zaire	325	1	1	0	0.6%
	Sudan	333	7	4	2	3.8%
	Reston	316	17	3	1	6.2%
Rhesus	Zaire	435	15	11	0	5.6%
	Sudan	455	4	2	0	1.3%
	Reston	455	5	6	0	2.4%

The animals were healthy imported nonhuman primates in captivity in the United States. None of these animals had a history of filovirus disease.

Table 4. Experimental design to determine the ability of filoviruses to persist in nonhuman primates surviving virulent filovirus challenges

Source of infected monkey	Monkey species	Infecting virus	Primary diagnostic method	Biopsy (day)	Autopsy (day)
Experimental challenge	African green	Pennsylvania	Virus isolation	49	324
				52	69
					51
	Cynomolgus				89[a]
	African green	Reston	Virus isolation	50	51
				53	324
				53	29
					69
	Cynomolgus	Reston	Virus isolation	53	94[a]
				54	94[a]
				61	104[a]
	Cynomolgus	Sudan	Virus isolation	47	104[a]
	African green			48	69
					69
					29
					29
Outbreak: quarantine facilities	Cynomolgus	Reston	IFA titer rise 64 to 256	153	548
			Seroconversion to 1024	153	548
			High IFA titer 1024	153	548
			Seroconversion to 1024	153	548
			High IFA titer 1024	153	548
			High IFA titer 1024	153	548
Freshly caught in Tanzania	African Green	Sudan?	Seroconversion to 64	30	109
			High IFA titer 1024		600
			Seroconversion to 16		109
			Seroconversion to 64		109
			IFA titer rise 16 to 256		109

IFA, immunofluorescence assay.
[a]Rechallenged day 83 with Zaire virus.

monkey to human. Filoviruses, however, appear to be related, at least at the genomic sequence level of their glycoproteins, to respiratory pathogens of humans, such as respiratory syncytial virus (FELDMANN et al. 1993; RYABCHIKOVA et al. 1996). It may be that the natural host may experience a respiratory infection. It is also clear that despite the vast number of monkeys on protocol over many years in many countries, many of which have been immunosuppressed as part of the experimental protocol, not a single case of filovirus disease in a laboratory primate has been reported in these circumstances. Lack of persistence in survivors may not be compatible with the concept of the monkey as the natural reservoir of filoviruses. Healthy monkeys, even with high filovirus antibody titers, do not threaten other monkeys nor their handlers.

Table 5. Comparison of filovirus viremia titers and polymerase chain reaction (PCR) assays for filovirus genome in experimentally infected nonhuman primates

Day postinfection	Blood				Liver		Kidney		Spleen	
	Virus isolation		RT-PCR		Virus isolation	RT-PCR	Virus isolation	RT-PCR	Virus isolation	RT-PCR
	Positive/ negative	Percent positive	Positive/ negative	Percent positive	Positive/ negative	Positive/ negative	Positive/ negative	Positive/ negative	Positive/ negative	Positive/ negative
3	22/23	49	0/7	0					3	
5	27/4	87	6/9	40						
6	2/2	100	11/4	73	3/3		3/3		3/3	
7	25/1	96	11/4	73	1/1		1/1		1/1	
9	17%5	77	6/8	43	2/0		2/0		2/0	
11	9/10	47	3/0	100	1/0		0/2		1/1	
12	2/2	100	0/7	0	1/1		1/0		1/0	
13	1/10	0	1/5	17						
14	3/4	43	0/7	0			1/0		1/0	
15	0/11	0	0/7	0						
18	0/11	0								
19	0/5	0					1/1		1/1	
20–29	0/27	0	0/4				0/1		0/3	
30–39	0/6	0								
40–49					0/3	0/1	0/3	0/1	0/3	0/1
50–59	0/4	0			0/6	0/2	0/6	0/3	0/6	0/3
60–69	0/4	0			0/3		0/3	0/2	0/3	
70–79										
80–89	0/5	0								
90–99										
100–199	0/6	0			0/12	0/2	0/2	0/2	0/3	0/2
200–299					0/2	0/6	0/2	0/2	0/2	0/2
300–399	0/1	0				0/2		0/2		0/2
400–499										
500–600					0/7	0/7	0/7	-0/7	0/7	0/7

12 Protection Studies in Nonhuman Primates

Though the market for an effective filovirus vaccine for human use is likely to be small, and many of the most appropriate potential recipients people with little economic or political clout, the absence of effective treatment, the alarm the disease arouses, and the severity of the threat of infection to medical and laboratory personnel demand treatment and prevention measures, including vaccine development. Clearly a first step would be to see if monkeys surviving infection with the Sudan virus, which had developed high levels of antibody detectable by IFA, could survive challenge with the Zaire virus. This experiment was first performed in 1980. Both challenged monkeys died with high virus titers in blood (BOWEN et al. 1980). Similar experiments using guinea pigs gave variable results. Furthermore, in vitro estimations of neutralizing antibody failed to detect neutralizing activity in survivors except using a fixed serum dilution, varying virus dilution technique (FISHER-HOCH et al. 1992a). The ability of antibody from survivors to neutralize filoviruses was therefore not demonstrated, and prospects for a simple, killed vaccine not encouraging.

Attempts to protect nonhuman primates by vaccinating with irradiated whole Zaire virus preparations were unsuccessful (J. MCCORMICK, unpublished data). In a later experiment, five cynomolgus monkeys which had survived initial filovirus challenges were rechallenged 83 days later with Ebola (Zaire) (Table 4) (FISHER-HOCH et al. 1992a). Three had survived Reston virus infection, one Pennsylvania virus and one Sudan virus. The immune response in these animals was studied using radioimmunoprecipitation (RIP) and neutralizing techniques. By RIP it was shown that antibody to the filovirus glycoprotein was virus-specific, but that Reston and Pennsylvania viruses cross-reacted strongly, indicating that they are very similar. In contrast, serum from the Sudan virus-infected survivor was unable to recognize the Zaire or the Asian filovirus glycoproteins. In all survivors, virus-specific anti-glycoprotein antibody could be detected by day 14 post-challenge, and was maximal on day 27. Virus-specific neutralizing antibody could be detected in all five animals. The only animal in which cross-reacting neutralizing antibody was seen was the one which had survived Sudan virus. Serum from this monkey, which survived the rechallenge, was able to neutralize Zaire virus. This neutralizing activity was again only detected using a constant antibody-varying virus dilution technique as described above, and attempts to perform a classic virus neutralization assay using serum dilutions failed to detect any neutralizing activity.

The five animals were challenged with the Zaire strain of Ebola virus. Three died with classical Ebola disease and high viremias. Two survived the rechallenge without developing detectable viremia, one original Sudan virus-challenged animal and one Reston virus-challenged animal. Neither survivor developed viremia. There was a marked drop in filovirus-specific IFA antibody titer after rechallenge, particularly in those that died; two animals had no demonstrable IFA antibody to African filovirus viruses on the day of death. Neither survivor had detectable antibody to the Zaire virus glycoprotein by RIP analysis before challenge, but neither developed Zaire virus viremia nor clinical illness, though mild elevations of LDH and AST and drop in platelet counts were observed. Antibody titers to the Zaire antigen did not rise in the two survivors.

The conflicting results of this study show that protection against fatal challenge with a Zaire virus can be achieved, but the conditions that are required for development of immunity are unexplained. It is clear, however, that since the survivors never developed viremia, they are protected against reinfection and replication of the virus, not just against development of overt disease. It is possible that a live homologous vaccine, expressing appropriate antigens, may be effective, since an association was observed between protection and possession of preexisting antibody to filovirus at the facility in Manila during the 1989–1990 epizootic of Reston virus. In this situation protection correlated with antibody titer. Even then a small number of animals (six) with preexisting antibody became infected (HAYES et al. 1992).

13 Smaller Laboratory Animals in Filovirus Studies

The outcome in smaller laboratory animals is variable and is also virus strain-, inoculum- and passage-dependent. Immediately following the Marburg outbreak, monkeys, mice, hamsters and guinea pigs were experimentally infected with the virus. Passage of the Marburg virus in guinea pigs and the pathology of the infection were described (SHU et al. 1969; ROBIN et al. 1971; KORB and SLENCZKA 1981) along with methods for the detection of filovirus antigen in the liver of guinea pigs by immunofluorescence (SLENCZKA et al. 1968). This established the use of the guinea pig for propagation of high-titer virus in the laboratory and titration of stock inoculae. Experimental infection of hamsters was also described, and this animal has been used for antibody production, particularly for Marburg virus (ZLOTNIK and SIMPSON 1969; ZLOTNIK 1969; KISSLING et al. 1970; SLENCZKA et al. 1971). In hamsters higher inoculae are required to produce overt disease than in nonhuman primates, and serial passage is required (SIMPSON et al. 1968; SIMPSON 1969).

Following the appearance of Ebola virus in 1976, guinea pigs were again used (JOHNSON et al. 1977; BOWEN et al. 1977). Primary isolation of the Sudan virus has been difficult in tissue culture, and guinea pigs have proved to be sensitive for isolation from clinical material. They were inoculated intraperitoneally to titrate virus and prepare stocks, for which purpose they remain a mainstay of laboratory preparation of virulent inoculae (ELLIS et al. 1978). At the same time comparison of the virulence of the Zaire and Sudan viruses of Ebola virus in guinea pigs confirmed the higher pathogenicity of the Zaire virus (BOWEN et al. 1980).

Mice do not develop disease with Marburg virus, but the virulence of Ebola virus in newborn mice has been demonstrated (VAN DER GROEN et al. 1979). Rabbits, mice, chick embryos, and hamsters have been used for reagent production, particularly production of polyclonal and monoclonal antibodies (HOFMANN and KUNZ 1968; SHEVTSOVA 1970; SLENCZKA et al. 1971; PATTYN et al. 1977; VAN DER GROEN et al. 1979; VAN DER GROEN and ELLIOT 1982; CHEPURNOV et al. 1994; SANCHEZ et al. 1996). Suckling mice have been used in unsuccessful attempts to detect neutralizing activity in humans (MCCORMICK et al. 1983).

Guinea pigs have also been used for studying the use of immunomodulators in both Ebola and Marburg virus disease processes in guinea pigs infected via the respiratory route (SERGEEV et al. 1995). Protection of guinea pigs using a killed vaccine has also been reported, but not confirmed (LUPTON et al. 1980). The variability of outcome in filovirus-infected guinea pigs has made some of these studies difficult to interpret. More recently a report describes efficacy of plasmids encoding Ebola virus proteins injected into outbred guinea pigs (XU et al. 1998). The authors state that the animals were killed on day 10 following challenge and that there were only five control animals. It is well known both that mortality is highly variable in outbred guinea pigs and that many will die of Ebola infection well after day ten. This report underlines the need to pay careful attention to biological variability in animal studies. Ideally guinea pig studies are best performed in inbred animals, and more controls than test animals will be needed to provide statistical significance. Extended observation periods are essential, particularly for survivors.

Guinea pigs have even been implicated in the ecology of filoviruses, but these observations were more likely due to the lack of specificity of the antibody assay than to real events (STANSFIELD et al. 1982). There are no reports of successful propagation of filoviruses in arthropods with the exception of a single report of propagation of filoviruses in *Aedes aegypti* which remains unconfirmed (KUNZ et al. 1968).

It is possible that several species lower on the phylogenetic tree than primates may be involved in silent persistence of the virus in nature. Information from animal experimentation regarding the natural history of filoviruses in their ecological niche has been hard to obtain through experimental infections, but some indicators of potential mechanisms for maintenance of the virus in nature may have been glimpsed.

Attempts to use small animals as models for human disease have generally not been successful. This is for several reasons. The outcome of infection is variable, even with high dose inoculae of the most pathogenic viruses. Adult mice, for instance, do not develop disease even when infected with the Zaire virus. This means that extrapolation to human disease may be uncertain. Blood samples for measurements of hematological and biochemical parameters are very small. Assay systems developed for human subjects may not work as well with rodent proteins as well as they do with nonhuman primate specimens, where in general, reagents for human use work very well. The advantages of a biological system with larger numbers than can be provided using nonhuman primates still leaves small laboratory animals their place in several types of careful studies, and with new molecular techniques they may regain importance in studies of pathophysiology and immunopathology (RYABCHIKOVA et al. 1996b).

In summary, variable pathogenicity is seen in smaller laboratory animals. Zaire virus kills guinea pigs more consistently after several adaptive passages, whereas the Sudan and Marburg viruses do not. Only the Zaire virus was found to be lethal for suckling mice (McCormick et al. 1983). Older mice and some other animals are resistant (Swanepoel et al. 1996). Currently the most use of small animals in laboratory studies of filoviruses is in pilot experiments asking simple questions, such as the production of a specific cytokine, or the ability to protect of a new vaccine candidate, and in routine tasks, such as raising antisera.

14 Biosafety in Experimental Filovirus Infections

There are problems associated with use of all animals for modeling human filovirus disease, especially nonhuman primates. All experiments with live virus must be carried out in BSL4 facilities; a constraint which is essential for the African filoviruses, but possibly questionable for the Asian viruses. This means that any assay which requires biologically active material, that is all bioassays and some enzymatic studies, have to be performed entirely within BSL4. The team handling animals therefore has to acquire a wide range of skills. In practice, every effort is made to harvest material and inactivate at the earliest possible step, so that assays can be performed in BSL2 laboratories, where quality control is more easily assured. Nevertheless, all animal experiments are expensive, difficult, strenuous, and potentially dangerous if experiments are not well-planned, staff well-trained, or if mistakes are made. Studies are only safe in a well-equipped, maintained and staffed space-suit laboratory, managed in such a way as to reduce stress on staff to the minimum. Infected animals are the most potent and unpredictable source of laboratory infection, and the handlers must all be very well aware at all times that their entire bodies must be completely protected against any penetrating injury. Above all, direct injury, such as by needle stick, must be avoided (Emond et al. 1977). At the same time, the handlers must be sufficiently comfortable to be able to work in a calm and relaxed manner both to avoid accidents to themselves and further distress to the animals.

15 Challenges for the Future

Nonhuman primates will continue to be important in the study of filovirus infections. Given the scarcity and difficulty of studying human disease, experimentally infected animals are the only way in which we may gain some understanding of the processes involved. The main problem, outside obvious expense, is that the biosafety restriction necessarily placed limits these studies to just a few laboratories. At present the available BSL4 laboratories are not focused on pathophysiology and treatment, since most current efforts are directed toward ecology and molecular studies. Detailed studies of disease processes in such a rapid infection are difficult and demanding to conduct, though newly developed continuous remote monitoring techniques would clearly make them easier and provide more complete data. Without considerable infusion of new, up-to-date expertise in the understanding of the underlying biology of disease, the prospects are currently not good for major advances in therapy or even prevention. This is worth pursuing, because understanding the intracellular molecular cascade of events in disease is essential to understanding the collapse of the microcirculation which precedes death. Furthermore, detailed understanding of the disease processes and immunology of survivors, and the true nature of protection, is essential for tailored development of effective vaccines.

Efforts to return to studies of pathophysiology using newly developed molecular techniques which allow precise measurements from small volume specimens would be well worthwhile. Large new arenas of understanding of the mechanisms of cytokine induction, interferon activities, lymphokines and leukotrienes applied to fresh studies of Ebola virus infection would not only help us to understand, and possibly treat, this disease, but might very well cast fresh light on a number of fulminating infectious diseases, since, at least in the final cascade of events, common pathways are likely to be involved. The nonhuman primate allows careful, controlled and sequential study of rapidly evolving events in a way which is extremely difficult to follow in patients even in the most well equipped and well run facilities.

It might be equally useful to develop better nonhuman primate models for Ebola infection. A smaller, less expensive monkey would allow larger numbers, but smaller specimen volumes might limit investigations. Probably the most relevant single exercise at present would be careful titration of dose and route of infection in order to develop a model which more closely resembles the mortality seen in human infections the Zaire strain of Ebola virus. African green monkeys experimentally inoculated with Marburg virus soon after the 1967 outbreak all died, regardless of route of inoculation, but progression to death was slowed if the inoculum titer was reduced. Reduction in inoculum, delivered by a more natural route than a needle, would provide an opportunity to study the disease during the acute and, most importantly, the convalescent phase, when the deleterious processes are rapidly reversed. These are key data. It would also provide a more realistic model for vaccine efficacy studies. It may be that the disappointing results in rechallenge experiments are related to the overwhelming parenteral dose delivered at rechallenge.

Finally, if we are to make real progress in understanding filovirus disease, we need to consider the use of the nonpathogenic filoviruses under less stringent safety conditions. Reston virus has infected humans, but has never caused disease. The species specificity of these viruses in terms of pathogenicity needs to be reviewed critically with a view to using these viruses to advance our knowledge more rapidly.

References

Baron RC, McCormick JB, Zubeir OA (1983) Ebola virus disease in southern Sudan: hospital dissemination and intrafamilial spread. Bull World Health Organ 61:997–1003

Baskerville A, Bowen ET, Platt GS, McArdell LB, Simpson DI (1978) The pathology of experimental Ebola virus infection in monkeys. J Pathol 125:131–138

Bowen ET, Lloyd G, Harris WJ, Platt GS, Baskerville A, Vella EE (1977) Viral haemorrhagic fever in southern Sudan and northern Zaire. Preliminary studies on the aetiological agent. Lancet 1:571–573

Bowen ET, Platt GS, Lloyd G, Raymond RT, Simpson DI (1980) A comparative study of strains of Ebola virus isolated from southern Sudan and northern Zaire in 1976. J Med Virol 6:129–138

Bowen ET, Platt GS, Simpson DI, McArdell LB, Raymond RT (1978) Ebola haemorrhagic fever: experimental infection of monkeys. Trans R Soc Trop Med Hyg 72:188–191

Centers for Disease Control (1990a) Update: Ebola-related filovirus infection in nonhuman primates and interim guidelines for handling nonhuman primates during transit and quarantine. MMWR Morb Mortal Wkly Rep 39:22–4,29–30

Centers for Disease Control (1990b) Update: evidence of filovirus infection in an animal caretaker in a research/service facility. MMWR Morb Mortal Wkly Rep 39:296–297

Centers for Disease Control (1990c) Update: filovirus infections among persons with occupational exposure to nonhuman primates. MMWR Morb Mortal Wkly Rep 39:266–267

Centers for Disease Control (1990d) Update: filovirus infection in animal handlers. MMWR Morb Mortal Wkly Rep 39:221

Centers for Disease Control (1990e) Update: filovirus infection associated with contact with nonhuman primates or their tissues. MMWR Morb Mortal Wkly Rep 39:404–405

Centers for Disease Control (1995) Update: outbreak of Ebola viral hemorrhagic fever–Zaire 1995. MMWR Morb Mortal Wkly Rep 44:468–9, 475

Chepurnov AA, Merzlikin NV, Chepurnova TS, Vorob'eva MS (1994) Preparation of rabbit antiserum to Ebola virus. Vopr Virusol 39:286–288

Dalgard DW, Hardy RJ, Pearson SL, Pucak GJ, Quander RV, Zack PM, Peters CJ, Jahrling PB (1992) Combined simian hemorrhagic fever and Ebola virus infection in cynomolgus monkeys. Lab Anim Care 42:152–157

Ellis DS, Bowen ET, Simpson DI, Stamford S (1978) Ebola virus: a comparison, at ultrastructural level, of the behaviour of the Sudan and Zaire strains in monkeys. Br J Exp Pathol 59:584–593

Emond RT, Evans B, Bowen ET, Lloyd G (1977) A case of Ebola virus infection. Br Med J 2:541–544

Feldmann H, Klenk HD, Sanchez A (1993) Molecular biology and evolution of filoviruses. Arch Virol Suppl 7:81–100

Fisher-Hoch SP, Brammer TL, Trappier SG, Hutwagner LC, Farrar BB, Ruo SL, Brown BG, Hermann LM, Perez Oronoz GI, Goldsmith CS, et al (1992a) Pathogenic potential of filoviruses: role of geographic origin of primate host and virus strain. J Infect Dis 166:753–763

Fisher-Hoch SP, Khan JA, Rehman S, Mirza S, Khurshid M, McCormick JB (1995) Crimean Congo-haemorrhagic fever treated with oral ribavirin. Lancet 346:472–475

Fisher-Hoch SP, Lloyd G, Platt GS, Simpson DI, Neild GH, Barrett AJ (1983) Haematological and biochemical monitoring of Ebola infection in rhesus monkeys: Implications for patient management. Lancet ii:1055–1058

Fisher-Hoch SP, Mitchell SW, Sasso DR, Lange JV, Ramsey R, McCormick JB (1987) Physiological and immunologic disturbances associated with shock in a primate model of Lassa fever. J Infect Dis 155:465–474

Fisher-Hoch SP, Perez Oronoz GI, Jackson EL, Hermann LM, Brown BG (1992b) Filovirus clearance in nonhuman primates. Lancet 340:451–453

Fisher-Hoch SP, Platt GS, Neild GH, Southee T, Baskerville A, Raymond RT, Lloyd G, Simpson DI (1985) Pathophysiology of shock and hemorrhage in a fulminating viral infection (Ebola). J Infect Dis 152:887–894

Gear JS, Cassel GA, Gear AJ, Trappler B, Clausen L, Meyers AM, Kew MC, Bothwell TH, Sher T, Miller GB, Schnider J, Koornhof HJ, Gomperts ED, Isaacson M and Gear, JH (1975) Outbreak of Marburg virus disease in Johannesburg. Br Med J 4:489–493

Geisbert TW, Jahrling PB, Hanes MA, Zack PM (1992) Association of Ebola-related Reston virus particles and antigen with tissue lesions of monkeys imported to the United States. J Comp Pathol 106:137–152

Georges AJ, Renault AA, Bertherat E, Baize S, Leroy E, Le Guenno B, Lepage J, Amblard J, Edzang S, Georges-Courbot MC (1996) Recent Ebola virus outbreaks in Gabon from 1994 to 1996: Epidemiologic and control issues. International Colloquium on Ebola virus Research, 4–7 September 1996, Antwerp, Belgium (Abstr)

Haas R, Maass G (1971) Experimental infection of monkeys with the Marburg virus. In: Martini GA, Siegert R (eds) Marburg virus disease. Springer, Berlin Heidelberg New York, pp 136–143

Hayes CG, Burans JP, Ksiazek TG, Del Rosario RA, Miranda ME, Manaloto CR, Barrientos AB, Robles CG, Dayrit MM, Peters CJ (1992) Outbreak of fatal illness among captive macaques in the Philippines caused by an Ebola-related filovirus. Am J Trop Med Hyg 46:664–671

Henderson BE, Kissling RE, Williams MC, Kafuko GW, Martin M (1971) Epidemiological studies in Uganda relating to the 'Marburg' agent. In: Martini GA, Siegert R (eds) Marburg virus disease. Springer, Berlin Heidelberg New York, pp 166–176

Hendricks K, Pearson S, McCormick JB, Fisher-Hoch SP (1992) Epizootic of Reston virus in an animal holding facility in Texas. Abstracts of the 42nd Annual Meeting of the American Society of Tropical Medicine and Hygiene, Boston, Massachusetts (Abstr)

Hofmann H, Kunz C (1968) "Marburg virus" (Vervet monkey disease agent) in tissue cultures (in German). Zentralbl Bakteriol Orig 208:344–347

Ivanoff B, Duquesnoy P, Languillat G, Saluzzo JF, Georges A, Gonzalez JP, McCormick JB (1982) Haemorrhagic fever in Gabon. I. Incidence of Lassa, Ebola and Marburg viruses in Haut-Ogooue. Trans R Soc Trop Med Hyg 76:719–720

Jaax N, Jahrling P, Geisbert T, Geisbert J, Steele K, McKee K, Nagley D, Johnson E, Jaax G, Peters C (1995) Transmission of Ebola virus (Zaire strain) to uninfected control monkeys in a biocontainment laboratory. Lancet 346:1669–1671

Jahrling PB, Geisbert TW, Dalgard DW, Johnson ED, Ksiazek TG, Hall WC, Peters CJ (1990) Preliminary report: isolation of Ebola virus from monkeys imported to USA. Lancet 335:502–505

Johnson BK, Gitau LG, Gichogo A, Tukei PM, Else JG, Suleman MA, Kimani R (1981) Marburg and Ebola virus antibodies in Kenyan primates [letter]. Lancet 1:1420–1421

Johnson BK, Gitau LG, Gichogo A, Tukei PM, Else JG, Suleman MA, Kimani R, Sayer PD (1982) Marburg, Ebola and Rift Valley Fever virus antibodies in East African primates. Trans R Soc Trop Med Hyg 76:307–310

Johnson E, Jaax N, White J, Jahrling P (1995) Lethal experimental infections of rhesus monkeys by aerosolized Ebola virus. Int J Exp Pathol 76:227–236

Johnson KM, Lange JV, Webb PA, Murphy FA (1977) Isolation and partial characterisation of a new virus causing acute haemorrhagic fever in Zaire. Lancet 1:569–571

Kissling RE, Murphy FA, Henderson BE (1970) Marburg virus. Ann NY Acad Sci 174:932–945

Kissling RE, Robinson RQ, Murphy FA, Whitfield S (1968) Agent of disease contracted from green monkeys. Science 160:888

Korb G and Slenczka W (1981) Light microscopy study of Ebola virus hepatitis in guinea pigs (in German). Verh Dtsch Ges Pathol 65:100–102

Kunz C, Hofmann H, Aspock H (1968) Propagation of "Marburg virus" (Vervet monkey disease agent) in Aedes aegypti (in German). Zentralbl Bakteriol Orig 208:347–349

Lupton HW, Lambert RD, Bumgardner DL, Moe JB, Eddy GA (1980) Inactivated vaccine for Ebola virus efficacious in guinea pig model [letter]. Lancet 2:1294–1295

Martini GA (1969) Marburg agent disease in man. Trans R Soc Trop Med Hyg 63:295–302

Martini GA (1971) Marburg Virus disease, clinical syndrome. In: Martini GA, Siegert R (eds) Marburg virus disease. Springer, Berlin Heidelberg New York, pp 1 9

Martini GA, Knauff HG, Schmidt HA, Mayer G, Baltzer G (1968) A hitherto unknown infectious disease contracted from monkeys: "Marburg-virus" disease. Ger Med Mon 13:457–470

McCormick JB, Bauer SP, Elliott LH, Webb PA, Johnson KM (1983) Biologic differences between strains of Ebola virus from Zaire and Sudan. J Infect Dis 147:264–267

Miranda ME, White ME, Dayrit MM, Hayes CG, Ksiazek TG, Burans JP (1991) Seroepidemiological study of filovirus related to Ebola in the Philippines [letter]. Lancet 337:425–426

Pattyn S, van der Groen G, Courteille G, Jacob W, Piot P (1977) Isolation of Marburg-like virus from a case of haemorrhagic fever in Zaire. Lancet 1:573–574

Robin Y, Bres P, Camain R (1971) Passage of Marburg virus in guinea pigs. In: Martini GA, Siegert R (eds) Marburg virus disease. Springer, Berlin Heidelberg New York, pp 117–122

Rollin P (1996) Clinical virology of human infection. International Colloquium on Ebola Virus Research, 4–7 September 1996, Antwerp, Belgium 33 [Abstr]

Ryabchikova E, Kolesnikova LV, Smolina MP, Tkachev VK, Pereboeva LA, Baranova SG, Grazhdantseva AA, Rassadkink Y (1996a) Ebola virus infection in guinea pigs: presumable role of granulomatous inflammation in pathogenesis. Arch Virol 141:909–921

Ryabchikova E, Strelets L, Kolesnikova LV, Pyankov O, Sergeev AN (1996b) Respiratory Marburg virus infection in guinea pigs. Arch Virol 141:2177–2190

Sanchez A, Trappier SG, Mahy BW, Peters CJ, Nichol ST (1996) The virion glycoproteins of Ebola viruses are encoded in two reading frames and are expressed through transcriptional editing. Proc Natl Acad Sci USA 93:3602–3607

Sergeev AN, Lub MI, P'iankova OG, Kotliarov LA (1995) The efficacy of the emergency prophylactic and therapeutic actions of immunomodulators in experimental filovirus infections (in Russian). Antibiot Khimioter 40:24–27

Shevtsova ZV (1970) The Marburg virus (Rhabdovirus simiae) (in Russian). Vopr Virusol 15:643–646

Shu HL, Siegert R, Slenczka W (1969) The pathogenesis and epidemiology of the "Marburg-virus" infection. Ger Med Mon 14:7–10

Siegert R, Shu HL, Slenczka W (1968) Isolation and identification of the "Marburg virus". Ger Med Mon 13:514–518

Simpson DI (1969) Vervet monkey disease transmission to the hamster. Br J Exp Pathol 50:389–392

Simpson DI, Zlotnik I, Rutter DA (1968) Vervet monkey disease: experimental infection of monkeys with the causative agents and antibody studies in wild-caught monkeys. Br J Exp Pathol 49:458

Slenczka W, Shu HL, Piepenberg G, Siegert R (1968) Detection of the antigen of the "Marburg virus" in the organs of infected guinea-pigs by immunofluorescence. Ger Med Mon 13:524–529

Slenczka W, Wolff G, Siegert R (1971) A critical study of monkey sera for the presence of antibody against the Marburg virus. Am J Epidemiol 93:496–505

Stansfield SK, Scribner CL, Kaminski RM, Cairns T, McCormick JB, Johnson KM (1982) Antibody to Ebola virus in guinea pigs: Tandala, Zaire. J Infect Dis 146:483–486

Stojkovic LJ, Bordjoski M, Gligic A, Stefanovic Z (1971) Two cases of cercopithecus-monkeys-associated haemorrhagic fever. In: Martini GA, Siegert R (eds) Marburg virus disease. Springer, B' ...n Heidelberg New York, pp 24–33

Swanepoel R, Leman PA, Burt FJ, Braack LE (1996) Studies on the ecology of filoviruses in southern Africa. International Colloquium on Ebola Virus Research 99 [Abstr]

van der Groen G, Elliot LH (1982) Lack of cross reactivity of rhabdovirus antibodies with Marburg and Ebola antigens in the indirect immunofluorescent antibody test. Ann Soc Belg Med Trop 62:67–68

van der Groen G, Jacob W, Pattyn SR (1979) Ebola virus virulence for newborn mice. J Med Virol 4:239–240

World Health Organization (1978a) Ebola haemorrhagic fever in Zaire 1976. Bull World Health Organ 56:271–293

World Health Organization (1978b) Ebola haemorrhagic fever in Sudan. 1976 Report of a World Health Organisation International Study Team. Bull World Health Organ 56:247–270

Zlotnik I (1969) Marburg agent disease: pathology. Trans R Soc Trop Med Hyg 63:310–327

Zlotnik I, Simpson DI (1969) The pathology of experimental vervet monkey disease in hamsters. Br J Exp Pathol 50:393–399

Animal Pathology of Filoviral Infections

E.I. Ryabchikova, L.V. Kolesnikova, and S.V. Netesov

1	Introduction	145
2	Filoviral Infections in Rodents	146
2.1	Adaptation of Filoviruses to Rodents	146
2.2	Marburg Infection in Guinea Pigs	148
3	Filoviral Infections in Nonhuman Primates	150
3.1	Organ Pathology	150
3.1.1	Blood Cells and Hematopoiesis	150
3.1.2	Blood Vessels	151
3.1.3	Liver	152
3.1.4	Spleen	154
3.1.5	Kidney	155
3.1.6	Lung	156
3.1.7	Gastrointestinal Tract	157
3.1.8	Reproductive System	158
3.1.9	Endocrine Glands	158
3.1.10	Heart	159
3.1.11	Central Nervous System	159
3.2	Species-Specific Differences	160
4	Pathogenetic Events	161
4.1	Entry of Filoviruses into Blood	161
4.2	Target Cells and Filovirus Spread	162
4.3	Evidence of Damage to the Immune System	164
4.4	Hemostatic Impairment in Filoviral Infections	166
4.4.1	Blood Clotting	166
4.4.2	Changes in Vascular Permeability	168
4.5	Implied Cause(s) of Death	169
5	Concluding Remarks	170
References		171

1 Introduction

Current concepts of filoviral pathology have been mainly based on experimental data obtained soon after discovery of the viruses. The relevant studies have been designed to identify and characterize the new hazardous pathogens and to describe

---

State Research Center of Virology and Biotechnology "Vector," Research Institute of Molecular Biology, Koltsovo, Novosibirsk region, 633159, Russia

the macroscopic and histopathologic changes in the virus affected animals. In comprehensive reviews (MURPHY 1985; PETERS et al. 1991; SIEGERT 1972; WULFF and CONRAD 1977; ZLOTNIK 1969), the features of fatal MBGV (Marburg virus) and EBOV (Ebola virus) infections at the terminal stage in monkeys have been highlighted:

- Absence of visual symptoms of disease, except for skin petechiae in occasional animals
- Hemostatic impairment, sporadic visceral and mucosal hemorrhages
- Hepatic and renal dysfunction
- Lymphoid depletion, little or no inflammatory response at infection sites
- Shock and rapid death of virtually all monkeys
- Absence of obvious cause(s) of fatal outcome

It should be stipulated that the terminal pattern of filovirus infections results from interplay of many factors and the contribution of each is still not known exactly. Furthermore, the high infectious doses of filoviruses used in most studies (BASKERVILLE et al. 1985; JOHNSON et al. 1995; MURPHY et al. 1971) can obscure the "natural" course of infection, especially at its early stage.

Modern virological methods have made it possible to investigate the genetic basis of virulence and to identify the genes and their expression products affecting severity of the course of viral disease (DANIEL et al. 1996; ISHIKAWA et al. 1995). Judgments concerning the intricate pathogenetic mechanisms of viral infections cannot rely only on integral parameters, such as mortality, temperature curve, blood cell count, and serum enzymes. There is an obvious need for studies on pathogenetic mechanisms at the levels of the cell and interacting cellular systems. For this purpose, cells maintaining viral reproduction and functional changes in infected cells, particularly those interfering with homeostasis, must be investigated.

Other approaches to an understanding of pathogenetic mechanisms underlying filoviral infections would be provided by observations on conversion of the viral infection from lethal to nonlethal forms, with the use of small doses of infectious viruses, followed by time course studies of postinfection changes, and comparison of species-specific features of filoviral infections. With reference to these approaches, we made an attempt to review recent and pertinent literature and our own data concerning filoviral pathology in animals.

2 Filoviral Infections in Rodents

2.1 Adaptation of Filoviruses to Rodents

Early studies have shown that infection caused by blood or organ suspensions from patients with Ebola or Marburg hemorrhagic fever is not lethal for guinea pigs and hamsters. Both viruses were adapted, however, to guinea pigs and hamsters after sequential passages, and infections then became lethal. Based on light microscopic observations distinctive damage to visceral organs, notably liver and spleen, became more conspicuous with increasing number of MBGV passages in guinea pigs and hamsters. Lymphoid depletion, necrotic changes, and impaired hemostasis developed in spleen during the course of the passages. Hepatocellular necrosis, activation of Kupffer cells, and occasional neutrophil infiltration became evident in liver. Virus-specific eosinophil inclusions appeared in increasing numbers in hepatic and splenic cells. Data on inflammatory infiltration in liver of MBGV infected guinea pigs are controversial. Brain lesions are distinguishing features of infection caused by adapted MBGV in hamsters (ROBIN et al. 1971; SIEGERT and SLENCZKA 1971; ZLOTNIK 1969, 1971; ZLOTNIK and SIMPSON 1969). It has been demonstrated that EBOV also becomes adapted after sequential passages in guinea pigs, however, the associated damage in organs has not been investigated (BOWEN et al. 1980).

Our time course studies on nonlethal infection have revealed that EBOV reproduces only in the cells of the mononuclear phagocyte system (MPS) of guinea pigs, producing pronounced focal granuloma-like liver inflammation (RYABCHIKOVA et al. 1993, 1996b). Interactions of EBOV with susceptible MPS cells are of two types: (1) infected cells are recognized by immune effector cells with resulting formation of an inflammatory (granuloma-like) focus around the infected cells; (2) infected cells are not recognized by immune effector cells, the inflammatory focus is not formed, the produced EBOV is freely released into the blood stream. The amount of granuloma-like foci by far surpasses that of "free" infected MPS cells in liver. This strongly suggested that the unadapted EBOV population predominantly contains virions showing type (1) interaction with MPS cells resulting in trapping of virus-infected cells and their progeny inside inflammatory foci. Attempts to detect infectious virus in the blood of guinea pigs by inoculation of newborn mice and the PFU method were unsuccessful to day 17 of the infection. Granuloma-like inflammation is a salient feature of nonlethal Ebola infection in guinea pigs. The ultrastructural findings include destruction of almost a half of the stromal and macrophagal cells in spleen and lymph nodes and a decrease in mitotic frequency. The structural changes in kidney are evidence of an increase in its functional activity.

During the course of nine sequential passages in guinea pigs, the concentration of EBOV in blood rose, disruptive changes in organs were enhanced, the focal pattern of hepatic damage became diffuse, inflammatory responses were sharply reduced, and signs of immunosuppression became exaggerated. Hepatocytes, adrenal cells, fibroblasts and endothelial cells became increasingly involved in EBOV reproduction during the passages. The pattern of histopathologic changes in liver and the other examined organs was similar to the one described for monkeys infected with EBOV (see Sect. 3.1), except for hemorrhages and signs of disseminated intravascular coagulation (DIC) syndrome. Based on analysis of our data, it may be concluded that selection and accumulation of type (2) interaction virions, showing no reaction with the immune effector cells, result from sequential passages of EBOV in guinea pigs. A population of EBOV isolated from an affected human and passaged twice in monkeys was heterogeneous with respect to interaction with guinea pig MPS cells. Virions causing an infection of MPS cells followed by development of the effector leukocyte response and inflammation made up the bulk of the population (PEREBOEVA et al. 1993; RYABCHIKOVA et al. 1993, 1996a). Leukocytes did not react with infected cells in monkeys inoculated with the same EBOV preparation (see Sect. 4.3), thereby reflecting species differences in the properties of MPS cells between guinea pigs and monkeys.

No electron microscopic data are available for adaptation of MBGV to guinea pigs and hamsters. This restricts comparisons of the changes we observed for EBOV adaptation with those previously reported for MBGV adaptation. The features common to both viruses during their passages are enhanced disruptive changes in organs and, what is noteworthy, lymphoid depletion and necrosis of lymphoid tissue (ROBIN et al. 1971; SIEGERT and SLENCZKA 1971; ZLOTNIK 1969, 1971; ZLOTNIK and SIMPSON 1969).

2.2 Marburg Infection in Guinea Pigs

Marked changes in liver and spleen occur in guinea pigs intraperitoneally (i.p.) infected with adapted MBGV. Engorgement of sinusoids and small vessels, accumulation of neutrophils in their lumen, and rarefaction of hepatocyte cytoplasm were observed on days 3–4 postinfection. By days 5–6, congestion increased, erythrocytes sludged, and echynocytes (irregularly shaped erythrocytes) formed. MBGV-infected hepatocytes were arranged in foci whose sizes increased with developing infection. Lipid dystrophy in hepatocytes was a frequent occurrence. Accumulation of neutrophils proceeded, but they did not migrate from the vascular bed. By days 7–8, just prior to death of the animals, liver lesions were focal due to small areas of necrosis containing infected and noninfected hepatocytes. The concomitant changes concerned endothelium of sinusoids and the majority of Kupffer cells. Many of the sinusoidal neutrophils were subject to lysis. There were no marked inflammatory changes in liver of MBGV-infected guinea pigs (BECHTELSHEIMER et al. 1971; RYABCHIKOVA et al. 1994a).

The first signs of damage in liver after aerosol and i.p. infection with MBGV were the same. Only accumulations of neutrophils in liver sinusoids were noted in intraperitoneal infection, while a clear-cut leukocyte response to the infected cells developed in aerosol challenge. In the case of aerosol infection, neutrophils and mononuclear leukocytes migrated into the perisinusoidal spaces and liver parenchyma; the periportal zones were increasingly infiltrated with leukocytes. MPS cells and lymphocytes made close contacts with the infected hepatocytes without disrupting them. At the terminal stage of airborne infection, the layer of sinusoids was disrupted, infected fibroblasts and endotheliocytes occurred in large numbers (RYABCHIKOVA et al. 1996b).

Damage of splenic red pulp was largely determined by severity of vascular lesions. In i.p. MBGV-infected guinea pigs, blood vessels were congested, macrophages were swollen, and neutrophils and platelets accumulated in the blood stream. Many of the neutrophils were lysed. At the terminal stage, swelling of endothelium and blood vessel thrombosis were additionally observed. In aerosol MBGV-infected guinea pigs, damage was more pronounced, sinus endothelium was disrupted, and necrotic foci appeared (RYABCHIKOVA et al. 1994a, 1996b).

In lymphoid tissue of spleen, lymph nodes and tonsils, the size of follicles and the number of cells they contained were reduced. Infected and uninfected stromal and MPS cells became necrotic, starting from day 4 after aerosol infection. Germinal centers were hardly discerned in splenic white pulp shortly before death. There were no signs of mitotic figures and lymphoblast differentiation (RYABCHIKOVA et al. 1996b). Changes in lymphoid tissue in aerosol MBGV-infected guinea pigs were similar to those described in i.p. infection (ZLOTNIK 1969). In our i.p. infection study animals exhibited weak signs of a specific immune response, including rare centers of lymphoid cell division with a few mitoses, large lymphoblasts and differentiating plasma cells (RYABCHIKOVA et al. 1994a). The different immune responses reported for MBGV-infected guinea pigs may be explained by the different passage levels of the virus used.

No visible changes were detected in the kidney of i.p. MBGV-infected guinea pigs with the light microscope (ZLOTNIK 1969). From day 4 postinfection, damage of tubular epithelium was revealed by electron microscopic examination. The basal infoldings lost their regular arrangement; mitochondria and cytoplasm became partly swollen. Tubular damage progressed to necrosis of some cells; blood cells clogged intertubular and glomerular capillaries; there was swelling and necrosis of many endothelial cells with developing infection. Glomerular cells were not appreciably altered (RYABCHIKOVA et al. 1994a). The same pattern of kidney changes, although more prominent at the terminal stage, was observed in aerosol MBGV-infected guinea pigs. There was cell necrosis in the proximal and distal tubules, diapedesis of erythrocytes, shrinking of a few glomeruli, small focal necrosis, and hemorrhages into interstitium near large blood vessels (RYABCHIKOVA et al. 1996b).

There were no obvious inflammatory changes in lung of MBGV-infected guinea pigs. Changes in microcirculation were observed during the course of infection by both the i.p. and aerosol routes; neutrophils and macrophages filled the lumen of capillaries, and slight diapedesis of erythrocytes into interstitial tissue and, to a lesser extent, alveolar spaces occurred (RYABCHIKOVA et al. 1994a, 1996b).

Damage of adrenals and intestine caused by MBGV in guinea pigs was studied only in aerosol infection. By 4–9 days postinfection, capillaries were clogged with erythrocytes and neutrophils and cellular debris were observed. Scarce small necrotic foci and small hemorrhages were seen in adrenal parenchyma and interstitium. Established contacts were seen between mononuclear leukocytes and virus-infected adrenal cells not associated with disruption of cell structure. There were no disruptive changes in the epithelium of the large and small intestines. Small necrotic foci occurred in the lamina propria at sites of infected MPS cells and fibroblasts. Congestion and accumulation of neutrophils in blood vessels were observed in the lamina propria (RYABCHIKOVA et al. 1996b).

Thus, when lethal for guinea pigs, Marburg infection was associated with disturbed microcirculation, lesions of liver and spleen, and suppression of the effector response of the immune system. Pathological changes started to appear later in aerosol challenge and they were more severe than in parenteral infection. A distinguishing feature of aerosol infection was leukocyte attack on the infected cells as a component of the nonspecific immune response.

3 Filoviral Infections in Nonhuman Primates

3.1 Organ Pathology

We obtained light and electron microscopic data on pathological changes in visceral organs and also on cell involvement in viral reproduction in green monkeys. Animals were inoculated with MBGV (Popp strain) i.p. and subcutaneously (s.c.) with EBOV (Zaire strain) in two experimental series. The infectious dose was 100 LD_{50}. Animals were killed at daily intervals over a period of 9 days postinfection. To determine the level of filoviral reproduction in organs by electron microscopy, we introduced the term "infection degree of an organ" denoting the total number of infected cells and viral particles in an organ.

3.1.1 Blood Cells and Hematopoiesis

Neutrophilia, pronounced lymphopenia and thrombocytopenia were the consistent features of Ebola infection in monkeys. Hematologic changes appeared 5–7 days postinfection with Zaire and Sudan strains, and later, at 9–10 days postinfection, with Reston and Pennsylvania isolates. The relative proportions of T and B lymphocytes remained unaltered in the background of progressing lymphopenia; erythrocytes, nonspecific activation of phagocytes and the phagocytic index did not change significantly in the course of Ebola infection in baboons (LUCHKO et al. 1995; FISHER-HOCH et al. 1985, 1992; JAAX et al. 1996; JOHNSON et al. 1995). Lymphopenia, thrombocytopenia and neutropenia were detected in MBGV-infected monkeys (SIEGERT et al. 1968). Lymphopenia had also been observed in MBGV-infected monkeys and guinea pigs (LUB et al. 1995; RYABCHIKOVA et al. 1996b).

An obvious cause of the hematologic changes was impairment of hematopoiesis. In rhesus monkeys, necrosis, abundant virions and hemorrhages were observed in bone marrow at the terminal stage of Ebola infection; granulopoiesis persisted until the death of the animals (JAAX et al. 1996; JOHNSON et al. 1995).

Our light and electron microscopic observations demonstrated profound disruptive changes, namely cell necrosis, cell depletion, diapedesis of erythrocytes, hemorrhages, a sharp decrease in the number of dividing and differentiating cells, and swelling of endothelial cells in bone marrow of EBOV-infected baboons,

MBGV-infected rhesus monkeys and guinea pigs. Persisting granulopoiesis and erythropoiesis require special studies to determine whether normal cells can still be produced. Monocytes and lymphocytes and their precursors were extremely rare findings in sections of bone marrow. There was a breakdown of stromal and MPS cells, with some still showing signs of viral replication; intercellular spaces harbored viral particles.

Many agents have been implied as being causative in hematopoietic impairment. It is known that fibrin degradation products suppress lymphopoiesis (ROBSON et al. 1993). Destruction of stromal and MPS cells may also affect the microenvironment providing hematopoiesis (AKIRA et al. 1993; BOGDAN and NATHAN 1993; CAMPBELL and WICHA 1988). The destruction itself may result from the combined effects of filoviral reproduction in cytoplasm and certain humoral agents. It is also known that MPS cells produce mediators that can elicit damage of bone marrow (COOK 1996; SATO et al. 1995). The pattern of changes in megakaryocytes supports the suggestion that humoral agents may be involved in hematopoietic impairment. We observed no signs of MBGV or EBOV replication in megakaryocytes showing distinctive signs of damage, including edematous cytoplasm and mitochondria, and fragmentation of the Golgi complex. Some megakaryocytes were smooth-surfaced and did not produce platelets, others produced platelets devoid of specific granules suggestive of their functional impairment. Loss of specific granules has been associated with the inability of platelets to aggregate in EBOV-infected rhesus monkeys (FISHER-HOCH et al. 1985).

Thus, the pathological changes in bone marrow and blood counts correlate in experimental Marburg and Ebola infections. Further studies are needed to elucidate the mechanisms suppressing agranulopoiesis and thrombocytopoiesis. It is hoped that the obtained data will be helpful in timely correction of hemostatic derangements.

3.1.2 Blood Vessels

The pathological changes in microcirculatory vasculature were broadly similar in all the organs of filovirus-infected monkeys we examined. However, the organs differed in the extent of damage, which was most evident in liver and spleen. Sludging of erythrocytes, changes in their shape and a decrease in endothelial pinocytosis were the concomitant findings in Ebola and Marburg infections, evidence of interrupted transendothelial transport and oxygen supply to tissues. Capillaries and small blood vessels were engorged, and small platelet thrombi occurred within the first 5 days postinfection in both MBGV- and EBOV-infected green monkeys. At the terminal stage of Ebola infection, swollen leukocytes, cell debris, fibrin and fibrin-cellullar thrombi frequently obstructed small visceral vessels. The organs in filovirus-infected green monkeys were ranked according to decreasing endothelial damage as follows: spleen, liver, kidney, lung, adrenals, lymph nodes, intestine, heart. Endothelium was weakly, if at all, structurally altered in salivary glands, pancreas, and striated and smooth muscles. The more profound changes in vascular permeability and structure were observed for the abundantly blood supplied organs. Direct damage to endothelial cells by the replicating virus was observed only 1–2 days before death of the animals. Infected endothelial cells disintegrated at some sites along the capillaries, leaving only the basement membrane intact. Virus replication was found only in endothelial cells of capillaries and venules. Occasionally edema of the subendothelial layer was observed in large vessels.

In the course of EBOV infection, pericytes swelled in the visceral organs and skin, and many of the pericytes had disintegrated by the time of death of the monkeys. There are data indicating that pericytes are involved in regulation of blood vessel permeability (COURTOY and BOYLES 1983; LONIGRO et al. 1996). There was thus reason to assume that pericyte damage may be implied in development of erythrocyte diapedesis, hemorrhages and also impairment of ionic permeability.

3.1.3 Liver

In necropsied EBOV-infected monkeys, the liver was enlarged, friable, and congested, with yellowish foci (2–3 cm in diameter) that often coalesced (BASKERVILLE et al. 1978; LUCHKO et al. 1995). In necropsied MBGV-infected monkeys, the liver was also enlarged, congested, friable and had fatty changes (SIEGERT 1972; ZLOTNIK 1969).

Biochemical data indicated that the detoxification and synthetic hepatic functions were impaired in Ebola and Marburg infections. In EBOV-infected rhesus monkeys, aminotransferases were elevated on days 4–5 postinfection; LDH, alkaline phosphatase and creatinine-phosphokinase rose at the terminal stage of infection (FISHER-HOCH et al. 1985, 1992; JAAX et al. 1996; JOHNSON et al. 1995). Elevated endotoxin levels also confirmed that hepatic detoxification in EBOV-infected rhesus monkeys is impaired (JAAX et al. 1996).

In baboons infected with EBOV (20–50 LD_{50} for this species), the values obtained from bilirubin and thymol tests were elevated on day 3 postinfection, evidencing of deranged permeability of hepatic vessels and hepatocellular dysfunction (Table 1). The levels of total protein, circulating immune complexes and malonic dialdehyde did not appreciably change by day 6 postinfection. The biochemical changes on days 5–6 demonstrated damage of liver and kidney, with the appearance of acute phase proteins and also activation of the complement system. The level of C-reactive protein increased fourfold 2 days postinfection and 4.5-fold 7 days postinfection. There was a threefold, 2.4-fold and 1.4-fold increase in the contents of $\alpha 1$-, $\alpha 2$- and γ-fractions of globulins, respectively; β-lipoproteins increased threefold. The albumin/globulin ratio decreased from 1.2 ± 0.1 to 0.8 ± 0.1 on day 5 and to 0.2 ± 0.1 on day 9 postinfection. The observed biochemical changes were not specific, but were mainly concomitants of fever onset; consequently, they were of little diagnostic significance. While illustrative of infection, the changes themselves do not provide insight into the pathogenetic mechanisms (LUCHKO et al. 1995).

There was a broad similarity between hepatocellular changes observed in green monkeys during the course of experimental infection with MBGV and EBOV. On

Table 1. Biochemical parameters of serum samples during the time course of Ebola virus infection in baboons[a] (Luchko et al. 1995)

Parameters	Measurement units	Days postinfection					
		0	1	3	5	7	8	9
ALT	AU	0.6 ± 0.1	0.7 ± 0.1	0.9 ± 0.1	1.1 ± 0.1*	1.8 ± 0.2*	2.2 ± 0.3*	2.7 ± 0.8*
AST	AU	0.6 ± 0.1	0.9 ± 0.1	1.0 ± 0.2	0.9 ± 0.2	2.7 ± 0.5*	2.8 ± 0.4*	2.7 ± 0.8*
Thymol probe	S/HO	2.1 ± 0.2	2.5 ± 0.3	4.4 ± 0.8*	2.6 ± 0.4	3.0 ± 0.7	3.9 ± 0.8	2.8 ± 0.8*
Bilirubin	mkmol/l	4.9 ± 0.5	7.2 ± 0.5*	10.4 ± 1.7*	6.9 ± 1.2	9.1 ± 0.9*	28.4 ± 8.3*	20.0 ± 4.0*
Urea	mmol/l	5.3 ± 0.4	5.6 ± 0.3	5.4 ± 0.4	6.8 ± 0.6	11.9 ± 0.9*	17.6 ± 3.6*	28.4 ± 5.5*
Creatinine	mmol/l	156.4 ± 6.6	154.1 ± 7.5	149.1 ± 9.2	167.0 ± 6.9	256.7 ± 27	335.0 ± 67	363.0 ± 88*

AU, activity units; *$p < 0.05$ compared with 0 day.
[a]The mean value for 13 monkeys.

day 2 postinfection, sinusoidal congestion developed; vacuoles appeared in some hepatocytes and Kupffer cells swelled. On days 3–4 postinfection, swelling of endothelium of sinusoids and small vessels and necrosis of scarce hepatocytes complicated the morphological pattern. Infected MPS cells were identified in blood, mostly in portal tract areas, from day 3. Hepatocytes were involved in filoviral replication a day later. An "avalanche" of pathological alterations occurred starting at day 5 postinfection: the cell lining of sinusoids and small vessels almost completely broke down, fibrin fibers were deposited in sinusoids, and hepatocytes became increasingly necrotic.

The morphological features of the infection at the terminal stage may be summarized as follows: the liver contains numerous infected hepatocytes, MPS cells and viral particles; infected fibroblasts and endotheliocytes are rarely encountered; there are scattered necrotic foci, and both infected as well as many uninfected cells disintegrate. The pattern of changes in blood vessels is virus species-dependent. In EBOV-infected monkeys, hepatic vessels contain numerous fibrin and fibrin-erythrocyte thrombi; there are also deposits of fibrin in Disse's spaces and interstitium. As for Marburg infection, deposition of fibrin is insignificant and restricted to necrotic areas. There is a remarkable absence of hepatic inflammatory infiltration in the background of severe organ damage in green monkeys infected with EBOV or MBGV.

The pattern of histopathologic changes in liver at the terminal stage of filoviral infections is similar in different monkey species (BASKERVILLE et al. 1978, 1985; FISHER-HOCH et al. 1992; GEISBERT et al. 1992; JAAX et al. 1996; JOHNSON et al. 1995; MURPHY et al. 1971; SIEGERT 1972).

3.1.4 Spleen

The spleen was enlarged, soft and brittle and often devoid of follicles in necropsied monkeys. Lymph nodes were enlarged, inconsistently, however, and hemorrhages were seen in capsules (SIEGERT 1972; SIMPSON 1968; ZLOTNIK 1969).

Splenic red pulp was more damaged than all the other organs studied in EBOV- or MBGV-infected green monkeys. In MBGV-infected green monkeys, pronounced changes in the red pulp appeared quite late, on days 4–5 postinfection, and included engorgement of sinuses and necrotic foci. The first MBGV-infected cells were revealed on day 5 postinfection in blood and interstitium; their number increased as infection developed. By days 6–7, sinus endothelium was almost completely disrupted. Tissue was virtually soaked with plasma. Small clumps of fibrin were deposited in lumen of blood vessels and intercellular spaces. Sinuses were tightly clogged with blood cells, fibrin and cellular debris. At the terminal stage of Marburg infection in green monkeys, red pulp had the appearance of a confluent thrombi made up of blood cells, disintegrating stromal and MPS cells, cellular debris whose fibrin layers were encrusted with viral particles. It was difficult to distinguish MBGV-infected cells in confluent thrombi in which MPS cells constituted the majority and fibroblasts and endotheliocytes the minority (MURPHY et al. 1971; RYABCHIKOVA et al. 1994a).

As for the EBOV-infected green monkeys, the red pulp showed damage on day 2 postinfection manifested as swelling and disruption of the MPS and stromal cells and pericytes. These changes became more severe as infection developed. The first infected MPS cells were found in blood vessels on day 3 postinfection. Fibrin deposited within and outside blood vessels, many erythrocytes left the vascular bed, and the sinus layer disintegrated from days 4–5. The architecture of the red pulp was completely damaged by necrosis on days 6–7 postinfection. The infection degree of spleen was not high at the terminal stage, presumably because the majority of EBOV-susceptible cells broke down at the preceding stages of infection. Cavities devoid of cells and occupied by fibrin were seen on sections as extensive areas containing fibrin bundles and rare erythrocytes. Aggregates of disintegrated cells, scarce virions and partly lysed neutrophils were interspersed between the areas. Taken together, all the signs of hemostatic impairment for splenic red pulp present a "synopsis" of what we observed for all the other examined organs.

Dramatic changes occurred in splenic red pulp in rhesus, green and cynomolgus monkeys infected with Zaire and Reston EBOV isolates, respectively (BASKERVILLE et al. 1978, 1985; GEISBERT et al. 1992; JAAX et al. 1996; JOHNSON et al. 1995). In EBOV Sudan strain-infected cynomolgus macaques, the degree of necrotic changes and fibrin deposition was lower than in animals infected with Zaire and Reston strains (FISHER-HOCH et al. 1992).The changes in splenic white pulp in filovirus-infected monkeys are described in Sect. 4.3.

3.1.5 Kidney

No appreciable lesions in kidney were, as a rule, observed in a study of autopsy material from filovirus-infected monkeys. Capsular and subcapsular hemorrhages have been occasionally noted (BASKERVILLE et al. 1978; LUCHKO et al. 1995).

Biochemical assays demonstrated impairment of kidney function from day 5 of Ebola infection in rhesus monkeys, and it was most severe at the terminal stage; blood levels of urea and creatinine and activities of lactate dehydrogenases, creatinine phosphokinase increased while Na, K, and Ca levels decreased (FISHER-HOCH et al. 1985; JAAX et al. 1996; JOHNSON et al. 1995). In EBOV-infected baboons, a significant increase in the concentration of creatinine and urea occurred on day 7 (Table 1) (LUCHKO et al. 1995).

The time course of histopathologic changes was the same in kidney of green monkeys inoculated with MBGV or EBOV. The tubules first showed lesions on day 3 postinfection. The pattern of changes was consistent with disturbed reabsorption, as evidenced by swelling of the cells of distal tubules and intertubular capillaries; disorganization of basal infoldings and mitochondrial cristae in the epithelium of the proximal tubules. Subsequently, these changes were accompanied by swelling of podocytes in glomeruli. Increasing capillary thrombosis, hemostasis, mild necrosis of tubular epithelium, deposition of mineral salt in tubules, and glomerular damage were observed on days 5–7. By the time the monkeys died, the capillary endothelium had broken down almost completely, perivascular spaces became edematous and a number of glomeruli necrotic. There were no associated inflammatory changes in kidney. These were the morphological changes unifying EBOV and MBGV infections in green monkeys. The additional features observed in the case of Ebola infection included fibrin thrombi, diapedesis, and interstitial hemorrhages. The progression of pathological processes in kidney was not directly related to filoviral replication in kidney epithelium. MBGV and EBOV were seen to replicate in rare intravascular MPS cells on days 5 and 6 postinfection, respectively. The infection degree of the kidney was low. Rare infected MPS cells and fibroblasts occurred in interstitium at the terminal stage of filoviral infections. Similar and more prominent changes have been observed in rhesus monkeys infected with high doses of EBOV; no marked changes were visualized in bladder, except for vascular thrombosis and rare hemorrhages (BASKERVILLE et al. 1978, 1985; JAAX et al. 1996).

At the terminal stage of infection caused by Reston virus in cynomolgus kidney, small scattered diffuse necrosis was present in the tubules of the cortical and medullar zones and glomerular endothelium. Fibrin deposition was seen in glomerular capillaries and rarely in interstitium. No changes in podocytes were found (GEISBERT et al. 1992).

3.1.6 Lung

Changes in respiratory function have not been studied in filovirus-infected monkeys. Cough and sneezing have not been noted. Shallow rapid respiration has only been reported for a small number of monkeys (JAAX et al. 1996). At postmortem examination, the lung was somewhat dense, appearing congested when sectioned; hemorrhages were occasionally encountered. The degree to which lung tissue was damaged varied a great deal, especially in Ebola infection. The variations may be due to individual susceptibility to the viruses and possible concurrent infections.

During the course of Marburg and Ebola infections, we observed development of microvasculature damage in the upper and lower respiratory tracts. At 4–5 days postinfection, the lungs were blood-congested. There was mild perivascular edema, and small aggregates of neutrophils and platelets were seen in blood vessels. In MBGV-infected green monkeys, the pattern remained unaltered up until the time of death; intravascular accumulations of neutrophils and diapedesis increased by the terminal stage. In EBOV-infected monkeys, monocytes and neutrophils accumulated in lung vessels; deposition of fibrin, diapedesis of erythrocytes, and thrombi composed of fibrin, necrotic debris and erythrocytes were seen. The terminal findings included dystrophic changes in endothelial cells, type II alveolocytes, and the epithelial cells of both glands and the lining of the conducting airways. There was desquamation of a few epithelial cells in trachea, bronchi and alveoli. Filoviruses usually did not reproduce in the epithelium of the respiratory tract. Reproduction of MBGV and EBOV was identified in blood MPS cells on day 5 and in interstitial MPS cells, alveolar macrophages, rare fibroblasts and endotheliocytes on days 6–7 postinfection.

Rhesus monkeys showed a wide range of individual differences in lung damage caused by high doses of EBOV. The lesions varied from mild (small necrosis of interalveolar septa and slight alveolar edema) to severe (extensive necrosis, collapsed alveoli, fibrin exudate and erythrocytes in alveoli). There were extensive hemorrhages and fibrin deposition in vessels and lung interstitium. The structure of the pulmonary airway cells was unaltered in most infected monkeys, and only some showed subacute inflammation. Many capillaries were clogged with necrotic monocytes, cellular debris, erythrocytes and platelets (BASKERVILLE et al. 1978, 1985; JAAX et al. 1996). In aerosol EBOV-challenged monkeys, interstitial pneumonia ranged from mild to moderate with foci surrounding terminal bronchioles and an increase in the number of alveolar macrophages (JOHNSON et al. 1995). In contrast, the number of alveolar macrophages decreased in s.c. EBOV-infected monkeys. It has been suggested that filovirus infection of monocytes may affect their subsequent differentiation into alveolar macrophages (KOLESNIKOVA et al. 1997). In all Reston virus-infected cynomolgus macaques, platelets accumulated in lung venules; most capillaries were obstructed by platelet aggregates, necrotic macrophages, monocytes and fibrin; pneumonia was observed in a few monkeys (GEISBERT et al. 1992).

Thus, lung damage specific to respiratory infection does not develop in filovirus-infected monkeys. Instead the observed damage is mainly due to hemostatic impairment; its association with edema of alveolar capillary cells suggests a reconsideration of the role of the lung in the pathogenesis of filoviral infection. The observed ultrastructural changes involved the respiratory pathways in all their entirety, evidence of disturbed gas exchange leading to deficient oxygenation.

3.1.7 Gastrointestinal Tract

Macroscopic lesions in filoviral infections were rare observations, being reduced to infrequent hemorrhages into the submucous and mucous layers of EBOV-infected monkeys (LUCHKO et al. 1995; JAAX et al. 1996).

In the course of our studies, we found no extensive discontinuities of the epithelium of the esophagus, stomach, or small and large intestines. The lesions of blood vessels specific to each virus contributed to the overall histopathologic pattern of changes in the intestine of EBOV- and MBGV-infected monkeys. Reproduction of MBGV and EBOV was rarely observed in a few circulating and interstitial MPS cells and fibroblasts at the terminal stage of infection.

The terminal histopathologic changes due to impaired hemostasis have been described in EBOV-infected rhesus monkeys. They showed hemorrhages, fibrin thrombi and single fibrin bundles in blood vessels and in mucous and submucous layers of the gastrointestinal tract. Erosions and necrosis of the intestinal epithelium presumably caused by deranged blood flow were also occasionally found (BASKERVILLE et al. 1978, 1985; JAAX et al. 1996).

In the Reston virus-infected cynomolgus macaques, no marked changes in the duodenal epithelium were observed. Viral replication was seen in monocytes and macrophages in blood, and lamina propria, fibroblasts and endothelium of blood vessels (GEISBERT et al. 1992). During the course of MBGV and EBOV infections, we found no epithelial cell lesions and inflammatory changes in salivary glands and pancreas. Only the microvasculature was altered. Filoviruses reproduced in rare MPS cells in blood, interstitium and also in single fibroblasts.

3.1.8 Reproductive System

In EBOV-infected adult female rhesus monkeys, no histopathologic changes were observed in ovaries, Fallopian tubes and vagina. In contrast, in males acute inflammation developed in the parietal and visceral layers of the tunica vaginalis and extended to the tunica albuginea and musculus cremaster from day 5 postinfection. The layers were abundantly infiltrated with lymphocytes and neutrophils, many were lysed. Testicular tissue was edematous. Prominent hemorrhages and deposition of fibrin between and within the canaliculi developed on days 6–7. There was terminal vacuolization and disintegration of some epithelial cells of the seminal tubules. Thinning of the spermatogonial layers was continuous, spermatocytes were rarely encountered, and there were no traces of sperm. A few fibrin-platelet thrombi occurred in veins of the tunica albuginea. EBOV was seen to reproduce in interstitial cells, endothelium and monocytes. However, in spite of marked degeneration, seminal tubules showed no signs of viral infection (BASKERVILLE et al. 1978, 1985). EBOV inclusions were found in stromal macrophages, fibroblasts and follicular cells in rhesus monkey ovary (JAAX et al. 1996).

3.1.9 Endocrine Glands

The endocrine glands have been scarcely studied in experimental filoviral infections. It should be emphasized that the hypothalamic-pituitary-adrenal axis has a significant influence on homeostasis and development of various diseases. Hormones are not only regulators of carbohydrate, protein and water-salt metabolism, but also suppressers of the production of cytokines, thereby exerting an anti-inflammatory effect (KAPCALA et al. 1995; PERRETTI and FLOWER 1994). Macroscopic examination of adrenals of EBOV-infected monkeys demonstrated that they retained their structural integrity, but with a thicker than normal cortical layer and hemorrhages in occasional cases (JAAX et al. 1996). There are no data concerning the hormone-producing function of endocrine glands in filoviral infections.

Our electron microscopic observations on changes in adrenals caused by EBOV in green monkeys revealed increased blood filling of the cortical layer, small hemorrhages, adhesions and small accumulations of platelets on blood vessel walls. The hemostatic disturbances were enhanced on day 5 postinfection. A deposition of slender fibrin bundles appeared around capillaries. The reproducing virus caused damage to a few adrenal cells in fascicular, glomerular and reticular zones and also to MPS cells on day 5 postinfection. Some cells were necrotic, the majority of the fascicular zone contained large mitochondria with partly lysed matrix and cristae; cytoplasm formed protrusions without organelles; the number of liposomes sharply decreased. In contrast to other organs, diffuse infiltration with neutrophils was observed in the cortical layer. On days 6–7 postinfection, the degree of adrenal infection and disintegration increased and numerous hemorrhages appeared; the number and size of necrotic areas in the parenchyma also increased. At the terminal stage, occasionally endothelial cells, in addition to adrenocortical cells and MPS cells, were involved in viral reproduction. Damage of infected adrenocortical cells suggestive of suppression of the hormone-producing function of the adrenals was noted. No appreciable changes were found in adrenal medulla of green monkeys during the course of EBOV infection.

It has been reported that small hemorrhages and diffuse necrosis are restricted to the adrenal fascicular zone at the terminal stage in rhesus and cynomolgus macaques infected with EBOV and Reston virus, respectively (BASKERVILLE et al. 1978; GEISBERT et al. 1992; JAAX et al. 1996). Little is known of the damage to other endocrine glands caused by deliberate filovirus infection. We found no disruptive lesions in islands of Langerhans scattered throughout the pancreas of green monkey during the course of EBOV infection. There was diffuse necrosis of thyroid and parathyroid glands in rhesus monkeys. Immunoreactivity was multifocal in thyroid interstitium, the number of immunopositive follicular epithelial cells being minimal. The chief cells of the parathyroid gland were occasionally immunopositive (JAAX et al. 1996; JOHNSON et al. 1995).

3.1.10 Heart

No damage to the heart was reported in filovirus- infected monkeys at autopsy. We observed abnormal blood filling of myocardial vessels in green monkeys during the first 2 days of infection with EBOV. Small accumulations of platelets formed in the capillaries 3 days postinfection. Congestion of blood vessels increased, small foci of diapedesis of erythrocytes, intravascular and perivascular deposition of fibrin appeared, a few endotheliocytes swelled. Later on, microcirculatory derangements were enhanced, fibrin deposition and diapedesis became more pronounced. Small foci of lysis of myofibrils, cristae and mitochondrial matrix presumably due to disturbed microcirculation were present in cardiomyocytes. Thus, the myocardium of EBOV-infected monkeys was little affected. Virus multiplication was found in circulating and interstitial MPS cells 4–5 days postinfection and in endothelial cells at the terminal stage of infection. Previous studies by light microscopy revealed no changes in the hearts of rhesus monkeys infected with EBOV (JAAX et al. 1996).

3.1.11 Central Nervous System

Data on brain damage caused by experimental infection of monkeys with filoviruses are fragmentary. In MBGV-infected green monkeys, blood congestion and small hemorrhages were found in brain (ZLOTNIK 1969). In EBOV-infected rhesus monkeys, perivascular edema and hemostasis were the only observed brain lesions (JAAX et al. 1996). In EBOV-infected baboons, we found no conspicuous hemorrhages, nor did we find evidence of erythrodiapedesis in brain cortex and medulla, in contrast to liver and kidney; infection with the virus was evident only in circulating MPS cells.

We now have shown, by both histopathologic and ultrastructural findings, that EBOV and MBGV have the most deleterious effects on liver, spleen, kidney and adrenals. However, the widespread impairment throughout the microcirculatory vasculature undoubtedly affects the function of all organ systems.

Even without supportive therapy, disease is protracted in filovirus-infected humans, as compared to monkeys. Consequently, monkey organs may be functionally impaired, although appearing morphologically spared. Thus, the course of disease in models of filovirus infection resembles the fulminant variant in humans not provided with adequate therapy. This must be taken into consideration in comparisons of experimental and clinical data.

3.2 Species-Specific Differences

Comparisons of the susceptibility of green monkeys and cynomolgus macaques to EBOV strains have demonstrated that the former are less susceptible to Pennsylvania, Reston and Sudan and that they have weaker symptoms of the disease, while Zaire is lethal for both monkey species. The infectious dose was 1000 $TCID_{50}$ (FISHER-HOCH et al. 1992).

We compared here the patterns of organ changes and the infection degree of organs at the terminal stage of infection with low doses of EBOV (1–10 LD_{50} for each species) in green and rhesus monkeys, baboons and cynomolgus macaques. Each species was represented by four monkeys. Specimens for electron microscopy were obtained promptly after the death of the monkeys on days 8–9 postinfection. EBOV reproduced in circulating and interstitial MPS cells, hepatocytes, adrenalcytes, fibroblasts and endotheliocytes in all four monkey species. The pattern of damage to visceral organs was the same in all the species. Disregarding differences within the monkey species, the morphological features characterizing filovirus infection in the four species are presented in this section. Green and rhesus monkeys differed from baboons in alterations in hemostasis (Table 2). There was extensive deposition of fibrin in the blood of all the visceral organs in green and rhesus monkeys, while fibrin deposited as rare clots or fibers mainly in necrotic areas in baboons. There were numerous and extensive hemorrhages in the visceral organs of baboons. Microcirculation was impaired without marked fibrin deposition and diapedesis of erythrocytes in cynomolgus macaques.

Table 2. The comparisons of viral reproduction, necrosis and hemostatic impairment in visceral organs in four monkey species infected by Ebola virus

Monkey species	Infection degree	Necrotic changes	Fibrin thrombi	Hemorrhages
Green monkeys	+ +	+ + +	+ + +	+ +
Rhesus monkeys	+	+ + + +	+ + + +	+ +
Baboons	+ +	+ + +	+ –	+ + + +
Cynomolgus macaques	+ + + +	+ +	+ –	+ –

Striking differences in the infection degree of visceral organs were found between cynomolgus macaques and the other three species. There were massive amounts of EBOV-infected cells and virions in all the visceral organs we examined in cynomolgus macaques. In addition to the predicted target cells, EBOV intensely replicated in numerous cells of bile canaliculi, endothelial cells of blood and lymph vessels and also in endocardial endothelium. The occurrence of necrotic lesions was unexpectedly low in a background of very high infection degree of visceral organs in cynomolgus macaques.

Data on damage to visceral organs in EBOV-infected monkey species are compiled in Table 2. Clearly, there is no direct relation between the level of viral replication in organs, severity of necrotic lesions and hemostatic impairment. The outcome is lethal for cynomolgus macaques in spite of lacking hemorrhages and manifestations of DIC syndrome.

Thus, EBOV attacks the same set of target cells and causes the same pattern of damage to lymphoid organs, most evidently liver, spleen, kidney and adrenals, in all the four monkey species. There was no inflammatory infiltration in the lesioned organs. All the monkey species showed the same signs of suppression of specific and nonspecific immune responses. Species-specific differences concerned only hemostatic impairment and the infectivity degree of organs.

Species-specific differences in fatal Ebola infection in monkeys may be accounted for in terms of the crucial role of EBOV-infected MPS cells in pathogenesis. We presume that infected MPS cells trigger immune suppression and microcirculatory and hemostatic impairment by release of mediators. Having taken into account heterogeneity and species specificity of MPS cells, we concluded from a survey of our data that, in fatal Ebola infection, either different functions of MPS cells or their different subsets responsible for regulation of specific and nonspecific immunity, microcirculation and hemostasis are impaired.

4 Pathogenetic Events

4.1 Entry of Filoviruses into Blood

Susceptible animals can be inoculated with infected cell culture fluid, organ suspensions and also blood samples from filovirus-affected subjects by the subcutaneous, intramuscular, intravenous, aerosol, intracerebral, or conjunctival route (HAAS et al. 1968; PETERS et al. 1991; SIEGERT 1972; SIMPSON 1969).

The infectious process can be arrested by inactivation of the invasive virus before it has gained access to blood stream. The question is, how long does a virus take to gain access? Judging by the scant data in the literature, access is dose-dependent. Thus, MBGV and EBOV were found to be present in blood 24 h after their parenteral injection at a dose higher than 100 LD_{50} to monkeys and guinea pigs (FISHER-HOCH et al. 1985; RYABCHIKOVA et al. 1994a). In guinea pigs aerosol-

bronchi (JOHNSON et al. 1995; KOLESNIKOVA et al. 1997; RYABCHIKOVA et al. 1994b), and Reston virus reproduced in rare alveolocytes, in epithelial cells of renal proximal tubules and in cells of adrenal medulla (GEISBERT et al. 1992).

Thus, filoviral infections in experimental animals are associated with active viral replication in MPS cells that provides not only production of virions, but also spread of infection to organ interstitium and parenchyma.

4.3 Evidence of Damage to the Immune System

It is important to recall that the host dies with generally no evidence of an immune response in fatal filoviral infections (PETERS et al. 1996; SIEGERT 1972). To identify morphological features of functional impairment of the immune system, we conducted time course studies on changes in lymphoid tissue of splenic white pulp, mesenteric, inguinal and peribronchial lymph nodes, lymphoid nodules and follicles of the respiratory and gastrointestinal tracts. Attention was also focused on leukocyte response to infected cells in parenchymal organs.

We found early involvement of lymphoid tissue in both MBGV- and EBOV-infected monkeys. Involvement was more severe in the latter during the course of infection because of stronger hemostatic impairment. Mitotic activities of lymphoid cells ceased on days 2–3 postinfection in all the anatomical locations of lymphoid tissue in green monkeys infected with EBOV or MBGV.

The mitosis-blocking agent may be of humoral nature. This appears quite plausible because: (1) filoviruses were not seen to replicate in lymphoid tissue at the early stage of infection; (2) mitosis was not observed at all the anatomical locations of lymphoid tissue. The action of the mitosis-blocking agent is specific with respect to lymphoid cells, judging by the continuing division of other cell type, particularly intestinal epithelium, throughout infection.

Swelling of MPS and stromal cells was observed from day 2 postinfection in all anatomical sites of lymphoid tissue. The first MBGV-infected MPS cells were recognized in blood vessels of lymphatic tissue in splenic white pulp and mesenteric lymph nodes at 5 days postinfection; in the case of EBOV infection, they were identified in white pulp on day 3 postinfection and lymph nodes on day 4 postinfection. The number of infected MPS cells increased and so did their lysis with progressing infection. No stromal and MPS cells remained unchanged by the terminal stage. It is noteworthy that both infected and uninfected cells disintegrated. Stromal and MPS cells disintegrated even where lymphoid tissue showed no obvious signs of viral infection. This was particularly true for lymphoid nodules and follicles of lung and intestine. The widespread disintegration of stromal and MPS cells may be of importance in filoviral pathogenesis because both cell types are directly involved in regulation of the immune response (CAMPBELL and WICHA 1988; TAUB and OPPENHEIM 1994; SATO et al. 1995). Another feature of lymphoid tissue lesions in MBGV-and EBOV-infected monkeys is disintegration of dendritic (antigen-presenting) cells on days 3–4 postinfection. These cells disintegrated at all sites of lymphoid tissue, as did MPS and stromal cells. Necrosis of dendritic cells undoubtedly aggravated the course of the disease, because precisely they are responsible for primary presentation of virus antigens to immunocompetent cells (KNIGHT and MACATONIA 1988; STEINMAN 1991). A better understanding of dendritic cell-filovirus interaction would be provided by the use of special, including in vitro, methods.

Involution of lymphoid follicles developed more rapidly in green monkeys infected with EBOV than MBGV. Rapid involution was manifest as a 69% reduction in the relative volume of the germinal centers of the follicles of the mesenteric lymph nodes during the first 2 days of Ebola infection. The reduction was mainly due to a decrease in the number of follicles and, to a lesser extent, in their size. The increasing lymphoid depletion became obvious starting at days 3–4 in both Marburg and Ebola infections. The lymphoid depletion was not caused by necrosis of lymphocytes on these days. Later in the course of infection, the few visible follicles of lymph nodes had a morphological appearance suggesting their waning function: small size, homogeneous germinal centers without blast cells. The B zones were more affected than the T zones in lymph nodes; however, the T zones showed no signs of functional activity. Splenic follicles underwent similar changes. Lymphoid tissue lost its appearance as a contiguous layer because of edema, necrosis, and lymphoid depletion at the terminal stage of Marburg and Ebola infections. All of the above changes became more pronounced by the time of death of the monkeys. The histopathologic pattern was complicated by additional damage to vascular endothelium, thrombosis and diapedesis of erythrocytes. Morphological evidence of filoviral reproduction was seen in MPS cells and a few endotheliocytes and fibroblasts of lymphoid tissue at the terminal stage.

The lesions described for lymphoid tissue of spleen and lymph nodes in EBOV-infected rhesus monkeys at the terminal stage (BASKERVILLE et al. 1978, 1985; JAAX et al. 1996; JOHNSON et al. 1995) are the same as those we observed in green monkeys.

Thus, lymphoid tissue is damaged during the course of MBGV and EBOV infections in monkeys. Its severity indicates that the function of the immune system is greatly affected. The changes are multicomponent and each component itself can have a significant impact on the function of the immune system. In lethal cases, the impairment is manifested in monkeys and guinea pigs at the morphological level as suppression of the cellular and humoral limbs of immunity: plasma cells are not formed and T cells fail to react with infected target.

The mechanism underlying suppression of the immune response in filoviral infections is unclear. The possibility cannot be ruled out that virus-specific factors may be inductive. The filoviral genome contains a region coding for surface glycoprotein which shows a high degree of homology with an immunosuppressive motif occurring in oncogenic retroviruses (BUKREEV et al. 1993; VOLCHKOV et al. 1992). It has been suggested that this glycoprotein may provide preferential binding of filoviruses to the MPS cells participating in the regulation of the immune response.

The nonstructural secreted glycoprotein (SGP) of EBOV abundantly released by EBOV-infected cells was believed to be involved in induction of immunosuppression (PETERS et al. 1996). However, this concept is not supported by observations that immunosuppression features of MBGV and EBOV infections are similar and that a SGP-like protein is not synthesized during MBGV replication.

Lymphoid tissue is affected before toxic products of other disintegrated cells enter the blood stream. Lymphoid tissue damage most likely occurs through specific mediators released early during infection by MPS cells. Based on the data in the literature on the functions of MPS mediators, it may be suggested that these mediators include tumor necrosis factor (TNF) and interleukins (BONE 1996; COOK 1996; TAUB and OPPENHEIM 1994).

The specific immunity responsible for the production of antibodies and the effector action of T cells and nonspecific defense are suppressed in EBOV- and MBGV-infected monkeys. Hypotheses concerning nonspecific defense were based on morphological manifestations such as contacts made between infected cells and leukocytes, followed by destruction of infected cells 5–6 days after inoculation. Neutrophils and large lymphocytes (natural killers) were engaged in these responses. Suppression of nonspecific defense is manifest as the inability of leukocytes to respond to infected cells, resulting in little or no inflammatory response at the infection site in monkeys. Guinea pigs also show signs of suppressed specific, but not nonspecific, immunity, even in nonlethal infection. Thus, there is reason to assume that specific and nonspecific immunity may be impaired independently of each other. Comparisons of aerosol and parenteral Marburg infections in guinea pigs suggest that circulating MPS cells may be engaged in suppression of nonspecific immunity. Leukocytes attack virus-infected cells in the case of aerosol infection. This is in contrast to the parenteral route, when they do not, perhaps due to the later infection of circulating MPS cells and, accordingly, later release of mediators halting leukocyte attack.

4.4 Hemostatic Impairment in Filoviral Infections

4.4.1 Blood Clotting

It is well known that viral hemorrhagic fevers are associated with defects, of different degrees, in blood clotting (COSGRIFF 1989). The hazards involved in handling filovirus-infected animals almost completely precludes the continuous monitoring required for adequate estimation of alterations in blood clotting. Development of DIC syndrome has not been reported for experimental Marburg infection (HAAS et al. 1968; LUB et al. 1995). In the case of Ebola infection, the factual data were inconsistent because different parameters were used from one study to another and also because animals were infected with high doses.

There are data showing an increase in thromboplastin and prothrombin time and a decrease in clotting factors I, VII and VIII in the course of Ebola infection in rhesus monkeys (FISHER-HOCH et al. 1985). In EBOV-infected baboons (Table 3), changes in certain parameters of the blood clotting system are evidence of rapid

Table 3. Blood clotting evidence of Ebola infection in baboons[a] (LUCHKO et al. 1995)

Days post-infection	Thrombin index (%)	Prothrombin index (%)	Fibrinogen (g/l)	Ethanol gel test	Fibrin degradation products
0	100.0 ± 0.0	100.0 ± 0.0	3.4 ± 0.4	0.0 ± 0.0	0.2 ± 0.1
1	105.8 ± 6.7	104.9 ± 6.8	5.3 ± 0.5	0.2 ± 0.1	0.3 ± 0.1
2	105.3 ± 10.5	139.5 ± 15.2*	5.0 ± 0.6	0.1 ± 0.1	n.d.
3	123.5 ± 12.8	142.7 ± 15.0*	4.0 ± 0.4	0.4 ± 0.2	0.1 ± 0.1
4	148.0 ± 15.5*	156.6 ± 16.9*	4.3 ± 0.4	0.5 ± 0.2	n.d.
5	137.9 ± 26.0	141.2 ± 14.7*	4.5 ± 0.4	0.3 ± 0.1	0.4 ± 0.1
6	103.5 ± 9.2	112.7 ± 12.1	3.74 ± 0.42	0.2 ± 0.1	0.3 ± 0.1
7	74.9 ± 7.2*	75.7 ± 6.9*	3.9 ± 0.6	1.0 ± 0.3*	1.7 ± 0.3*
8	63.8 ± 0.6*	62.8 ± 15.0*	3.0 ± 0.6	0.3 ± 0.2	3.0 ± 0.0*
9	20.1 ± 4.6*	32.3 ± 6.6*	1.8 ± 0.3*	0.0 ± 0.0	1.5 ± 0.5*

n.d, not determined; *$p < 0.05$ compared with 0 day.
[a] The mean value for 13 monkeys.

hypercoagulation, reaching a maximum by day 4 and followed by a decrease to hypocoagulation (LUCHKO et al. 1995).

In many monkeys, DIC syndrome developed at the terminal stage of Ebola infection, as demonstrated by the presence of fibrin degradation products in blood (FISHER-HOCH et al. 1985; JAAX et al. 1996). At the morphological level, DIC syndrome is evidenced by numerous fibrin thrombi at target locations in small visceral vessels of rhesus monkeys (BASKERVILLE et al. 1978, 1985; JAAX et al. 1996; JOHNSON et al. 1995). Widespread fibrin deposition is the key diagnostic feature of DIC syndrome. In fact, there is a view that diagnosis of DIC syndrome cannot rely on ambiguously interpreted biochemical data; rather, that it is valid under the condition that fibrin thrombi and masses are abundant at postmortem examination (HAMILTON et al. 1978). However, the absence of thrombi does not entirely exclude DIC syndrome because the time of observation and termination of fibrinolysis should be coincident.

In interpreting histopathologic data, deposition of fibrin in necrotic and inflammatory areas must be taken into account. In filovirus-infected animals, it is expedient to perform microscopic studies on blood vessels in the less damaged organs, such as pancreas, intestine, salivary gland. At the terminal stage of Ebola infection our studies revealed evidence of DIC syndrome in rhesus and green monkeys and no signs of widespread fibrin deposition in baboons quite possibly due to termination of fibrinolysis (Tables 2, 3). No marked deposition of fibrin was observed in cynomolgus macaques at this stage; however, biochemical data allowing us to entirely dismiss DIC syndrome were lacking.

There is, so far, no answer to the question what induces development of DIC syndrome in EBOV-infected monkeys? The presence of thromboplastin in blood provides favorable conditions for transformation of fibrinogen into fibrin. The primary sources of thromboplastin are activated blood monocytes and, to a lesser extent, damaged endothelial cells (RATNOFF and FORBES 1984). The monocytes are widely involved in EBOV reproduction from the start of infection. Quite possibly, mediators released by the infected monocytes initiate the clotting derangement that may result in DIC. Although lacking experimental support for filoviruses, the possibility that virus-infected monocytes may affect blood clotting has been demonstrated for other viral hemorrhagic fevers (COSGRIFF 1989; LEWIS et al. 1989).

Endothelial lesions became more severe as Ebola infection progressed. DIC syndrome may be induced by the combined effects of clotting factors released by the virus-infected monocytes and damaged endothelial cells. Greater damage to clotting and development of DIC syndrome distinguish Ebola from Marburg infection. Judging by recent molecular biological data, a cause of the differences in hemostatic impairment between the two infections may be the specific properties of low molecular weight protein (SGP), which is abundantly secreted by EBOV-, but not MBGV-, infected cells (PETERS et al. 1996). SGP may bind to cells or plasma factors involved in clotting regulation, thereby aggravating hemostatic impairment.

4.4.2 Changes in Vascular Permeability

All experimental filoviral infections are associated with varying degrees of impairment of vascular permeability. Symptoms including bleeding from injection site, blood vomiting, bleeding from rectum and vagina, and hemorrhages into skin and mucosa were observed sporadically in EBOV-infected monkeys (FISHER-HOCH et al. 1992; JOHNSON et al. 1995; LUCHKO et al. 1995) and infrequently in MBGV-infected monkeys (SIEGERT et al. 1968). Bleeding by diapedesis into visceral organs is a feature common to all filoviral infections, being more prominent in the case of Ebola.

Hemorrhages are one of the most consistently observed and striking symptoms in filovirus-affected humans. This raises the question whether hemorrhages are the consequences of blood vessel injury or perhaps diapedesis of erythrocytes. As visualized by light microscopy, the areas of hemorrhagic rash contained engorged capillaries and veins and small hemorrhages; outer epidermal layers were separated from the basal layers in EBOV-infected rhesus monkeys. There was no inflammatory infiltration of hemorrhagic areas (BASKERVILLE et al. 1978).

According to our electron microscopic observations, blood vessel walls were structurally unaltered; there was no inflammatory response in hemorrhagic skin areas in EBOV-infected baboons. Endothelial cells did not disintegrate to a greater extent in hemorrhagic skin than in other skin areas. Hemorrhages resulted from diapedesis. EBOV was seen to reproduce in circulating MPS cells and a few endothelial cells at sites distinct from those of erythrocyte passage through blood vessel walls. Stray infected MPS cells occurred in skin connective tissue also outside diapedesis sites. Erythrocytes which migrated from blood vessels to connective tissue were not hemolysed. Examination of hemorrhages in intestinal mucosa demonstrated that they resulted from erythrocyte passage through blood vessel walls that retained their morphological integrity.

4.5 Implied Cause(s) of Death

In analyzing the pathology of experimental filoviral infections, it is hard to pinpoint the direct cause of terminal shock and death. The cascade of pathogenetic responses triggered by infected MPS cells most likely culminates in terminal shock.

It will be recalled that the mononuclear phagocyte system includes diverse cells, such as blood monocytes, resident visceral and lymphoid macrophages, interstitial macrophages and dendritic cells of the immune system. MPS cells of all types interact and regulate a variety of functions via humoral mechanisms involving mediators, each having a wide range of functions in vivo (BONE 1996; RATNOFF and FORBES 1984; TAUB and OPPENHEIM 1994).

A scheme can be suggested for the cascade of pathogenetic events of filoviral infection in which each step exacerbates the severity of the successive step. First, MPS cells release mediators, which cause injury to the microcirculation and recruit neutrophils into the blood stream (MAYANSKY 1995; TAUB and OPPENHEIM 1994). Increasing disruption of the microcirculatory vasculature leads to disruption of organ blood supply, in turn causing impairment to the function of visceral organs, particularly liver. Hepatic detoxification is further impaired by involvement of hepatocytes in viral reproduction. General intoxication is aggravated by impairment of kidney filtration and detoxification due to alterations in hemodynamics and accumulation of toxic products in blood. Viral involvement of the adrenals provokes impairment of their hormone-producing functions, thereby contributing to the overall pathogenesis. The whole set of changes leads to severe intoxication and a disturbance of homeostasis. A suppressed immune response provides a favorable background for viral reproduction to become rampant.

Involvement of an increasing number of MPS cells in viral reproduction and progressing damage of the endothelium enhance the disturbance of hemostasis. Endothelial damage becomes more prominent by involvement of these cells in viral reproduction at the terminal stage.

It would appear that neutrophils greatly contribute to the pathogenesis of filovirus infections. It is known that their accumulation and destruction within blood vessels is a major cause of severe hemostatic impairment, multiple organ failure, and the adult respiratory distress syndrome. The impairment is, as a rule, fatal (BONE 1996; LEAVER et al. 1995).

To recapitulate, filovirus infections of lung were not coincident with inflammatory infiltration, conducting airways were not obstructed, marked pulmonary edema did not develop. We observed tight obstruction of alveolar capillaries by blood cells and accumulation of neutrophils in monkeys infected with both MBGV and EBOV. These changes, in conjunction with those in the endothelial lining of alveolar capillaries, must affect pulmonary gas exchange with resultant oxygen insufficiency of visceral organs and brain. It is conceivable that progressing hypoxia and the accumulation of toxic substances in blood act as inducers of terminal shock in filovirus-infected animals.

From our filoviral infection studies it can be inferred that events crucial for pathogenesis take place in the blood stream. Indeed, extravascular inflammatory responses were suppressed in lethal infection. The disease caused by MBGV or EBOV may be viewed as a variant of intravascular inflammation, or systemic inflammatory response syndrome. Current concepts of pathogenesis of intravascular inflammation assign a major role to interaction of: (1) neutrophils, (2) damage to endothelial cells, and (3) cytokines released by monocytes. Of the latter, TNF and interleukin-1 (IL-1) are the most important because of their wide range of physiological and pathophysiological effects (MAYANSKY 1995; BONE 1996; TAUB and OPPENHEIM 1994). The role of TNF and IL-1 in the pathogenesis of lethal filoviral infections certainly needs experimental confirmation. At this junction, only statements concerning indirect manifestations, including widespread hemostatic impairment, acute phase protein synthesis, and a distinct pattern of morphological changes are justified.

5 Concluding Remarks

In summary, the main features of experimental filoviral infections are:

- Primary replication of viruses in monocytes and macrophages
- Spread of infection by virions and infected macrophages
- Early impairment of microcirculation and immunity
- Involvement of hepatocytes, adrenocortical cells, endothelial cells and fibroblasts in filoviral reproduction during infection
- Development of damage in detoxification, hemostasis and regulatory responses in the course of infection

Based on the present results, it may be concluded that infection of MPS cells by filoviruses triggers a cascade of interrelated pathological responses. The fatal disease caused by filoviruses appears to be the cumulative result of inadequate responses to the invasive pathogen. Impairment of hemostasis, immunity, detoxification and regulatory systems becomes irreversible during the course of infection, leading to death of the animal.

The pathogenetic cascade proposed here is a broad outline of the pathogenic events still needing elaboration of the details. Infection of MPS cells is envisioned as the primary step in pathogenesis. It appears warranted to identify subsets of mononuclear phagocytes that are the earliest targets of EBOV and MBGV with a view to developing methods to ensure their abolition and to analyze the molecular biological features of their interaction with filoviruses.

Acknowledgements. Our experimental data presented here were obtained in accordance with the research Program of basic investigations supported by the Russian Ministry of Science and Technology. We are grateful to Dr. A.N. Sergeyev for collaboration during animal experiments and helpful discussions and to Mrs. A.N. Fadeeva for translating the manuscript from Russian to English.

References

Akira S, Tetsuya T, Kishimoto T (1993) Interleukin-6 in biology and medicine. Adv Immunol 54:1–5
Baskerville A, Bowen ETW, Platt GS, McArdell LB, Simpson DIH (1978) The pathology of experimental Ebola virus infection in monkeys. J Pathol 125:131–138
Baskerville A, Fisher-Hoch SP, Neild GH, Dowsett AB (1985) Ultrastructural pathology of experimental Ebola haemorrhagic fever virus infection J Pathology 47:199–209
Bechtelsheimer H, Korb G, Gedigk P (1971) Marburg virus hepatitis. In: Martini GA, Siegert R (eds) Marburg virus disease. Springer, Berlin Heidelberg New York, pp 62–67
Bogdan C, Nathan C (1993) Modulation of macrophage function by transforming growth factor b, interleukin-4 and interleukin-10. Ann NY Acad Sci 685:713–716
Bone RC (1996) Toward a theory regarding the pathogenesis of the systemic inflammatory response syndrome: what we do and do not know about cytokine regulation. Crit Care Med 24:163–172
Bowen ETW, Platt GS, Lloyd G, Raymond RT, Simpson DIH (1980) A comparative study of strains of Ebola virus isolated from Southern Sudan and Northern Zaire in 1976. J Med Virol 6:129–138
Bukreev AA, Volchkov VE, Blinov VM, Netesov SV (1993) The GP-protein of Marburg virus contains the region similar to the "immunosuppressive domain" of oncogenic retrovirus P15 E proteins. FEBS Lett 323:183–187
Campbell AD, Wicha MS (1988) Extracellular matrix and the hematopoietic microenvironment. J Lab Clin Med 112:140–146
Cook DN (1996) The role of MIP-1 alpha in inflammation and hematopoiesis. J Leukoc Biol 59:61–66
Cosgriff TM (1989) Viruses and hemostasis. Rev Infec Dis 11:672–688
Courtoy P, Boyles J (1983) Fibronectin in the microvasculature: localization in the pericyte-endothelial interstitium. J Ultrastr Res 83:258–273
Daniel AM, Swenson JJ, Mayreddey RPR, Khalili K, Frisque R (1996) Sequences within the early and late promoters of archetype JC virus restricted viral DNA replication and infectivity. Virology 216:90–101
Ellis DS, Bowen ETW, Simpson DIH, Stamford S (1978) Ebola virus: a comparison, at ultrastructural level, of the behavior of the Sudan and Zaire strains in monkeys. Br J Exp Pathol 59:584–593
Feldmann H, Bugany H, Mahner F, Klenk HD, Drenckhahn D, Schnittler HJ (1996) Filovirus-induced endothelial leakage triggered by infected monocytes/macrophages. J Virol 70:2208–2214
Fisher-Hoch SP, Platt GS, Neild GH, Southee T, Baskerville A, Raymond RT, Lloyd G, Simpson DIH (1985) Pathophysiology of shock and haemorrhage in fulminating viral infection (Ebola). J Infect Dis 152:887–894
Fisher-Hoch SP, Brammer TL, Trappier SG, Hutwagner LC, Farrar BB, Ruo SL, Brown BG, Hermann LM, Perez-Oronoz GI, Goldsmith CS, Hanes MA, McCormick JB (1992) Pathogenic potential of filoviruses: role of geographic origin of primate host and virus strain. J Infect Dis 166:753–763
Geisbert TW, Jahrling PB, Hanes MA, Zack PM (1992) Association of Ebola-related Reston virus particles and antigen with tissue lesions of monkeys imported to the United States. J Comp Pathol 106:137–152
Haas R, Maass G, Muller J, Oehlert W (1968) Experimentelle Infektionen von Cercopithecus aethiops mit dem Erreger des Frankfurt-Marburg, Syndroms (FMS) Z Med Mikrobiol Immunol J 154:210–220
Hamilton PJ, Stalker AL, Douglas AS (1978) Disseminated intravascular coagulation: a review. J Clin Pathol 31:609–619
Ishikawa K, Nagai H, Katayama K, Tsutsui M, Tanabayashi K, Takeuchi K, Hishiyama M, Saitoh A, Takagi M, Gotoh K, Muramatsu M, Yamada A (1995) Comparison of the entire nucleotide and deduced amino acid sequences of the attenuated hog cholera vaccine strain GPE5- ÿ0 and the wild-type parental strain ALD. Arch Virol 140:1385–1391
Jaax NK, Davis KJ, Geisbert TJ, Vogel P, Jaax GP, Topper M, Jahrling PB (1996) Lethal experimental infection of rhesus monkeys with Ebola-Zaire (Mayinga) virus by the oral and conjunctival route of exposure. Arch Pathol Lab Med 120:140–155
Johnson E, Jaax N, White J, Jahrling P (1995) Lethal experimental infections of rhesus monkeys by aerosolized Ebola virus. Int J Exp Path 76:227–236

Kapcala LP, Chautard T, Eskay RL (1995) The protective role of the hypothalamic-pituitary-adrenal axis against lethality produced by immune, infectious, and inflammatory stress. Ann NY Acad Sci 771:419–437

Knight SC, Macatonia SE (1988) Dendritic cells and viruses. Immunol Lett 19:177–182

Kolesnikova LV, Ryabchikova EI, Rassadkin YuN, Grazhdantseva AA (1997) Ultrastructural stereologic analysis of monkeys lung in experimental Ebola infection. Bull Exp Biol:123:205–208

Leaver HA, Yap PL, Rogers P, Wright I, Smith G, Williams PE, France AJ, Craig SR, Walker WS, Prescott RJ (1995) Peroxides in human leukocytes in acute septic shock: a preliminary study of acute phase changes and mortality. Eur J Clin Invest 25:777–783

Lewis RM, Morril JC, Jahrling PB, Cosgriff TM (1989) Replication of hemorrhagic fever virus in monocytic cells. Rev Infect Dis 11:736–742

Lonigro AJ, McMurdo L, Stephenson AH, Sprague RS, Weintraub NL (1996) Hypotheses regarding the role of pericytes in regulating movement of fluid, nutrients, and hormones across the microcirculatory endothelial barrier. Diabetes 45:38–43

Lub MYU, Sergeev AN, Pyankova OG, Pyankov OV, Petricshenko VA, Kotlyarov LA (1995) Certain pathogenetic characteristics of a disease in monkeys in infected with the Marburg virus by an airborne route. Probl Virol 4:158–161

Luchko SV, Dadaeva AA, Ustinova EN, Sizikova LP, Ryabchikova EI, Sandakhchiev LS (1995) Experimental study of Ebola hemorrhagic fever in baboon models. Bull Exp Biol Med 9:302–304

Mayansky AN (1995) Present-day evolution of the idea of II Metchnikoff on intravascular inflammation. Immunologia 4:8–14

Morens DM (1994) Antibody-dependent enhancement of infection and the pathogenesis of viral disease. Clin Infect Dis 19:500–512

Murphy FA (1985) Marburg and Ebola viruses. In: Pattyn SR (ed) Virology. Raven, New York, pp 1111–1118

Murphy FA, Simpson DIH, Whitefield SC, Zlotnik I, Carter GB (1971) Marburg virus infection in monkeys: ultrastructural studies. Lab Invest 24:279–291

Nossal GJV, Abbot A, Mitchell J (1968) Antigens in immunity. XIV. Electron microscopic radioautographic studies of antigen capture in the lymph node medulla. J Exp Med 127:263–276

Pereboeva LA, Tkachev VK, Kolesnikova LV, Krendeleva LYA, Ryabchikova EI, Smolina MP (1993) Ultrastructural changes of guinea pigs organs in sequential passages of Ebola virus. Probl Virol 4:179–182

Perretti M, Flower RJ (1994) Cytokines, glucocorticoids and lipocortins in the control of neutrophil migration. Pharmacol Res 30:53–59

Peters CJ, Johnson ED, McKee KT (1991) Filoviruses and management of viral hemorrhagic fevers. In: Belshe RB (ed) The textbook of human virology. Mosby Year Book, St. Louis, pp 699–712

Peters CJ, Sanchez A, Rollin PE, Ksiazek TG, Murphy FA (1996) Filoviridae: Marburg and Ebola Viruses. In: Fields BN, Knipe DM, Howley PM et al (eds) Field's Virology, 3rd edn. Lippincott-Raven, Philadelphia, pp 1161–1176

Pyankov OV, Sergeev AN, Pyankova OG, Chepurnov AA (1995a) Experimental Ebola fever in Macaca mulatta. Problems Virol 3:113–115

Pyankov OV, Sergeev AN, Pyankova OG, Kolesnikova LV (1995b) Pathologic changes in the organism of primates infected with Ebola virus through respiratory tract. J Aerosol Medicine 8:76

Ratnoff OD, Forbes ChD (1984) Disorders of haemostasis. New York, pp 577

Robin J, Bres P, Camain R (1971) Passage of Marburg virus in guinea pigs. In: Martini GA, Siegert R (eds) Marburg Virus Disease. Springer, Berlin Heidelberg New York, pp 117–122

Robson SC, Saunders R, Purves LR, deJager C, Corrigall A, Kirsch RE (1993) Fibrin and fibrinogen degradation products with an intact D-domain C-terminal gamma chain inhibit an early step in accessory cell-dependent lymphocyte mitogenesis. Blood 81:3006–3014

Ryabchikova EI, Baranova SG, Tkachev VK, Grazhdantseva AA (1993) Morphological changes in Ebola infection of guinea pigs. Probl Virol 4:177–179

Ryabchikova EI, Vorontsova LA, Skripchenko AA, Shestopalov AM, Sandakhchiev LS (1994a) The peculiarities of internal organs damage in experimental animals infected with Marburg disease virus. Bull Exp Biol Med 4:430–434

Ryabchikova EI, Kolesnikova LV, Tkachev VK, Pereboeva LA, Baranova SG, Rassadkin YuN (1994b) Ebola infection in four monkey species. In: Frontiers of viral pathogenesis: Ninth International Conference of Negative Strand Viruses. Estoril, Portugal, 2–7 October, 1994, p 164

Ryabchikova E, Kolesnikova L, Smolina M, Tkachev V, Pereboeva L, Baranova S, Grazhdantseva A, Rassadkin Yu (1996a) Ebola virus infection in guinea pigs: presumable role of granulomatous inflammation in pathogenesis. Arch Virol 141:909–921

Ryabchikova E, Strelets L, Kolesnikova L, Pyankov O, Sergeev A (1996b) Respiratory Marburg virus infection in guinea pigs. Arch Virol 141:2177–2189

Sato T, Selleri C, Young NS, Maciejewski JP (1995) Hematopoietic inhibition by interferon-gamma is partially mediated through interferon regulatory factor-1. Blood 86:3373–3380

Schnittler HJ, Mahner F, Drenckhahn D, Klenk HD, Feldmann H (1993) Replication of Marburg virus in human endothelial cells: a possible mechanism for the development of viral hemorrhagic disease. J Clin Invest 91:1301–1309

Sergeev AN, Lub MYu, Kotlyarov LA, Pyankov OV, Vorontsova LA (1995) Some aspects of Marburg virus pathogenesis in guinea pigs infected through respiratory tract. J Aerosol Medicine 8:76

Siegert R, Shu HL, Slenczka W (1968) Zur Diagnostik und Pathogenese der Infection mit Marburg-Virus. Deutsch Ärztebl Ärztl Mitt 65:1827–1830

Siegert R, Slenczka W (1971) Laboratory diagnosis and pathogenesis. In: Martini GA, Siegert R (eds) Marburg Virus Disease. Springer, Berlin Heidelberg New York, pp 157–160

Siegert R (1972) Marburg virus. In: Virology monographs. vol. 11. Springer, Berlin Heidelberg New York, pp 98–153

Simpson DIH, Bowen EJW, Bright WF (1968) Vervet monkey disease: experimental infection of monkeys with causative agent and antibody studies in wild caught monkeys. Lab Anim Handb 68:75–81

Simpson DIH (1969) Marburg agent disease. Trans Roy Soc Trop Med Hyg 63:303–309

Steinman RM (1991) The dendritic cell system and its role in immunogenicity. Annu Rev Immunol 9:271–283

Skripchenko AA, Ryabchikova EI, Vorontsova LA, Shestopalov AM, Vjazunov SA (1994) Marburg virus and mononuclear phagocytes: study of interaction. Prob Virol 5:215–219

Taub DD, Oppenheim JJ (1994) Chemokines, inflammation and the immune system. Ther Immunol 1:229–246

Volchkov VE, Blinov VM, Netesov SV (1992) The envelope glycoprotein of Ebola virus contains an immunosuppressive-like domain similar to oncogenic retroviruses. FEBS Lett 305:181–184

Wulff H, Conrad JL (1977) Marburg virus disease. In: Kurstak E, Kurstak Ch (eds) Comparative diagnosis of viral disease. Academic, New York, pp 3–33

Zlotnik I (1969) Marburg agent disease: pathology. Trans Roy Soc Trop Med Hyg 63:310–323

Zlotnik I, Simpson DIH (1969) The pathology of experimental vervet monkey disease in hamsters. Brit J Exp Path 50:393–399

Zlotnik I (1971) Marburg disease: The pathology of experimentally infected hamsters. Martini GA, Siegert R (eds) Marburg Virus Disease. Springer, Berlin Heidelberg New York, pp 129—135

Molecular Pathogenesis of Filovirus Infections: Role of Macrophages and Endothelial Cells

H.J. SCHNITTLER[1] and H. FELDMANN[2]

1	Introduction	175
2	Proposed Course of Filovirus Infection	177
3	Pathogenesis of Filovirus Infection	179
3.1	Role of Monocytes/Macrophages During Filovirus Infection	179
3.1.1	Replication	180
3.1.2	Activation	180
3.1.3	Spreading	184
3.2	Immunosuppression Associated with Filovirus Infection	187
3.3	Role of Endothelial Cells During Filovirus Infection	188
3.3.1	Replication	188
3.3.2	Hemorrhage	190
3.3.3	Changes in Permeability	191
3.3.3.1	Effects of Mediators	191
3.3.3.2	Structural Components	192
3.3.3.3	Signaling Pathways	195
3.3.4	Further Alterations	195
4	Conclusions	196
References		197

1 Introduction

Filoviruses are enveloped, nonsegmented negative-stranded RNA viruses and constitute a separate family within the Order Mononegavirales (KILEY et al. 1982; FELDMANN et al. 1993; MURPHY et al. 1995). The family consists of a single genus, *Filovirus*, which can be separated into two species, Marburg and Ebola. Species Marburg comprises five independent strains with Musoke (MUS) being designated the prototype virus of the family. Species Ebola is subdivided into four distinct subspecies; Cote d'Ivoire, Reston, Sudan, and Zaire (FELDMANN and KLENK 1996; SANCHEZ et al. 1996). The genomes of Marburg virus (MBGV) and Ebola virus (EBOV) as well as the structural protein-encoding genes and their products have

---

[1]Institut für Physiologie, Westfälische Friedrich-Wilhelms-Universität, Robert-Koch-Str. 27a, 48149 Münster, Germany
[2]Institut für Virologie, Philipps-Universität, Robert-Koch-Str. 17, Marburg, Germany

been studied extensively during the past years (for review see FELDMANN et al. 1993; FELDMANN and KLENK 1996; PETERS et al. 1996a). Detailed information on the molecular biology of filoviruses are given in the Chapter by FELDMANN and KILEY this volume.

Infections with filoviruses often cause a fulminant hemorrhagic disease in humans and monkeys (SMITH et al. 1967; MARTINI 1971; PATTYN 1978). Among all viral hemorrhagic fevers (HFs), Marburg and Ebola HF are characterized as the most severe forms, associated with hemorrhagic manifestations, marked hepatic involvement, coagulation disorders, generalized shock, and the highest case-fatality rate (22%–90%) (MARTINI and SIEGERT 1971; PATTYN 1978; FELDMANN and KLENK 1996; PETERS et al. 1996a). The viruses are pantropic, but no single organ shows sufficient damage to account for either the onset of the severe shock syndrome or the bleeding tendency (ISHAK et al. 1982). The clinical symptoms and the pathology of filovirus infections are described in the Chapters by ZAKI/GOLDSMITH and FISHER-HOCH/MCCORMICK this volume.

The pathophysiological changes, however, that make filovirus infections so devastating are just beginning to be unraveled. As in some other viral HFs (hemorrhagic fever with renal syndrome (HFRS), Dengue HF, Lassa fever), filovirus infections are associated with generalized fluid distribution problems, hypotension, coagulation disorders and bleeding tendency finally resulting in fulminate shock (PETERS et al. 1991). Filovirus HF can be compared to a syndrome provoked by systemic treatment with cytokines which is also known from endotoxin-induced shock (BROUCKAERT and FIERS 1996; DINARELLO 1996). These cytokine-induced symptoms are designated as the systemic inflammatory response syndrome (SIRS), which describes a surplus reaction of the host triggered by pathogens or their products (BROUCKAERT and FIERS 1996).

Morphological studies on filovirus-infected humans and monkeys suggested that monocytes/macrophages and fibroblasts are preferred sites of filovirus replication in early stages of the disease (GEISBERT et al. 1992; RYABCHIKOVA et al. 1994; ZAKI and PETERS 1997). Human monocytes/macrophages in culture were highly sensitive to filovirus infection resulting in massive production of infectious particles and finally cell lysis (FELDMANN et al. 1996). Most importantly, infected monocytes/macrophages became activated early in infection resulting in the release of several proinflammatory cytokines including tumor necrosis factor-α (TNF-α) (FELDMANN et al. 1996; WEST et al. 1996). Furthermore, endothelial cells are targeted by cytokines and respond by increased permeability, procoagulant activity, expression of cell adhesion molecules and loss of blood pressure regulation. These pathophysiological reactions have been described for endotoxin shock and SIRS. Endothelial cells are also suitable targets for filoviruses in tissue and organ cultures, and virus replication caused extensive cell lysis of these cells (SCHNITTLER et al. 1993). Post-mortem observations of human cases clearly demonstrated viral antigen in endothelial cells of several organs (ZAKI and PETERS 1997). Therefore, it is likely that the mononuclear phagocyte system (MPS) and the endothelium play a central role in the pathogenesis and in the development of the disease-decisive symptoms of filoviral HF.

2 Proposed Course of Filovirus Infection

Person-to-person transmission by intimate contact with skin and secretion products of infected patients is the main route of infection for humans. Virus seems to enter through minute lesions of the skin and mucosae. From there, viral particles can either get direct access to the vascular system or indirect access via the lymphatic system. The latter would presuppose a primary viral replication in local lymph nodes followed by an ascending infection finally reaching the venous vascular system leading to viremia (FENNER 1948; JAHRLING et al. 1996a; PETERS et al. 1996b) (Fig. 1). Viremia is accompanied by the initial symptoms, including fever, headache and asthenia. Around day 3 post-onset of disease additional less specific symptoms, such as severe watery diarrhea, abdominal pain, sore throat, nausea, vomiting, cough, arthralgia, and pharyngeal and conjunctival injections are observed. Patients are dehydrated, apathetic, disoriented, and may develop a characteristic, nonpruritic, maculopapular centripetal rash associated with varying degrees of erythema and desquamate by day 5–7 of illness. Hemorrhagic manifestations develop during the peak of the illness and are of prognostic value for the disease. Bleeding into the gastrointestinal tract is most prominent beside petechiae and hemorrhages from puncture wounds and mucous membranes. Death occurs between days 7 and 16 (mean days 8–9) and is associated with a hypovolemic shock syndrome. In post-mortem material multiple focal necroses can be observed in lymph nodes, liver, spleen, gastrointestinal tract, mucosae, kidneys, lung, heart, brain and skin.

Primary sites of virus replication during filovirus infections are cells of the MPS (GEISBERT et al. 1992; RYABCHIKOVA et al. 1994; ZAKI and PETERS 1997; FELDMANN et al. 1996) that are located in several organs, such as the liver (Kupffer, cells), spleen, lymph nodes, lung (alveolar macrophages), serous cavities (pleural and peritoneal macrophages), and nervous system (microglia) (WEISS 1983). Lymph nodes, liver and spleen, however, are known as preferred organs for filovirus replication, whereas other organ involvement is irregular and does not follow a characteristic scheme (FELDMANN and KLENK 1996; PETERS et al. 1996a). These differences in organ tropism can most likely be explained by the direct access of viral particles to sessile cells of the MPS without penetration of cellular or tissue barriers as has been postulated recently (SCHNITTLER and FELDMANN 1998). In lymph nodes macrophages typically occur in the lymph sinus and are surrounded by lymphatic fluid allowing infection by filoviruses that were taken up by lymph vessels from small lesions (Fig. 1). Following infection and primary replication in these macrophages, virus particles get access to secondary lymph nodes and finally to the vascular system (viremia). The subsequent step during the course of infection seems to be mediated by the presence of macrophages in the liver sinuses and spleen. In contrast to most other organs, the spleen possesses an open blood circulation with sheathed capillaries (macrophages) and sessile macrophages present along the venous sinuses (Fig. 1) (WEISS 1983). Thus, during viremia these macrophages can directly be infected without penetration of cellular or tissue barriers,

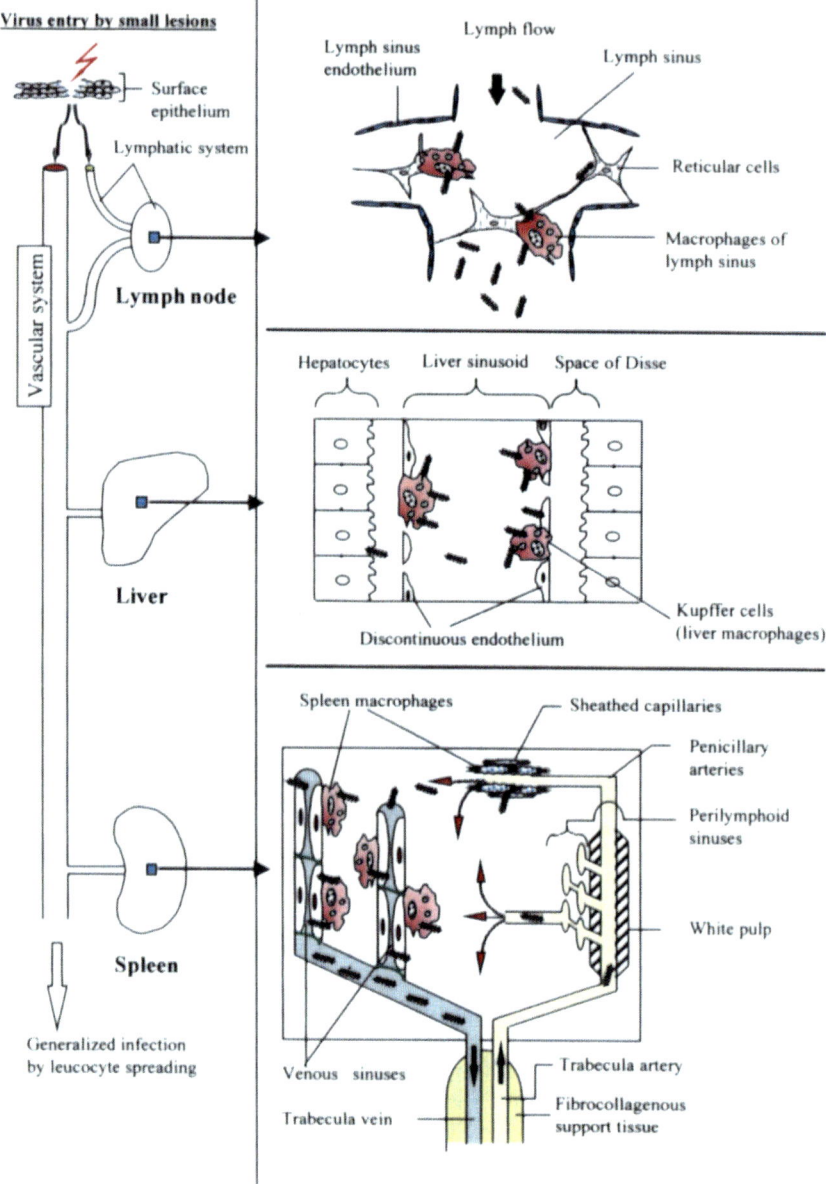

Fig. 1. Hypothesized course of virus entry and primary virus replication during filovirus infection. Virus replication in/and activation of lymphatic macrophages (*upper right*), Kupffer cells (*middle right*), and spleen macrophages (*lower right*) may be important for primary organ tropism during filovirus infection. The generalized infection seems to be caused by an extravasation of infected leukocytes, as discussed in the text. (From SCHNITTLER and FELDMANN 1998)

similar to the situation in lymph nodes. The same holds true for the liver, where virus can get direct access to the liver Kupffer cells (sessile macrophages) that are frequently located at the venous portal sinuses (Fig. 1). Therefore, in contrast to most other organs, sessile macrophages of these preferred organs can directly be infected during viremia and may substantially contribute to primary virus replication and spread (SCHNITTLER and FELDMANN 1998).

In addition, the discontinuous portal liver sinus endothelium does not rest on a regular basement membrane and displays large gaps and transcellular holes allowing macromolecules and water as well as larger particles, such as chylomicrons and virus particles, to directly enter the space of Disse from the blood (Fig. 1). This allows the virus to infect hepatocytes without passing cellular or tissue barriers during viremia (SCHNITTLER and FELDMANN 1998). Filoviral particles in the space of Disse have been found in post-mortem liver material from infected humans and monkeys (GEISBERT et al. 1992; ZAKI and PETERS 1997). Since hepatocytes are less frequently infected than macrophages (ZLOTNIK 1969; GEISBERT et al. 1992), the liver tropism in filovirus infection might be primarily mediated by Kupffer cells. The endothelium of several organs (e.g., kidney, circumventricular organs of the brain) is also discontinuous; however, the basement membranes and the blood-liquor barrier in those organs may prevent the direct attack of virus particles on parenchymal cells. These organs are most likely infected later on by extravasation of infected circulating monocytes, other leukocytes and/or endothelial cells that subsequently contribute to the pantropic manifestation of filovirus infections.

3 Pathogenesis of Filovirus Infection

3.1 Role of Monocytes/Macrophages During Filovirus Infection

The MPS, consisting of monocytes and tissue macrophages, is a dynamic cellular system with representation in all tissues and with the potential to exert a modulatory role in tissue homeostasis and in local immunological and inflammatory responses. Macrophages mediate a unique tissue response to external stimuli by: (1) interaction with extracellular molecules which they internalize and process, (2) secretion of many products, such as cytokines, proteases, complement proteins, and peroxide, that are important for the inflammatory response, (3) interaction with B and T lymphocytes which allows intervention with the specific immunological responses, (4) being located close to the microvasculature and surrounding epithelial and mesenchymal cells, and (5) expression of surface receptors for lymphokines which lead to macrophage activation (UNANUE and ALLEN 1987).

Activated macrophages are highly microbicidal and thus, in general, the MPS is an important part of the host defense mechanism. In most cases macrophages confer protection to viral infections by two different processes: (1) intrinsic, by acting as phagocytes or as nonpermissive host cells and (2) extrinsic, by retarding or ablating virus multiplication in neighboring cells through destruction of those cells. In addition, activated macrophages participate in virus inhibition by producing cytokines, mediating antibody-dependent cytotoxic cell lysis and triggering the specific humoral immune response. However, phagocytosis can also lead to persistence, e.g., human immunodeficiency virus (HIV), or multiplication, e.g., HIV, herpes viruses, MBGV, of viruses. Thus macrophages may aid transmission to other cells and distant sites or enhance infection by depression of the immune defenses.

3.1.1 Replication

In contrast to lymphocytes, eosinophils and plasma cells, circulating monocytes and macrophages in various tissues are preferred sites of filovirus replication (GEISBERT et al. 1992). Virus replication in monocytes/macrophages has been described in humans, monkeys, guinea pigs, and hamsters (ZLOTNIK 1969; GEISBERT et al. 1992; RYABCHIKOVA et al. 1994; ZAKI and PETERS 1997). More recently, MBGV replication has been investigated in cultured human peripheral blood monocytes/macrophages (FELDMANN et al. 1996) (Fig. 2). The mode of virus entry into macrophages, phagocytosis or receptor-mediated, is still unknown. Synthesis of viral-specific subgenomic RNAs was detectable 6h post-infection; evidence for replication was obtained earliest by 12h post-infection (WEST et al. 1996). Examination of infected monocytes/macrophages by transmission electron microscopy revealed numerous viral particles in intracellular cytoplasmic vacuoles. Virus budding was observed at membranes surrounding the vacuoles and at the plasma membrane (Fig. 2). Within 24–32 hours post-infection, the first infectious virus was released, as detected by plaque assays. At 5–6 days post-infection, cytolysis and early signs of cytopathogenic effects, with cells rounding off, were seen (FELDMANN et al. 1996).

3.1.2 Activation

Cultured human monocytes/macrophages became activated upon filovirus infection resulting in an increase of transcription of several multifunctional cytokines such as TNF-α, and interleukin (IL)-1β, IL-6, and IL-8 (FELDMANN et al. 1996; WEST et al. 1996). The time course of cytokine expression and release was studied with TNF-α, the prototype cytokine and important mediator released by activated monocytes/macrophages (Fig. 3a). Approximately 4–6h post-infection with MBGV a significant increase in release was observed and peak values (ca. 3ng/ml) were reached 12–24h post-infection (FELDMANN et al. 1996). This was confirmed by the detection of TNF-α-specific mRNA transcripts around 6h post-infection (WEST et al. 1996). To date, it remains uncertain if activation depends on virus replication or viral antigen taken up by monocytes/macrophages. The exact molecular mechanism of virus-induced activation is unknown as well.

Beside their other functions, cytokines are also involved in the development of shock and SIRS that can be provoked by administration of cytokines (BROUCK-

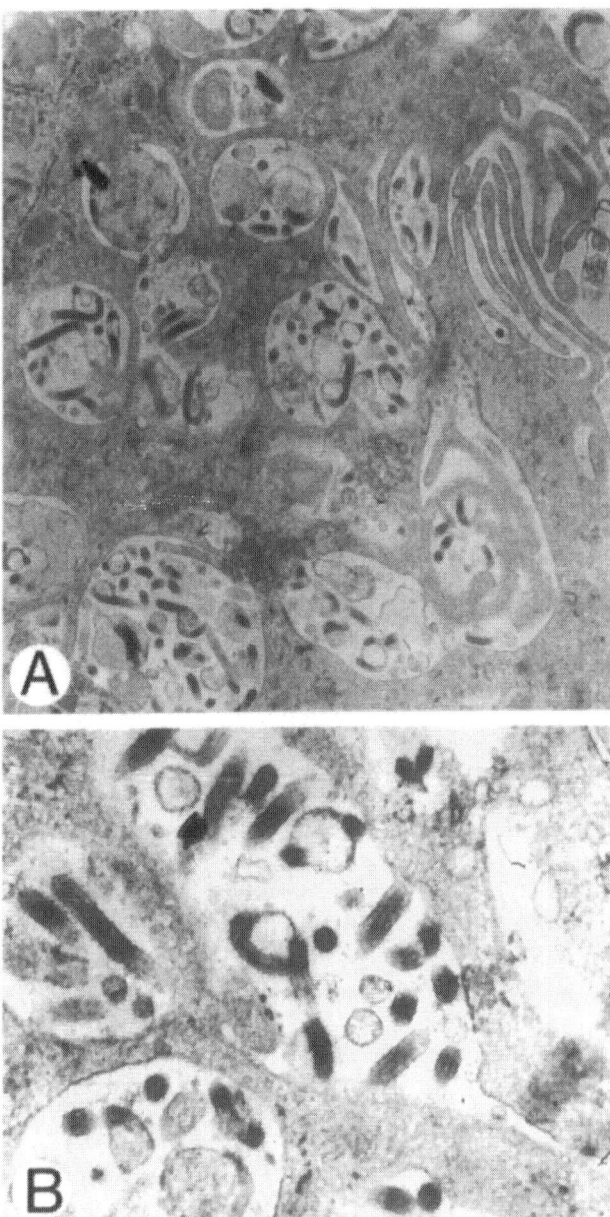

Fig. 2A,B. Electron microscopy of Marburg virus (MBGV)-infected cultured human peripheral blood monocytes/macrophages. Virus replication in cultured human peripheral blood monocytes/macrophages resulted in an accumulation of large numbers of particles in intracellular vacuoles **A**. Virus budding was observed from intracellular membranes surrounding vacuoles **B** as well as from the plasma membrane of infected cells. (From FELDMANN et al. 1996)

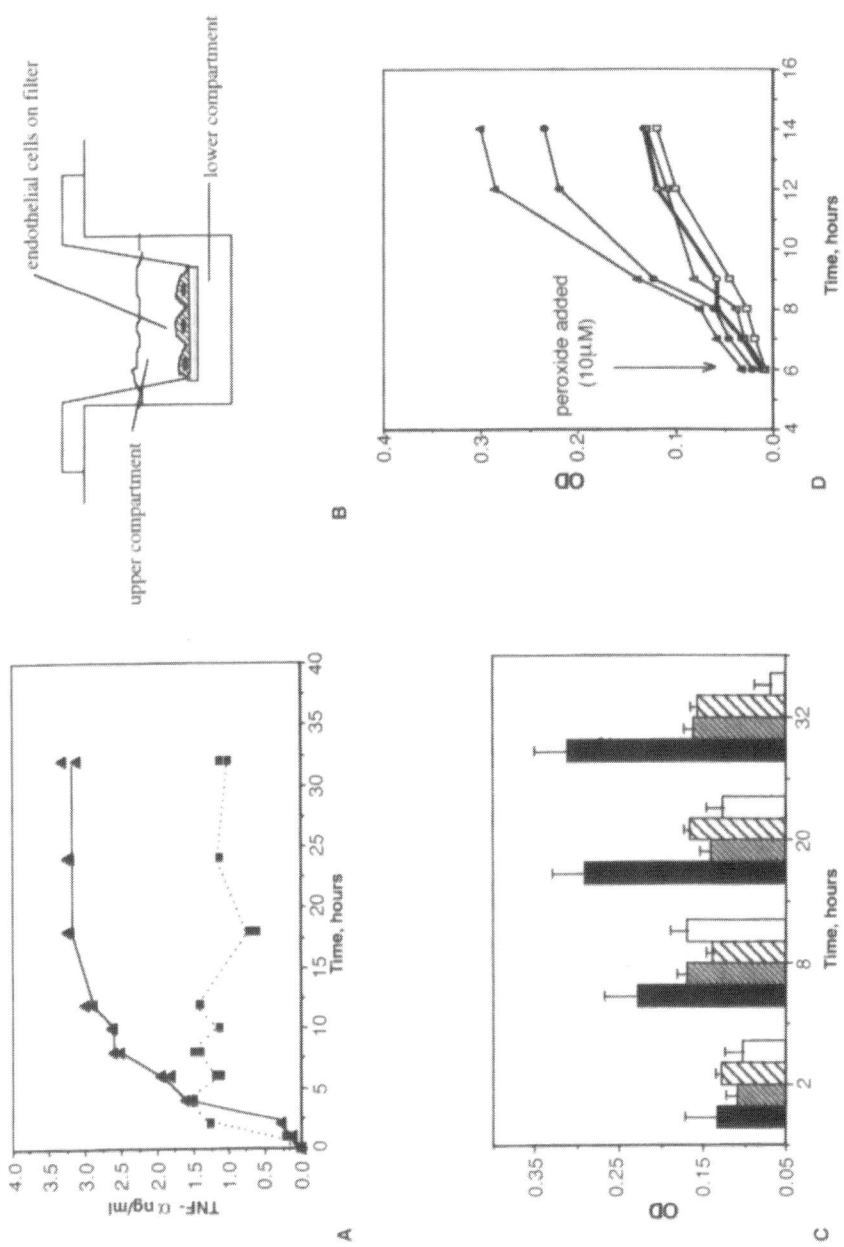

AERT and FIERS 1996). Shock and SIRS can be caused by multiple trauma, ischemic reperfusion injury, acute transplant rejection, and acute inflammatory stages, such as hepatitis and pancreatitis, but also by bacterial (e.g., endotoxin) (DINARELLO 1996) and viral infections. The initial symptoms observed in filoviral HF (e.g., fever and headache) are unspecific and most likely triggered by cytokines (PETERS et al. 1996a). Therefore, the rapid onset of symptoms might correlate with infection and/ or activation of circulating monocytes/macrophages and sessile macrophages. Recently, elevated levels of several mediators, including TNF-α, were detected in acute sera of several EBOV-infected patients from Gabon (GEORGES et al. 1996) and Zaire (VILLINGER et al. 1998) as well as from infected monkeys (IGNATYEV et al. 1996). Elevated TNF-α values have also been reported from patients with Argentine HF caused by Junin virus, a member of the family Arenaviridae, that correlated with the severity of the disease (HELLER et al. 1992). One can assume that filovirus-infected monocytes/macrophages do not only react by the release of cytokines, but also of other proinflammatory mediators (e.g., histamine, serotonin), proteases (e.g., elastase) and peroxide (H_2O_2). The latter factors might be of increasing significance at final stages of the disease when infected monocytes/macrophages undergo cell lysis (FELDMANN et al. 1996). Secreted macrophage products act on multiple organs, but the endothelium and the vascular wall seem to be the main targets.

Apart from effects on endothelial cells described below (see Sect. 3.3.3), cytokines produced and secreted from activated monocytes/macrophages also play an important role in the host defense mechanism against viral infections (ADERKA et al. 1986; GONG et al. 1991; JOHNSTON et al. 1995; KAWANISHI et al. 1995). Many cytokines inhibit viral replication either directly or indirectly as has been described for interferon (IFN), IL-1, and TNF α. IL-1ß and TNF-α are needed for a number of immunological functions, such as the proliferative and functional T lymphocyte responses, including T lymphocyte-monocyte interaction during antigen presentation (LE and VILCEK 1987; UNANUNE and ALLEN 1987). TNF-α inhibits virus replication both directly and indirectly by inducing IFN and other cytokines and

---

Fig. 3A–D. Role of monocyte/macrophage activation on endothelial permeability. **A** Time-dependent increase in TNF-α release from MBGV-infected culture human monocytes/macrophages. *Triangle*, MBGV infection; *square*, mock infection. Note, TNF-α increase is observed within hours post-infection. **B** Transwell filter system used to assay the effect of supernatants derived from MBGV-infected monocytes/macrophages on endothelial permeability. (From FELDMANN et al., 1996). **C** Increased permeability of HUVEC cultures in response to culture supernatants of MBGV-infected monocytes/macrophages using the transwell filter system shown in **B**. Culture supernatants of MBGV-infected monocytes/macrophages prior to (*solid bars*) and after (*shaded bars*) incubation with a neutralizing anti-TNF-α antibody (60 µg/ml; Boehringer Mannheim, Germany); culture supernatants of mock-infected monocytes/macrophages prior to (*hatched bars*) and after (*open bars*) incubation with a neutralizing anti-TNF-α antibody (60 µg/ml; Boehringer Mannheim, Germany); OD, optical density at 470 nm. **D** Effect of recombinant TNF-α and H_2O_2 on endothelial permeability using the transwell filter system shown in **B**. *Open circle*, no treatment; *open square*, no TNF-α and 10 µM H_2O_2; *open triangle*, 1 ng TNF-α per ml and 10 µM H_2O_2; *solid circle*, 10 ng TNF-α per ml and 10 µM H_2O_2; *solid triangle*, 100 ng TNF-α per ml and 10 µM H_2O_2; OD, optical density at 470 nm

augmenting inflammation, phagocytosis and cytotoxic activity. Nevertheless, some cytokines, such as TNF-α, possess a protean nature, with both beneficial and adverse effects for the host. Thus, TNF-α has been shown to activate replication of cytomegalovirus (HAAGMANS et al. 1994) and HIV (HAN et al. 1996). It also stimulates the adenovirus E3 promotor resulting in a reduction of cell surface expression of HLA class I molecules, thus allowing the virus to take advantages of an immune mediator to escape immunosurveillance (KÖRNER et al. 1992). The effect of cytokines on filovirus replication is currently unknown.

3.1.3 Spreading

Infected macrophages found in various tissues probably originate from infected circulating monocytes/macrophages that extravasate following infection (GEISBERT et al. 1992; RYABCHIKOVA et al. 1994; FELDMANN et al. 1996). Thus, intracellular budding and accumulation of viral particles in vacuoles (Fig. 2) might provide a mechanism by which virus hidden from the host's immune surveillance may spread. Virus could be passed to other cells by exocytosis of the vacuoles or by release after cell lysis (FELDMANN et al. 1996). A similar role of monocytes/macrophages has been discussed for the spread of other virus infections such as HIV (ORENSTEIN et al. 1988) and Visna virus (PELUSO et al. 1985).

The proposed extravasation of infected monocytes/macrophages might be triggered by a virus-induced cytokine release (see Sect. 3.1.2) which activates the endothelium. In addition, infected endothelial cells may undergo autoactivation, as described later on in this chapter (see Sect. 3.3.1). It has been shown that cytokines, such as TNF-α and IL-1β, H_2O_2, and various proinflammatory mediators increase the expression of various cell adhesion molecules on endothelial cells such as intracellular adhesion molecule-1 (ICAM-1), vascular cell adhesion molecule-1 (VCAM-1), and E- and P-selectin (endothelial cell activation) (BEVILAQUA et al. 1987; PATEL et al. 1992; BRADLEY et al. 1993, 1995; HAHNE et al. 1993; GOTSCH et al. 1994). E and P-selectin function as Ca^{2+}-dependent lectins and mediate the binding of leukocytes to the surface of endothelial cells and, thus, initiate the extravasation process (SHIMIZU et al. 1992; VESTWEBER 1992, 1993; LENTER et al. 1994). Molecular cross-bridges between cell adhesion molecules of endothelial cells and leukocytes slow leukocytes down and initiate rolling along the vascular endothelium where cytokines may be presented to their receptors on the leukocyte surface resulting in stimulation (PIZCUETA and LUSCINSKAS 1994). Firm adhesion to endothelium is mediated by activated integrins (HYNES 1992) which are glycosylated heterodimeric cellular adhesion receptors (SONNENBERG 1993). Leukocyte integrin activation is transient and thus permits cells to move along the endothelial surface until appropriate signals for translocation are received. The subsequent diapedesis into the tissue seems to be mediated via the interaction of lymphocyte function-associated antigen (LFA-1) and ICAM-1, and the movement is directed via a chemotactic gradient (HOGG 1993).

In order to achieve diapedesis of leukocytes interendothelial junctions have to be opened (Fig. 4). Leukocyte transmigration through the endothelial layer further seems to be dependent on the platelet endothelial cell adhesion molecule 1 (PECAM-1) (DeLisser et al. 1994) and the cadherin/catenin-complex (DelMaschio et al. 1996 Moll et al., 1998) which are both localized at endothelial junctions (Figs. 4, 5). PECAM-1 is a member of the immunoglobulin superfamily and is also expressed at surfaces of neutrophils, monocytes, some lymphocytes subsets, and platelets (DeLisser et al. 1994). A change in the distribution of PECAM-1 in situ occurs after systemic application of histamine (Leach et al. 1995). Evidence has been provided that PECAM-1 is involved in recruitment of leukocytes into inflammatory sites through the interendothelial junctions (DeLisser et al. 1994) and the extracellular matrix (Liao et al. 1995). Furthermore, a disorganization of the cadherin/catenin complex (see Sect. 3.3.3.2; Fig. 4) has been described following adhesion of polymorphonuclear leukocytes to endothelial cells (DelMaschio et al. 1996 Moll et al., 1998). In this way, monocytes are enabled to enter sites of inflammation (Shanley et al. 1995). In necrotic foci of filovirus-infected organs, however, only very few inflammatory cells are found (Peters et al. 1996a) suggesting a deficient immunoreaction due to inhibition of leukocyte activation and transmigration or lack of chemotactants. This may be supported by recent data

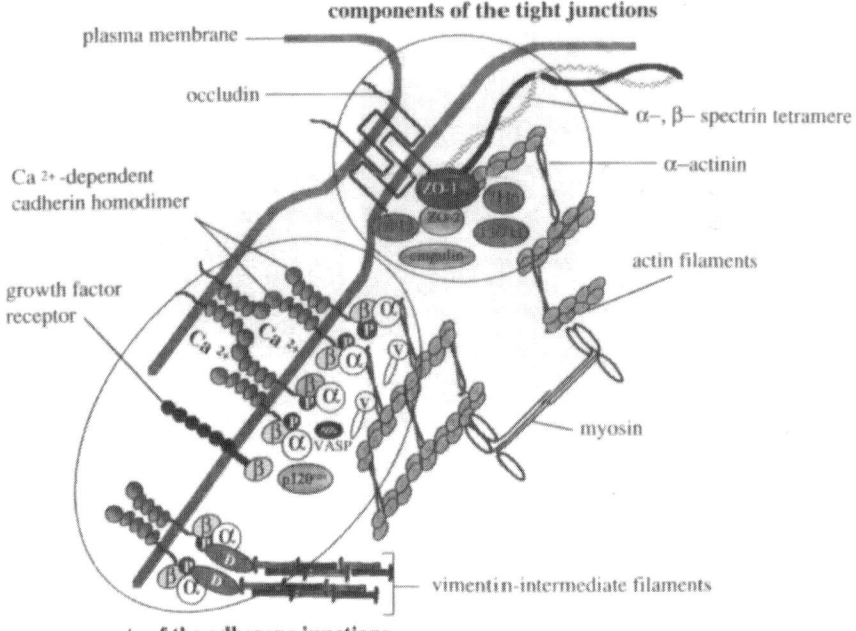

Fig. 4. Organization of interendothelial adherens and tight junctions. The endothelium displays an extended adherens junctional zone in which gap (not shown) and tight junctions are morphologically inserted. The structure and organization of the interendothelial junctions are described in detail in Sect. 3.3.3.2. α, α-catenin; β, β-catenin; D, desmoplakin; P, plakoglobin; V, vinculin; $VASP$, vasodilator stimulated phosphoprotein; ZO-1, ZO-2, zonula occludens protein-1 and -2

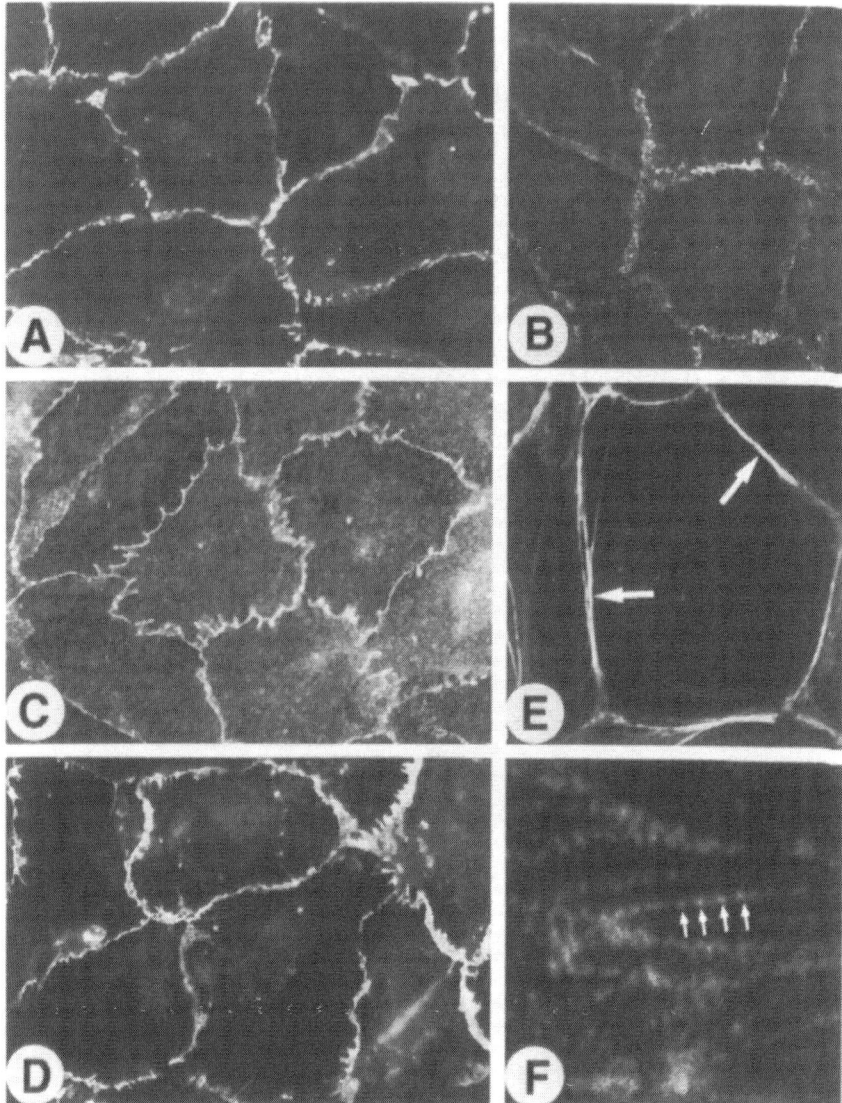

Fig. 5A–F. Components of interendothelial tight and adherens junctions in cultured human umbilical vein endothelial cells (HUVEC). Proteins were visualized by indirect immunofluorescence using either phalloidin **E** or antibodies (**A–D, F**) directed to: **A** Ca^{2+}-dependent vascular endothelial cell adhesion molecule VE-cadherin, a component of adherens junctions: **B** cadherin-associated protein plakoglobin (γ-catenin): **C** tight junction-associated protein occludin: **D** platelet endothelial cell adhesion molecule-1 (PECAM-1): **E** actin filament, and **F** myosin. Note the staining pattern along the cell-to-cell contacts in **A–D**. The actin filaments of the dense peripheral band (DPB) (**E**, *white arrows*) are stained by rhodamine-labeled phalloidin which specifically reacts with polymerized but not monomeric actin. Myosin, a further component of the DPB, displays a sarcomeric pattern (**F**, *white arrows*) which is a predictor of endothelial contractility. For details on the organization of interendothelial junctions see Sect. 3.3.3.2 and Fig. 4

indicating binding of the soluble glycoprotein of EBOV (sGP) to neutrophils (YANG et al. 1998).

3.2 Immunosuppression Associated with Filovirus Infection

Fatal filovirus infections usually end with high viremia and no evidence of an effective immune response (PETERS et al. 1996a). EBOV Reston infection in monkeys is an exception in that it shows a rise in nonproductive antibodies shortly before death (JAHRLING et al. 1996a). Neutralizing antibody titers in human convalescent sera can, if at all, only barely be detected in laboratory tests, but a hyperimmune horse serum against EBOV displayed a fairly high neutralizing titer (KRASNYANSKII et al. 1994; JAHRLING et al. 1996b). Beside anecdotal case reports suggesting the potential benefit of passive immunization against EBOV infections (BOWEN et al. 1978), recent reports from the 1995 Zairian outbreak about effective treatment of acutely ill patients with whole blood transfusions from convalescent patients suggested that quantities of antibody, predicted to be marginally effective in laboratory tests, may still be protective (JAHRLING et al. 1996b; MUPAPA et al. 1998). A study in which EBOV-infected monkeys were treated with equine hyperimmune serum indicated also that passively acquired antibody can have a beneficial effect in reducing the viral burden (JAHRLING et al. 1996b). Nevertheless, data available today do not support an important role of neutralizing antibodies (humoral immune response) in virus clearance.

In humans and monkeys circulating monocytes/macrophages have been described as primary target cells of filoviruses (MURPHY et al. 1971; BASKERVILLE et al. 1985; GEISBERT et al. 1992; RYABCHIKOVA et al. 1994), and filovirus replication in cultured monocytes/macrophages is cytolytic (FELDMANN et al. 1996). Furthermore, an extensive disruption of the parafollicular regions in the spleen and lymph nodes has been observed resulting in destruction of antigen-presenting dendritic cells (GEISBERT et al. 1992; ZAKI and PETERS 1997; JAHRLING et al. 1996a; PETERS et al. 1996a). Thus, cellular immunity is also affected during filoviral HF. Altogether, filoviruses induce immunosuppression in the infected host which may contribute to the rapid generalization and the severity of the disease.

The mechanisms leading to the immunosuppressed status of the hosts are unknown and currently being investigated. The single surface glycoprotein (GP) is assumed to be the major antigenic molecule of virion particles. Its interaction with the host immune system may be modulated by the high content of carbohydrates which accounts for approximately 50% of the molecular weight (see Chap. 1; FELDMANN et al. 1991; GEYER et al. 1992; PETERS et al. 1994). For EBOV, it has been demonstrated that the small secreted glycoprotein sGP interacts with the host immune response by binding to neutrophils (YANG et al. 1998). This protein is the primary expression product of the glycoprotein gene and thus bears the NH_2-terminal 300 amino acids of the full-length structural GP which is expressed by an editing mechanism of the same gene. In addition, relatively high amounts of GP_1, the NH_2-terminal proteolytic cleavage fragment of GP, are released into the medium of filovirus-infected cells, and it is thought that this form of GP could interfere with the host immune system as well (VOLCHKOV et al. 1998a,b). In contrast to sGP, all filovirus full-length GP molecules possess a sequence close to the COOH-terminus resembling a presumptive immunosuppressive domain found in retrovirus glycoproteins (VOLCHKOV et al. 1992; WILL et al. 1993). Peptides synthesized according to this 26 amino acid long region inhibited the blastogenesis of lymphocytes in response to mitogens, inhibited production of cytokines, and decreased proliferation of mononuclear cells in vitro (IGNATYEV et al. 1993). It is not yet known if the immunosuppressive domain on the GP is functional on mature molecules, but evidence for GP mediation of the above mentioned effects has been reported (AGAFANOV et al. 1993).

3.3 Role of Endothelial Cells During Filovirus Infection

The endothelium lines the inner surface of the heart and the vessels. Under physiological conditions it provides an anti-thrombogenic surface and establishes a cellular barrier that is involved in controlling the traffic of molecules, water and leukocytes between the intravascular and extravascular spaces. Furthermore, blood pressure regulation is largely mediated by endothelial cells which release vasoactive agents. These important endothelial functions are changed dramatically in systemic inflammation as observed in various forms of shock. Release of cytokines, chemokines, histamine, peroxides and other proinflammatory agents during the development of septic shock seem to be primarily responsible for profound alterations in endothelial morphology and functions. This includes increase in endothelial permeability, expression of cell adhesion molecules, reduction of vascular tone and increase in procoagulant activity (BROUCKAERT and FIERS 1996; DINARELLO 1996). Numerous endothelial dysfunctions and finally a shock syndrome can be evoked by systemic administration of cytokines. TNF-α has been shown to significantly contribute to a SIRS, a condition which basically mimics shock symptoms. Therefore a crucial role of cytokines in endothelial dysfunction and shock is evident (BROUCKAERT and FIERS 1996). Increased TNF-α levels were observed in patients infected with EBOV (GEORGES et al., 1996; VILLINGER et al., 1998) and could be experimentally provoked in cell cultures of human monocytes/macrophages after MBGV infection (FELDMANN et al., 1996). These data suggest an important role of TNF-α in inducing circulatory shock during filovirus infections.

3.3.1 Replication

Significant injury in microvasculature was observed in different tissues of filovirus-infected humans and viral particles were localized in pericytes and endothelial cells (ZAKI and PETERS 1997). Comparable observations have also been described from experimentally infected monkeys in late stages of the disease (RYABCHIKOVA et al. 1994). Endothelial cells are sufficient targets for filoviruses as demonstrated in cultured human umbilical vein endothelial cells (HUVEC) (Fig. 6) and the endo-

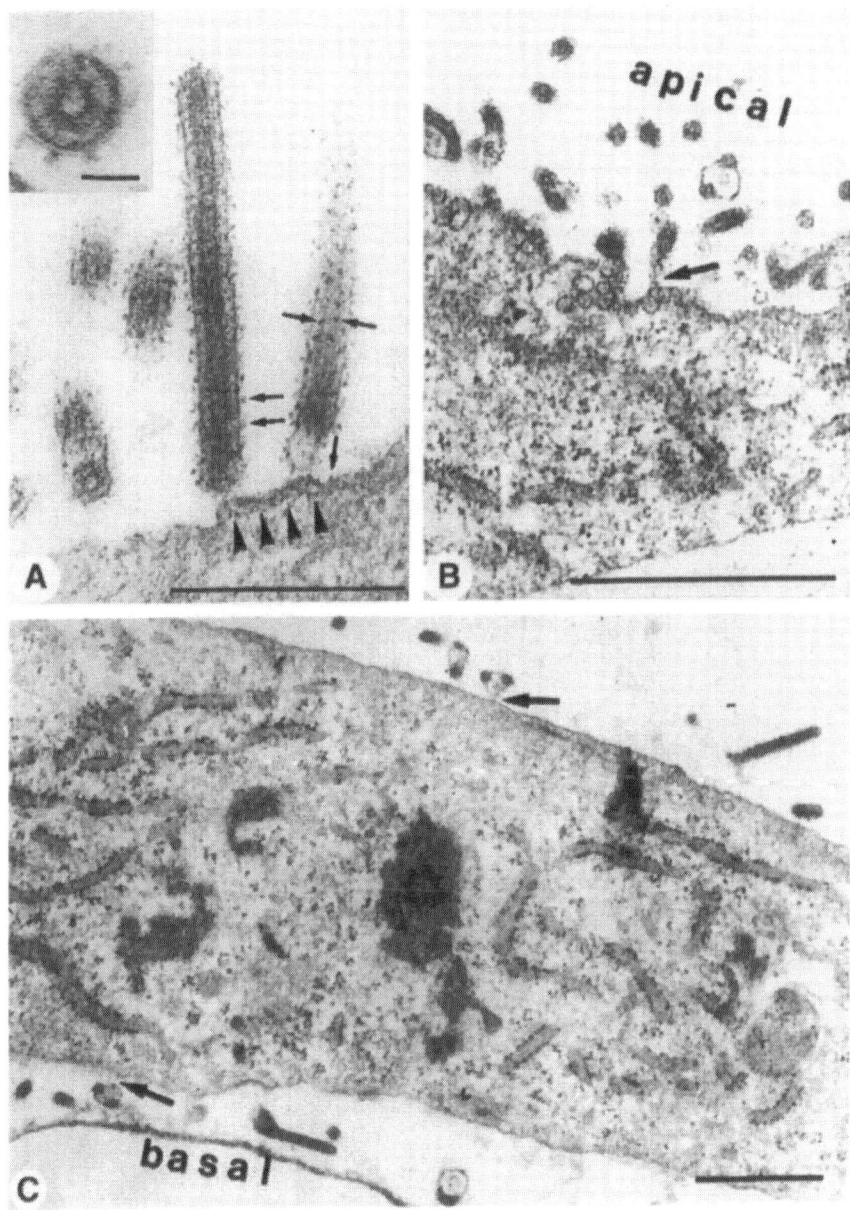

Fig. 6A–C. Virus budding from cultured human umbilical vein endothelial cells (HUVEC). **A** Particles consist of a central channel surrounded by the nucleocapsid (*inset*). The nucleocapsid is further surrounded by a membrane envelope in which spikes are inserted (*arrows*). The plasma membrane is thickened at locations where virus budding occurs (*arrowheads*). **B** Virus particles preferentially budded from the apical plasma membrane. **C** Following destruction of the interendothelial contacts due to lytic replication, virus particles were also observed on the basolateral sides of infected cells. *Bars* in **A, B, C**, 0.5 µm; *bar inset* **A**, 50 nm

thelium of organ-cultured human umbilical cord veins (SCHNITTLER et al. 1993). Recently it has been shown that GP mediates the binding to endothelial cells (YANG et al. 1998). The replication of MBGV in HUVEC was cytolytic, and first budded virus was detected approximately 12 h post-infection. Budding of mature filamentous virions occurred preferentially from the apical surface. Plasma membranes appeared to be thickened where virus budding occurred, which might be caused by an incorporation of viral proteins (Fig. 6A). The vectorial budding in endothelial cells was lost after prolonged incubation due to a destruction of endothelial cells (SCHNITTLER et al. 1993). The vectorial budding may be of strategical importance for virus spread in the infected host. Firstly, replication in endothelial cells could maintain and reinforce the viremic phase during infection. This is supported by the preferential budding from the apical plasma membrane early post-infection (Fig. 6B) (SCHNITTLER et al. 1993). Secondly, virus-induced cytolysis, which occurred later on, may support virus spreading into the underlying tissue (Fig. 6C). This is in line with observations made in organ cultured umbilical cord veins (SCHNITTLER et al. 1993) as well as in filovirus-infected monkeys (FISHER-HOCH et al. 1985; RYABCHIKOVA et al. 1994) and humans (ZAKI and PETERS 1997).

3.3.2 Hemorrhage

The destruction of endothelial cells by virus-induced cytolysis contributes to hemorrhages which typically occur late in infection. Small defects of endothelial cells caused by filovirus replication might be initially covered by the spread of adjacent intact cells, a mechanism known as small wound healing (WONG and GOTTLIEB 1988). However, following extended virus-induced damage this repair mechanism fails, and blood extravasates into the underlying tissue.

The importance of coagulation pathway alterations regarding the bleeding tendency in filovirus infections is less clear. Endothelial cells are involved in the regulation of procoagulant and anticoagulant intravascular activities. Experimental data support the occurrence of a disseminated intravascular coagulation (DIC) in experimentally infected monkeys (FISHER-HOCH et al. 1983, 1985), but it is still not clear that it is crucial or central in the overall pathogenesis of filoviral HFs of humans (PETERS et al. 1996a). The intrinsic clotting pathway seems to be altered whereas the extrinsic one is mainly undisturbed (FISHER-HOCH et al. 1983, 1985). Endothelial-dependent procoagulation can be initiated by tissue factor release that is triggered by thrombin, platelet-derived growth factor, lipopolysaccharide and cytokines, such as IL-1 and TNF-α (MANN et al. 1992). An increase in TNF-α has been demonstrated in filovirus-infected cultured macrophages (FELDMANN et al. 1996), and elevated levels were detected in acute sera of EBOV-infected patients (GEORGES et al. 1996; VILLINGER et al. 1998). A role of tissue factor in the development of DIC during filovirus infections has been postulated (GROB 1995). Furthermore, DIC is associated with an increase in fibrinogen-fibrin degradation products by which the fibrinogen fragment D is able to disorganize endothelial monolayers and thus might contribute to endothelial damage and vascular leakage (DANG et al. 1984). Thus, the bleeding tendency in filovirus infections has to be considered as a symptom caused by multiple factors. The cytolytic destruction of endothelial cells and several alterations in the regulation of the coagulation pathways seem to be important.

3.3.3 Changes in Permeability

3.3.3.1 Effects of Mediators

Infected hosts pre-agonally develop shock and generalized edematous swelling which seem to be caused by a reduced endothelial barrier function due to an increase in endothelial permeability. Endothelial barrier function is ensured by intercellular junctional complexes that mediate cell-cell adhesion and sealing between adjacent cells. Stimulation of cultured endothelial cells by supernatants of filovirus-activated macrophages caused a dissociation of the interendothelial junctions (interendothelial gap formation) resulting in a breakdown of the barrier function (Figs. 3C, 7B) (FELDMANN et al. 1996). This phenomenon is designated as paraendothelial permeability and is in general a common response of the endothelium to local and systemic inflammation. Changes in endothelial permeability can be induced by certain agents, such as thrombin, histamine, serotonin, cytokines, proteolytic enzymes, and peroxides (MAJ and PALADE 1961; MAJNO et al. 1969; SACHS et al. 1978; WEISS and REGIANI 1984; GARCIA et al. 1986; HENSON and JOHNSTON 1987; CYBULSKY et al. 1988; DOWNIE et al. 1992; ISHII et al. 1992; LUM et al. 1993; SUTTORP et al. 1993a,b; SEEGER et al. 1995; FELDMANN et al. 1996). In contrast to histamine and thrombin, which cause a quick and short increase in paraendothelial permeability, polypeptide mediators, such as cytokines cause long-lasting effects. It should be noted that an increase in permeability mediated by cytokines seems to be the result of various mediators acting in concert. Supernatants of filovirus-infected cultured human macrophages were shown to contain elevated levels of TNF-α, and the data indicated that TNF-α plays an important, but not exclusive, role in the increase in permeability (Fig. 3) (FELDMANN et al. 1996). At comparable concentrations which had no direct permeability-increasing effect in vitro, TNF-α was capable of priming the effect in the presence of other factors or agents, such as IL-2 and H_2O_2 (Fig. 3D) (ISHII et al. 1992; FELDMANN et al. 1996). This is in line with a transcriptional activation of several cytokine mRNAs detected in filovirus-infected monocytes/macrophages (see Sect. 3.1.2) (WEST et al. 1996). Finally, filovirus infections cause lysis of macrophages by which an uncontrolled release of proinflammatory mediators has to be assumed.

The different mediators finally leading to interendothelial gap formation act by different signaling pathways. Experiments done in several cell culture models and in situ provide strong evidence that all these substances and substance groups target the endothelial actin filament system and most likely the proteins of the interendothelial junctions. The actin filament system is part of the endothelial cytoskeleton and can be localized in endothelial cells in situ and in culture as a component of stress fibers (FRANKE et al. 1984; DRENCKHAHN and WAGNER 1986;

WHITE et al. 1983), interendothelial junctions (WHITE et al. 1983; GABBIANI et al. 1983; SCHNITTLER et al. 1990), and the membrane cytoskeleton HÜTTNER et al. 1985; KETIS et al. 1986). The junction-associated actin filaments, designated as the dense peripheral band (DPB) (Figs. 4, 5), seem to be an important subcellular structure to maintain morphology and function of intercellular adherens and tight junctions. Apart from actin, the DPB and the stress fibers in endothelial cells additionally consist of myosin, tropomyosin and α-actinin (SCHNITTLER et al. 1990; DRENCKHAHN and WAGNER 1986; GABBIANI et al. 1983) which are also characteristic components of the contractile apparatus in muscles. These proteins are arranged in a sarcomere-like periodic pattern providing the apparatus for ATP- and Ca^{2+}-dependent contractility in nonmuscle cells (Figs. 4, 5). Increased paraendothelial permeability caused by interendothelial gap formation seems to depend on actin/myosin filament contractions that require Ca^{2+}, ATP and myosin light chain phosphorylation (SCHNITTLER et al. 1990; WYSOLMERSKI and LAGUNOFF 1990; GOECKELER and WYSOLMERSKI 1995). Phosphorylation of myosin light chain is a requirement for actin/myosin nonmuscle and smooth muscle contractilities. A TNF-α-triggered increase in endothelial permeability is partly caused by a reduction of the antioxidant capacity in endothelial cells (ISHII et al. 1992). This is important since elevated peroxide levels lead to a rise in intracellular Ca^{2+} (SHASBY et al. 1985), which is a basic requirement for actin/myosin-mediated endothelial contraction. In addition, evidence has been provided that other proinflammatory mediators, such as thrombin and histamine, also act by a contractile mechanism leading to increased permeability (GARCIA et al. 1986; STASEK et al. 1992; MOY et al. 1993; PATTERSON et al. 1994).

A breakdown of the barrier function due to interendothelial gap formation can also be induced by drugs, such as cytochalasins and *Clostridium botulinum* toxin, which directly perturb the actin-filament system (SUTTORP et al. 1991; STEVENSON and BEGG 1994, 1992–1994; AKTORIES et al. 1992). Furthermore, interendothelial gap formation mediated by proinflammatory mediators requires the interaction of stimulus recognition, signal transduction and at least a regulated dissociation of interendothelial junctions.

3.3.3.2 Structural Components

Interendothelial junctions consist of proteins that are also common in adherens type junctions, tight junctions and gap junctions of epithelial cells. Compared with epithelial cells, the endothelium does not display a specific order of junctions, but rather consists of an extended adherens junctional zone in which gap and tight junctions are morphologically inserted (FRANKE et al. 1988). Tight and adherens junctions primarily consist of integral membrane proteins that are linked to the actin and intermediate filaments of the cytoskeleton via a chain of interposed proteins (Figs. 4, 5). These proteins are important for cell-to-cell and cell-to-substrate adhesion and are the targets of diverse signaling cascades which probably regulate opening and closing of cell-to-cell junctions (CITI 1993; DEJANA et al. 1995; JOKUSCH and RÜDIGER 1996; MACHESKY and HALL 1996; SCHNITTLER 1998). It

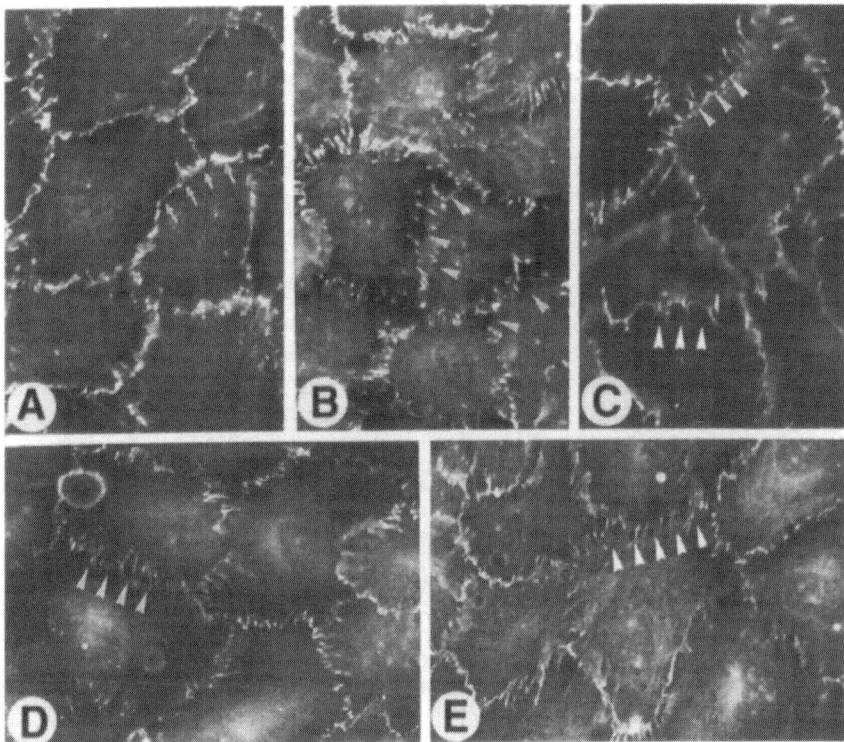

Fig. 7A–E. Interendothelial gap formation and redistribution of junction-associated proteins in confluent monolayers of human umbilical vein endothelial cells (HUVEC) following treatment with different proinflammatory agents. Cells were treated with **A** supernatants of MOCK-infected human monocytes/macrophages; **B** supernatants of Marburg virus (MBGV)-infected human monocytes/macrophages; **C** 10 ng tumor necrosis factor (TNF)-α per ml and 10 μM H_2O_2; **D** histamine; **E** 1 μM Ca^{2+}-ionophore A23187. Note, interendothelial junctions are maintained following treatment with supernatants of mock-infected human monocytes/macrophages (**A**, *white arrows*). In contrast, interendothelial gap formation is associated with a remarkable redistribution of junction-associated proteins, here demonstrated for plakoglobin (**B–E**, *white arrowheads*), and is observed in all cases of increased permeability

remains to be determined whether these proteins of the junctional complexes respond specifically to diverse stimuli. However, interendothelial gap formation in filovirus infections and other inflammatory reactions resulting in increased permeability is associated with a dissociation of tight and adherens junctions and requires the dissociation of interendothelial cell adhesion molecules (FELDMANN et al. 1996) (Fig. 7B).

Occludin is an integral membrane protein exclusively appearing at tight junctions in epithelial and endothelial cells (FURUSE et al. 1993). This 65 kDa protein possesses four hydrophobic domains with two nearly uncharged extracellular loops (Figs. 4, 5). It is believed that these hydrophobic loops are responsible for the sealing features of tight junctions by interaction with hydrophobic loops of homotypic occludin of the opposite cell (FURUSE et al. 1993). At the cytoplasmic side several tight junction-associated proteins have been identified, the zonula occludens protein-1 (ZO-1) and -2 (ZO-2) (STEVENSON et al. 1986; GUMBINER et al. 1991), cingulin (CITI et al. 1988), 7H6 antigen (ZHONG et al. 1993), rab 13 (ZAHRAOUI et al. 1994), and an uncharacterized 130 kDa protein (BALDA et al. 1993). ZO-1 binds spectrin tetramers (ITOH et al. 1991) which are components of the membrane cytoskeleton (Fig. 4), and together with actin filaments seems to be important for tight junction stability in epithelial and endothelial cells. In contrast to epithelial cells, endothelial cells express a ZO-$1^{\alpha-}$ isoform, lacking 80 amino acids at the COOH-terminus (WILLOT et al. 1992). This isoform seems to be more plastic in attaching to the underlying actin filament cytoskeleton than epithelial ZO-1. Such a plasticity might influence the physiologically dynamic nature of endothelial tight junctions in response to pathological stimuli such as increased permeability and transmigration of leukocytes.

Since interendothelial junctions are interdigitated complexes of tight and adhesion junctions, formation of interendothelial gaps requires dissociation of both. Adherens junctions in endothelial cells are characterized by proteins of the Ca^{2+}-dependent cadherin family which consists of various subtypes (LIAW et al. 1990; RUBIN et al. 1991; LAMPUGNANI et al. 1992, 1995; SALOMON et al. 1992). Evidence has been provided that cadherins of adjacent cells form a zipper-like structure (SHAPIRO et al. 1995) and bind via α-, β-catenin and plakoglobin (γ-catenin) to DPB facing the interendothelial junctions (Figs. 4, 5). Increased endothelial permeability by culture supernatants of filovirus-activated monocytes/macrophages was associated with a remarkable reorganization of plakoglobin (Fig. 7) (FELDMANN et al. 1996), vascular endothelial cadherin (VE-cadherin), β-catenin, and a dissociation of α-catenin (SCHNITTLER and FELDMANN 1997). Redistribution of the cadherin/catenin complex can also be evoked in cultured human endothelial cells by treatment with TNF-α/H_2O_2 (FELDMANN et al. 1996), histamine and the Ca^{2+} ionophore A23187 (Fig. 7), and was further associated with a redistribution of the junction-associated actin filament system. The detailed function of each of the junctional proteins is largely unknown. Plakoglobin seems to be important for maturation of adherens junctions (LAMPUGNANI et al. 1995) and has been shown to be essential to mechanical strength between interendothelial junctions (SCHNITTLER et al. 1997) as well as to stabilization of myocardial desmosomes (RUIZ et al. 1996). Desmoplakin, another protein associated with interendothelial junctions and known to be a characteristic component of epithelial desmosomes, has recently been described (VALIRON et al. 1996). This protein provides the link of intermediate filaments to interendothelial junctions and may therefore contribute to junctional stability (VALIRON et al. 1996). It seems reasonable to assume that a local loss of homotypic binding strength between both occludin (tight functions) and cadherin (adherens junctions) occurs during interendothelial gap formation. This seems to be dependent on the binding of junctional proteins to the actin filament system by which plakoglobin may play a crucial role. The local loss of actin filaments from junctional proteins during inflammation in combination with the actin/myosin-mediated contraction and redistribution seems to be an important mechanism in interendothelial gap formation.

3.3.3.3 Signaling Pathways

As already mentioned above, the proteins specifically targeted by diverse intracellular signal cascades are largely unknown, but several signaling events have been identified suggesting the involvement of several kinase and phosphatase activities. cAMP has been suggested to play a role in preventing interendothelial gap formation and permeability. Increases in cAMP reduced edema during acute lung injury (FARRUKH et al. 1987; KOBAYASHI et al. 1987) as well as interendothelial gap formation and permeability (LAPOSATA et al. 1983; KILLACKEY et al. 1986; CARSON et al. 1989; CASNOCHA et al. 1989; STELZNER et al. 1989; BUCHAN and MARTIN 1992; SUTTORP et al. 1993b; PATTERSON and GARCIA 1994). In addition, increased cAMP, mediated by cholera toxin, which ADP-ribosylates stimulatory G-proteins and, in turn, protects against the thrombin-induced increase in paraendothelial permeability, also attenuates myosin light chain (MLC) activation and abrogates contractile processes subsequent to protein kinase C (PKC) activation, but independent of MLC activation (PATTERSON et al. 1994). In contrast, phospholipase C-dependent activation of G-protein with a subsequent inositol trisphosphate (IP3)-mediated increase in Ca^{2+} and diacylglycerol (DAG) is an initial event in the thrombin-provoked endothelial increase in permeability (DE MICHELE et al. 1990; GARCIA et al. 1990, 1991, 1992). The intracellular rise in Ca^{2+}, however, seems to be an important mediator leading to intercellular gaps probably by actin/myosin contraction. This has also been shown to occur following treatment with calcium ionophores, *E.coli* hemolysin, *Staphylococcus aureus* α-toxin and bradykinin (SUTTORP et al. 1985, 1990; SCHILLING et al. 1988; SCHNITTLER et al. 1990). Recently it was shown that inactivation of actin-regulating Rho-proteins by *Clostridium difficile* toxin B in cultured endothelial cells cause an increase in permeability (HIPPENSTIEL et al. 1997). This is of special importance, since Rho seems to be the top of several signal cascades involved in actin filament regulation (MACHESKY and HALL 1996). In addition, junction-associated proteins, such as VE-cadherin, catenins, p120^{cas} and vinculin, are substrates for protein kinases and phosphatases (HAMAGUCHI et al. 1993; REYNOLDS et al. 1994; BRADY-KALNAY et al. 1995; JOKUSCH and RÜDIGER 1996; KRYPTA et al. 1996). Evidence has been provided that protein modifications, such as phosphorylation and dephosphorylation, change binding affinities between junction-associated proteins. Therefore, it seems reasonable to assume that such modifications of junction-associated proteins are involved in regulation of paraendothelial permeability in filovirus infections as well.

3.3.4 Further Alterations

Based on experimental data and clinical symptoms observed during filovirus infections it is reasonable to assume that cytokines released by virus-activated macrophages will cause endothelial cell activation resulting in changes in endothelial coagulation properties (see Sect. 3.3.2), an increased expression of cell adhesion molecules (see Sect. 3.3.3), and dysregulation of vascular tone.

Under physiological conditions, nitric oxide (NO) plays a key role in regulating myocardial contractility, vascular tone, platelet-endothelial interactions, leukocyte adhesion, and the immune response to infectious pathogens. NO occurs in a wide variety of cell types and tissues, including vascular endothelium, platelets, macrophages, and neuronal cells, and are produced by different nitric oxide synthetase (NOS) isoforms (LOSCALZO and WELCH 1995). Excess NO production has been implicated in the pathogenesis of hypotension and myocardial depression during various forms of shock (DINERMAN et al. 1993). Mice deficient in inducible NOS (iNOS) (iNOS$^{-/-}$) do not response to endotoxin by a fall in blood pressure compared with wild-type animals (MCMICKING et al. 1995). Along with many other reports, this study shows that, in general, the blood pressure fall in shock syndromes, including pathogen-induced syndromes, is dependent on NO that is produced via a mediator-triggered cascade.

4 Conclusions

The primary targets for filovirus infections seem to be the mononuclear phagocytic cells in which the viruses lytically replicate. Following infection via small lesions,

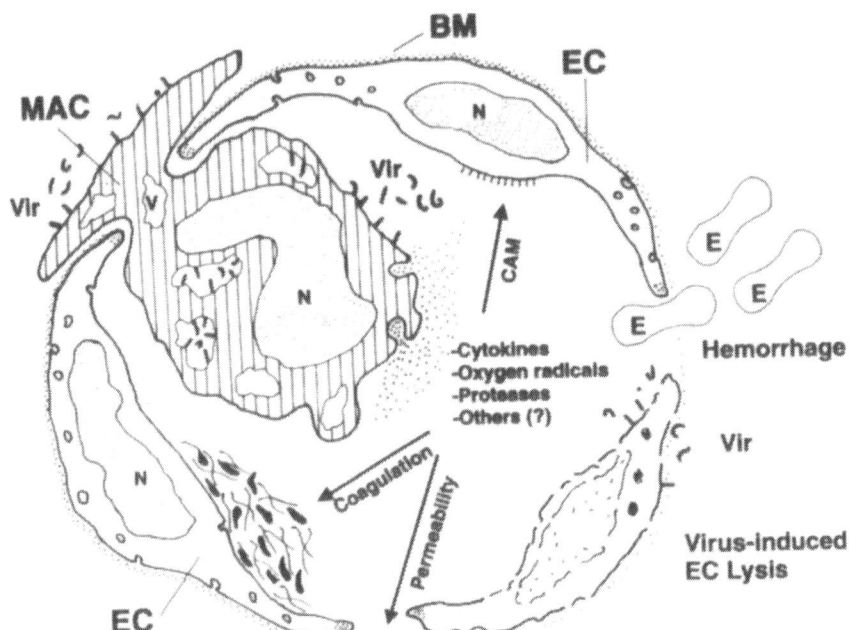

Fig. 8. Filovirus-induced effects on mononuclear phagocytic cells (MPS) and endothelial cells. *EC*, endothelial cell; *MAC*, macrophage; *Vir*, virus particle; *CAM*, cell adhesion molecule; *E*, erythrocyte; *BM*, basement membrane; *N*, nucleus; *V*, vacuole

virus particles enter lymph vessels or directly the vascular system. The primary organ tropism (lymph nodes, liver, spleen) may be explained by the direct access of viral particles to sessile cells of the MPS without penetration through cell or tissue barriers. This is the case in the lymph node, where macrophages are surrounded by lymphatic fluid, as well as in the spleen and the liver, where macrophages and hepatocytes have direct contact with the circulating blood. The pantropism in filovirus infections typically occurring at late stages of the disease seems to be due to cytolytic replication in endothelial cells and monocytes/macrophages and a subsequent extravasation of infected monocytes/macrophages resulting in spread of infectious viral particles. The disease-decisive symptoms during severe filovirus infections are most likely associated with a generalized vascular instability and dysregulation. Besides evidence for direct vascular involvement due to cytolytic replication in endothelial cells in vivo and in culture, the role of active mediator molecules in the pathogenesis of the disorders has to be considered. Filoviruses infect and activate monocytes/macrophages. Macrophage-derived mediators act on multiple organs, but the endothelium and the vascular wall seem to be the main targets as observed in various other forms of shock. The endothelium responds with an increase in permeability, dysregulation of vascular tone, expression of cell adhesion molecules and a change in the anticoagulatory phenotype. The bleeding tendency (hemorrhages) is probably caused by multiple factors, of which cytolytic virus replication in endothelial cells as well as cytokine-mediated processes and coagulation disorders are important. The above discussed pathophysiological events are illustrated in Figs. 1 and 8. Further studies have to be performed to characterize the role of other subsets of leukocytes in filovirus infections. In addition, special attention should be given to investigations into the immunosuppression associated with filovirus infections and interactions between viral and host cell proteins.

Acknowledgements. The authors are grateful to Hans-Dieter Klenk for helpful discussions. We further acknowledge financial supports over the years by grants from the Deutsche Forschungsgemeinschaft (Forschergruppe Kl 238/1–1; SFB 286, project A6; SFB 355, project B5, SFB 535, project A4, grant Fe 286/4–1), and the Kempkes-Stiftung (grant 21/95). H.F. held a fellowship of the National Research Council (NRC) at the Centers for Disease Control and Prevention, Special Pathogens Branch from 1992 to 1994.

References

Aderka D, Le JM, Vilcek J (1986) IL-6 inhibits lipopolysaccharide-induced tumor necrosis factor production in cultured human monocytes, U937 cells, and in mice. J Immunol 143:3517–3523
Agafanov AP, Ignatyev GM, Akimenko ZA, Volchkov VE (1993) Study of immunogenic and protective properties of Marburg virus GP, NP and VP40 proteins. IXth International Congress of Virology, Glasgow, Scotland, p 300
Aktories K, Wille M, Just I (1992) Clostridial actin-ADP-ribosylating toxins. Curr Top Microbiol Immunol 175:97–113
Balda MS, Gonzalez-Mariscal L, Matter K, Cereijido M, Anderson JM (1993) Assembly of tight junction: the role of diacylglycerol. J Cell Biol 123:193–302

Baskerville A, Fisher-Hoch SP, Neild GH, Dowsett AB (1985) Ultrastructural pathology of experimental Ebola haemorrhagic fever virus infection. J Pathol 147:199–209

Bevilacqua MP, Pober JS, Mendrick DL, Cotran RS, Gimbrone Jr MA (1987) Identification of an inducible endothelial-leukocyte adhesion molecule. Proc Natl Acad Sci USA 84:9238–9242

Bowen ETW, Lloyd G, Platt G, Mcardell LB, Webb PA, Simpson DIH (1978) Virological studies on a case of Ebola virus infection in man and monkeys. In: Pattyn SR (ed) Ebola virus haemorrhagic fever. Elsevier/North-Holland Biomedical, Amsterdam, pp 95–100

Bradley JR, Johnson D, Pober JS (1993) Endothelial activation by hydrogen peroxide: selective increase of intercellular cell adhesion molecule-1 and major histocompatibility complex I. Am J Pathol 142:1598–1609

Bradley JR, Thiru S, Pober JS (1995) Hydrogen peroxide-induced endothelial retraction is accompanied by a loss of the normal and spatial organization of endothelial cell-adhesion molecules. Am J Pathol 147:627–641

Brady-Kalnay SM, Rimm DL, Tonks NK (1995) Receptor protein tyrosine phosphatase PTPμ-associates with cadherins and catenins in vivo. J Cell Biol 130:977–986

Brouckaert P, Fiers W (1996) Tumor necrosis factor and the systemic inflammatory response syndrome. In: Rietschel ET, Wagner H (eds) Pathophysiology of septic shock (Curr Top Microbiol Immunol). Springer, Berlin Heidelberg New York, pp 167–187

Buchan KW, Martin W (1992) Modulation of barrier function of bovine aortic and pulmonary artery endothelial cells: Dissociation from cytosolic calcium content. Br J Pharmacol 107:932–938

Carson MR, Shasby SS, Shasby DM (1989) Histamine and inositol phosphate accumulation in endothelium: cAMP and a G protein. Am J Physiol 263:L664–L669

Casnocha SA, Eskin SG, Hall ER, McIntire LV (1989) Permeability of human endothelial monolayers: Effect of vasoactive agonist and cAMP. J Appl Physiol 67:1997–2005

Citi S (1993) The molecular organization of tight junctions. J Cell Biol 121:485–489

Citi S, Sabanay H, Jakes R, Geiger B, Kendrick-Jones J (1988) Cingulin, a new peripheral component of tight junctions. Nature 333:272–276

Cybulsky MI, Chan MKW, Movat HZ (1988) Biology of disease: acute inflammation and microthrombosis induced by endotoxin, interleukin-1, and tumor necrosis factor and their implication in gram-negative infection. Lab Invest 58:365–378

Dang CV, Bell WR, Kaiser D, Wong A (1984) Disorganization of cultured vascular endothelial cell monolayers by fibrinogen fragment D. Science 227:1487–1490

Dejana E, Corada M, Lampugnani MG (1995) Endothelial cell-to-cell junctions. FASEB J 9:910–918

DeLisser HM, Newman RJ, Albelda SM (1994) Molecular and functional characterization of PECAM-1/CD31. Immunol Today 15:490–495

DelMaschio A, Zanetti A, Corada M, Rival Y, Ruco L, Lampugnani MG, Dejana E (1996) Polymorphonuclear leucocytes adhesion triggers the disorganization of endothelial cell-to-cell adherens junctions. J Cell Biol 135:497–510

De Michele MA, Moon DG, Fenton JW, Minnear FL (1990) Thrombin's enzymatic activity increases permeability of endothelial monolayers. J Appl Physiol 69:1599–1606

Dinarello CA (1996) Cytokines as mediators in the pathogenesis of septic shock. In: Rietschel ET, Wagner H (eds) Pathophysiology of septic shock (Curr Topics in Microbiology and Immunology). Springer, Berlin Heidelberg New York, pp 133–165

Dinerman JL, Lowenstein CJ, Snyder SH (1993) Molecular mechanisms of nitric oxide regulation. Circ Res 73:217–222

Downie GH, Ryan U, Hayes BA, Friedmann M (1992) Interleukin-2 directly increases albumin permeability of bovine and human vascular endothelium in vitro. Am J Respir Cell Mol Biol 7:58–65

Drenckhahn D, Wagner J (1986) Stress fibers in the splenic sinus endothelium in situ: molecular structure, relationship to the extracellular matrix, and contractility. J Cell Biol 102:1738–1747

Farrukh IS, Gurtner GH, Michael JR (1987) Pharmacological modification of pulmonary vascular injury: possible role of cAMP. J Appl Physiol 62:47–54

Feldmann H, Bugany H, Mahner F, Klenk HD, Drenckhahn D, Schnittler HJ (1996) Filovirus-induced endothelial leakage triggered by infected monocytes/macrophages. J Virol 70:2208–2214

Feldmann H, Klenk HD (1996) Marburg and Ebola viruses. Adv Virus Res 47:1–52

Feldmann H, Klenk HD, Sanchez A (1993) Molecular biology and evolution of filoviruses. Arch Virol, Suppl 7:81–100

Feldmann H, Will C, Schikore M, Slenczka W, Klenk HD (1991) Glycosylation and oligomerization of the spike protein of Marburg virus. Virology 182:353–356

Fenner F (1948) The clinical features of mouse-pox (infectious ectromelia of mice) and the pathogenesis of the disease. J Pathol Bacteriol 60:529–552
Fisher-Hoch SP, Platt GS, Lloyd G, Simpson DIH (1983) Haematological and biochemical monitoring of Ebola infection in rhesus monkeys: implication for patient management. Lancet 1983:1055–1058
Fisher-Hoch SP, Platt GS, Neild GH, Southee T, Baskerville A, Raymond RT, Lloyd G, Simpson DIH (1985) Pathophysiology of shock and hemorrhage in a fulminating viral infection (Ebola). J Infect Dis 152:887–894
Franke RP, Gräfe M., Schnittler HJ, Seiffge D, Mittermayer C, Drenckhahn D (1984) Induction of human vascular endothelial stress fibres by fluid shear stress. Nature (London) 307:648–649
Franke WW, Cowin P, Grund C, Kuhn C, Kapprell HP (1988) The endothelial junction. The plaque and its components. In: Simionescu N, Simionescu M (eds) Endothelial cell biology in health and disease. Plenum, New York, pp 147–166
Furuse M, Hirase T, Nagafuchi A, Yonemura S, Tsukita S, Tsukita S (1993) Occludin: a novel integral membrane protein localizing at tight junctions. J Cell Biol 123:1777–1788
Gabbiani G, Gabbiani F, Lombardi D, Schwartz SM (1983) Organization of actin cytoskeleton in normal and regenerating arterial endothelial cells. Proc Natl Acad Sci USA 80:2361–2364
Garcia JGN, Domingues J, English D (1991) Sodium fluoride induced phosphoinositide hydrolysis, Ca^{2+} mobilization, and prostacyclin synthesis in cultured human endothelium: Further evidence for regulation by a pertusis toxin-insensitive guanine nucleotide-binding protein. Am J Respir Cell Mol Biol 5:113–124
Garcia JGN, Fenton JW, Natarajan V (1992) Thrombin stimulation of human endothelial cell phospholipase D activity: Regulation by phospholipase C, protein kinase C, and cyclic adenosine 3',5'-monophosphate. Blood 79:2056–2067
Garcia JGN, Painter RG, Fenton JW, English D, Callahan KS (1990) Thrombin-induced prostacyclin biosynthesis in human endothelium: Role of guanine nucleotide regulatory proteins in stimulus/coupling responses. J Cell Physiol 142:186–193
Garcia JGN, Siflinger-Birnboim A, Bizios R, Del Vecchio PJ, Fenton II JW, Malik AB. (1986) Thrombin-induced increase in albumin permeability acr₀ · the endothelium. J Cell Physiol 128:96–104
Geisbert TW, Jahrling PB, Hanes MA, Zack PM (1992) Association of Ebola-related Reston virus particles and antigen with tissue lesions of monkeys imported to the United States. J Comp Pathol 106:137–152
Georges AJ, Renaut AA, Bertherat E, Baize S, Leroy E, LeGuenno B, Lepage J, Amblard J, Edzang S, Georges-Courbot MC (1996) Recent Ebola virus outbreaks in Gabon from 1994 to 1996: epidemiologic and control issues. International Colloquium on Ebola Virus Research, Antwerp Belgium, p 47
Geyer H, Will C, Feldmann H, Klenk HD, Geyer R (1992) Carbohydrate structure of Marburg virus glycoprotein. Glycobiology 2:299–312
Goeckeler ZM, Wysolmerski RB (1995) Myosin light chain kinase-regulated endothelial cell contraction: The relationship between isomeric tension, actin polymerization, and myosin phosphorylation. J Cell Biol 130:613–627
Gong JH, Sprenger H, Hinder F, Bender A, Schmidt A, Horch S, Nain M, Gemsa D (1991) Influenza A virus infection of macrophages. Enhanced tumor necrosis factor-alpha (TNF-alpha) gene expression and lipopolysaccharide-triggered TNF-alpha release. J Immunol 147:3507–3513
Gotsch U, Jäger U, Dominis M, Vestweber D (1994) Expression of P-selectin on endothelial cells is upregulated by LPS and TNF-a in vivo. Cell Adhes Commun 2:7–14
Grob C (1995) Tissue factor initiation of disseminated intravascular coagulation in filovirus infection. Med Hypoth 45:380–382
Gumbiner B, Lowenkopf T, Apatira D (1991) Identification of a 160-kDa polypeptide that binds to the tight junction protein ZO-1. Proc Nat Acad Sci USA 88:3460–3464
Haagmans BL, Stals FS, van der Meide PH, Bruggeman CA, Horzinek MC, Schijns VE (1994) Tumor necrosis factor alpha promotes replication and pathogenicity of rat cytomegalovirus. J Virol 68:2297–2304
Hahne M, Jager U, Isenmann S, Hallmann R, Vestweber D (1993) Five tumor necrosis factor-inducible cell adhesion mechanisms on the surface of mouse endothelioma cells mediate the binding of leukocytes. J Cell Biol 121:655–664
Hamaguchi M, Matsuyoshi N, Ohnishi Y, Gotoh B, Takeichi M, Nagai Y (1993) $p60^{v-src}$ causes tyrosine phosphorylation and inactivation of the N-cadherin-catenin cell adhesion system. EMBO J 12:307–314

Han X, Becker K, Degen HJ, Jablonowski H, Strohmeyer G (1996) Synergistic stimulatory effects of tumor necrosis factor alpha and interferon gamma on replication of human immunodeficiency virus type I and on apoptosis of HIV-1-infected host cells. Eur J Clin Invest 26:286–292

Heller MV, Saavedra MC, Falcoff R, Maiztegui JI, Molinas FC (1992) Increased tumor necrosis factor-levels in Argentine hemorrhagic fever. J Infect Dis 166:1203–1204

Henson PM, Johnston RB (1987) Tissue injury in inflammation: oxidants, proteinases, and cationic proteins. J Clin Invest 79:669–674

Hippenstiel S, Tannert-Otto S, Vollrath N, Krüll M, Just I, Aktories K, vEichel Streiber C, Suttorp N (1997) Glucosylation of small-GTP binding proteins disrups endothelial barrier function. Am J Physiol 272:L38–L43

Hogg N (1993) A model of leukocyte adhesion to vascular endothelium. Curr Top Microbiol Immunol 184:79–86

Hüttner I, Walker C, Gabbiani G (1985) Aortic endothelial cell during regeneration. Remodeling of cell junctions, stress fibers, and stress fiber-membrane attachment domains. Lab Invest 53:287–302

Hynes RO (1992) Integrins: versatility, modulation, and signalling in cell adhesion. Cell 69:11–25

Ignatyev GM, Agafonov AP, Streltsova MA, Kashentseva EA (1996) Inactivated Marburg virus elicits a nonprotective immune response in Rhesus monkeys. J Biotechnol 44:111–118

Ignatyev GM, Blinov VM, Vochkov VE, Netesov SV (1993) New aspects in the phenomenon of immunosuppression caused by filoviruses. IXth International Congress of Virology, Glasgow, Scotland, p 299 (P 52-4)

Ishak KG, Walker DH, Coetzer JAW, Gardner JJ, Forelkin L (1982) Viral hemorrhagic fevers with hepatic involvement: pathologic aspect with clinical correlation. Prog Liver Dis 3:495–515

Ishii Y, Partridge CA, Del Vecchio PJ, Malik AB (1992) Tumor necrosis factor-α-mediate decrease in glutathione increases the sensitivity of pulmonary vascular endothelia cells to H_2O_2. J Clin Invest 189:794–802

Itoh M, Yonemura S, Nagafuchi A, Tsukita S, Tsukita S (1991) A 220 kD undercoat constitutive protein: its specific localization at cadherin-based cell-cell adhesion sides. J Cell Biol 115:1449–1462

Jahrling PB, Geisbert J, Swearengen JR, Jaax JP, Lewis ', Huggins JW, Schmidt JJ, LeDu JW Peters CJ (1996b) Passive immunization of Ebola virus-infected cynomolgus monkeys with immunoglobulin from hyperimmune horse. Arch Virol Suppl 11:135–140

Jahrling PB, Geisbert TW, Jaax NK, Hanes MA, Ksiazek TG, Peters CJ (1996a) Experimental infection of cynomolgus macaques with Ebola-Reston filoviruses from the 1989–1990 U.S. epizootic. Arch Virol Suppl 11:115–134

Johnston IC, Dunster IM, Schneider-Schaulies J, Schneider-Schaulies S (1995) Measles virus replication in neural cells. Trends Microbiol 3:361–365

Jokusch BM, Rüdiger M (1996) Crosstalk between cell adhesion molecules: vinculin as a paradigm for regulation by conformation. Trends Cell Biol 6:311–315

Kawanishi Y, Hayashi N, Katayama K, Ueda K, Takehara T, Miyoshi E, Mita E, Kasahara A, Fusamoto H, Kamada T (1995) Tumor necrosis factor-alpha and interferon-gamma inhibit synergistically viral replication in hepatitis B virus-replicating cells. J Med Virol 47:272–277

Ketis NV, Hoover RL, Karnovsky MJ (1986) Isolation of bovine aortic endothelial cell plasma membranes: identification of membrane-associated cytoskeletal proteins. J Cell Physiol 128:162–170

Kiley MP, Bowen ETW, Eddy GA, Isaäcson M, Johnson KM, McCormick JB, Murphy FA, Pattyn SR, Peters D, Prozesky OW, Regnery RL, Simpson DIH, Slenczka W, Sureau P, van der Groen G, Webb PA, Wulff H (1982) Filoviridae: a taxonomic home for Marburg and Ebola viruses? Intervirology 18:24–32

Killackey JJ, Johnston MG, Movat HZ (1986) Increased endothelial permeability of microcarrier-cultured endothelial monolayers in response to histamine and thrombin. Am J Pathol 122:50–61

Kobayashi H, Kobayashi T, Fukushima M (1987) Effects of dibutyryl cAMP on pulmonary air embolism-induced lung injury in awake sheep. J Appl Physiol 63:2201–2207

Körner H, Fritzsche U, Burgert HG (1992) Tumor necrosis factor alpha stimulates expression of adenovirus early region 3 proteins: Implications for viral persistence. Proc Natl Acad Sci USA 89:11857–11861

Krasnyanskii VP, Mikhailov VV, Borisevich IV, Gradoboev VN, Evseev AA, Pshenichnov VA (1994) Preparation of hyperimmune horse serum to Ebola virus. Vopr Virus 2:91–92

Krypta RM, Su H, Reichardt LF (1996) Association between a transmembrane protein tyrosine phosphatase and the cadherin-catenin complex. J Cell Biol 134:1519–1529

Lampugnani MG, Corada M, Caveda L, Breviario F, Ayalon O, Geiger B, Dejana E (1995) The molecular organisation of endothelial cell to cell junctions: Differential association of plakoglobin, β-catenin, and a-catenin with vascular endothelial cadherin (VE-cadherin) J Cell Biol 129:203–217

Lampugnani MG, Resnati M, Raiteri M, Pigott R, Pisacane A, Houen G, Ruco LP, Dejana E (1992) A novel endothelial-specific membrane protein is a marker of cell-cell contacts. J Cell Biol 118:1511–1522

Laposata M, Dovnarsky DK, Shin HS (1983) Thrombin-induced gap formation in confluent endothelial cell monolayers in vitro. Blood 62:549–556

Le JM, Vilcek J (1987) Accessory function of human fibroblasts in mitogen-stimulated interferon-gamma production by T-lymphocytes. Inhibition by interleukin 1 and tumor necrosis factor. J Immunol 139:3330–3337

Leach L, Eaton BM, Westcott EDA, Firth JA (1995) Effect of histamine on endothelial permeability and structure and adhesion molecules of the paracellular junctions of perfused human placental microvessels. Microvasc Res 50:323–337

Lenter M, Levinovitz A, Isemann S, Vestweber D (1994) Monospecific and common glycoprotein ligands for E- and P-selectin on myeloid cells. J Cell Biol 125:471–481

Liao F, Huynh HH, Eiroa A, Greene T, Polizzi E, Muller WA (1995) Migration of monocytes across endothelium and passage through extracellular matrix involve separate molecular domains of PECAM-1. J Exp Med 182:1337–1343

Liaw CW, Cannon C, Power MD, Kibonoka PK, Rubin LL (1990) Identification and cloning of two species of cadherins in bovine endothelial cells. EMBO J 9:2701–2708

Loscalzo J, Welch G (1995) Nitric oxide and ist role in the cardiovascular system. Prog Cardiovasc Disease XXXVIII No 2:87–104

Lum H, Andersen TT, Siflinger-Birnboim A, Tiruppathi C, Goligorsky MS, Fenton II JW, Malik AB (1993) Thrombin receptor peptide inhibits thrombin-induced increase in endothelial permeability by receptor desensitization J Cell Biol 120:1491–1499

Machesky LM, Hall A (1996) Rho: a connection between membrane receptor signalling and the cytoskeleton. Trends Cell Biol 6:304–310

Majno G, Palade GE (1961) Studies on inflammation; I. The c.`:ct of histamine and serotonin on vascular permeability: an electron microscopic study. J Biophys Biochem Cytology 11:571–626

Majno G, Shea SM, Leventhal M (1969) Endothelial contraction induced by histamine-type mediators. J Cell Biol 42:647–672

Mann KG, Krishnaswamy S, Lawson JH (1992) Surface dependent hemostasis. Sem Hematol 29:213–226

Martini GA (1971) Clinical syndrom. In: Martini GA, Siegert R (eds) Marburg virus disease, 1st edn. Springer, Berlin Heidelberg New York, pp 1–9

Martini GA, Siegert R (1971) Marburg virus disease, 1st edn. Springer, Berlin Heidelberg New York, pp 1–230

McMicking JD, Nathan C, Hom G, Chartrain N, Fletcher DS, Trumbauer M, Stevens K, Xie Q, Sokol K, Hutchinson N, Chen H, Mudgett JS (1995) Altered response to bakterial infection and endotoxic shock in mice lacking inducible nitric oxide synthase. Cell 81:641–650

Moll T, Dejana E. Vestweber D (1998) In vitro degradation of endothelial catenins by a neutrophil protease. J Cell Biol 140:403–407

Moy AB, Shasby SS, Scott BD, Shasby M (1993) The effect of histamine and cyclic adenosine monophosphate on myosin light chain phosphorylation in human umbilical vein endothelial cells. J Clin Invest 92:1198–1206

Mupapa K, Masamba M, Kibadi K, Kuvula K, Bwaka A, Kipasa M, Muyembe T (on behalf of the International Scientific and Technical Committee) (1998) Treatment of Ebola hemorrhagic fever with blood transfusions from convalescent. J Infect Dis (in press)

Murphy FA, Fauquet CM, Bishop DHL, Ghabrial SA, Jarvis AW, Martelli GP, Mayo MA, Summers MD (1995) Virus Taxonomy: classification and nomenclature of viruses. Sixth Report of the International Committee on Taxonomy of Viruses. Arch Virol, Suppl 10:265–292

Murphy FA, Simpson DIH, Whitfield SG, Zlotnik I, Carterm GB (1971) Marburg virus infection in monkeys. Lab Invest 24:279–291

Orenstein JM, Meltzer MS, Phipps T, Gendelman HE (1988) Cytoplasmic assembly and accumulation of human immunodeficiency virus types 1 and 2 in recombinant human colony-stimulating factor-1-treated human monocytes: an ultrastructural study. J Virol 62:2578–2586

Patel KD, Zimmerman GA, Prescott SM, McEver RP, McIntyre TM (1992) Oxygen radicals induce human endothelial cells to express GMP-140 and bind neutrophils. J Cell Biol 112:749–759

Patterson CE, Davis HW, Schaphorst KL, Garcia JGN (1994) Mechanisms of cholera toxin prevention of thrombin-and PMA-induced endothelial cell barrier dysfunction. Microvasc Res 35:308–315

Patterson CE, Garcia JGN (1994) Regulation of thrombin-induced endothelial cell activation by bacterial toxins. Blood Coagul Fibrinolysis 5:63–72

Pattyn SR (1978) Ebola virus hemorrhagic fever, 1st edn. Elsevier/North-Holland, Amsterdam, pp 1–436

Peluso R, Haase A, Stowring L, Edwards M, Ventura P (1985) A Trojan horse mechanism for the spread of Visna virus in monocytes. Virology 147:231–236

Peters CJ, Johnson ED, McKee KT (1991) Filoviruses and management and viral hemorrhagic fevers. In: Belshe RB (ed) Textbook of human virology. Mosby Year Book, St Louis, pp 699–712

Peters CJ, Sanchez A, Feldmann H, Rollin PE, Nichol ST, Ksiazek TG (1994) Filoviruses as emerging pathogens. Seminars in Virology 5:147–154

Peters CJ, Sanchez A, Rollin PE (1996a) Filoviridae: Marburg and Ebola viruses. In: Fields BN, Knipe DM et al (eds.) Virology, 3rd edn. Raven, Philadelphia, pp 1161–1176

Peters CJ, Jahrling PB, Khan AS (1996b) Patients infected with high-hazard viruses: Scientific basis for infection control. In: Schwarz TF, Siegel G (eds) Imported virus infections. Arch Virol Suppl 11:141–168

Pizcueta P, Luscinskas FW (1994) Monoclonal antibody blockade of L-selectin inhibits mononuclear leukocyte recruitment to inflammatory sites in vivo. Am J Pathol 145:461–469

Reynolds AB, Daniel J, McCrea PD, Wheelock MJ, Wu J, Zhang Z (1994) Identification of a new catenin: the tyrosine kinase substrate $p120^{cas}$ associates with E-cadherin complex. Mol Cell Biol 14:8333–8342

Rubin LL, Hall DE, Porter S, Barbu K, Cannon C, Horner HC, Janatpour M, Liaw CW, Manning K, Morales J, Tanner LI, Tomaselli KJ, Bard F (1991) A cell culture model of the blood-brain barrier. J Cell Biol 115:1725–1735

Ruiz P, Brinkmann V, Ledermann B, Behrend M, Grund C, Thalhammer C, Vogel F, Birchmeier C, Günthert U, Franke WW, Birchmeier W (1996) Targeted mutation of plakoglobin in mice reveals essential functions of desmosomes in the embryonic heart. J Cell Biol 135:215–225

Ryabchikova EI, Kolesnikova LV, Tkachev VK, et al. (1994) Ebola infection in four monkey species. Ninth International Conference on Negative Strand RNA Viruses, Estoril, Portugal, Abstract 246:164

Sachs T, Moldow CF, Craddock PR, Bowers TK, Jacob ``A (1978) Oxygen radicals mediate endothelial damage by complement-stimulated granulocytes: An in vitro model of imune complex vasculitis. J Clin Invest 61:1161–1167

Salomon D, Ayalon O, Patel-King R, Heynes RO, Geiger B (1992) Extrajunctional distribution of N-cadherin in cultured human endothelial cells. J Cell Sci 102:7–17

Sanchez A, Trappier S, Mahy BWJ, et al. (1996) Glycoprotein genes of Ebola viruses: unusual organization and mechanisms of gene expression. Proc Natl Acad Sci USA 93:3602–3607

Schilling WP, Ritchie AK, Navarro LT, Eskin SG (1988) Bradykinin-stimulated calcium influx in cultured bovine aortic endothelial cells. Am J Physiol 255:H219–H227

Schnittler HJ, Feldmann H (1997) Filovirus-induced increase in endothelial permeability. J Mol Med 75:B69

Schnittler HJ, Feldmann H (1998) Marburg and Ebola virus. Does the primary course of infection depend on the accessibility of organ-specific macrophages. Clin Infect Dis 27 (in press)

Schnittler HJ, Mahner F, Drenckhahn D, Klenk HD, Feldmann H (1993) Replication of Marburg virus in human endothelial cells: a possible mechanism for the development of viral hemorrhagic disease. J Clin Invest 91:1301–1309

Schnittler HJ (1998) Structural and functional aspects of intercellular junctions in vascular endothelium. Bas Res Cardiol (in press)

Schnittler HJ, Püschel B, Drenckhahn D (1997) Role of cadherins and plakoglobin in adhesion between endothelial cells under resting conditions and fluid shear stress. Am J Physiol 273:H2396–H2405

Schnittler HJ, Wilke A, Gress T, Suttorp N, Drenckhahn D (1990) Role of actin and myosin in the control of paracellular permeability in pig, rat, and human vascular endothelium. J Physiol (London) 431:379–401

Seeger W, Hansen T, Rössig R, Schmehl T, Schütte H, Krämer HJ, Walmrath D, Weissmann N, Grimminger F, Suttorp N (1995) Hydrogen peroxide-induced increase in lung endothelial and epithelial permeability – effect of adenylate cyclase stimulation and phosphodiesterase inhibition. Microvasc Res 50:1–17

Shanley TP, Warner RL, Ward PA (1995) The role of cytokines and adhesion molecules in the development of inflammatory injury. Mol Med Today 1:40–44

Shapiro L, Fannon AM, Kwong PD, Thompson A, Lehmann MS, Grubel G, Legrand F, Als Nielsen J, Colman DR, Hendrickson WA (1995) Structural basis of cell-cell adhesion by cadherins. Nature 374:327–337

Shasby DM, Lind SE, Shasby SS, Goldsmith JC, Hunninghake GW (1985) Reversible oxidant-induced increases in albumnin transfer across endothelium: alterations in cell shape and calcium homeostasis. Blood 65:605–614

Shimuzu Y, Newman W, Tanaka Y, Shaw S (1992) Lymphocyte interactions with endothelial cells. Immunol Today 13:106–112

Smith CEG, Simpson DIH, Bowen ETW (1967) Fatal human disease from vervet monkeys. Lancet II:1119–1121

Sonnenberg A (1993) Integrins and their ligands. Curr Top Microbiol Immunol 184:7–35

Stasek JE, Patterson CE, Garcia JGN (1992) Protein kinase C phosphorylates caldesmon$_{77}$ and vimentin and enhances albumin permeability across cultured bovine pulmonary artery endothelial cell monolayers. J Cell Physiol 153:62–75

Stelzner TJ, Weil JV, ÓBrien RF (1989) Role of cyclic adenosine monophosphate in the induction of endothelial barrier properties. J Cell Physiol 139:157–166

Stevenson BR, Begg DA (1994) Contraction-dependent effects of cytochalasin D on tight junctions and actin filaments in MDCK epithelial cells. J Cell Sci 107:367–375

Stevenson BR, Siliciano JD, Mooseker MS, Goodenough, DA (1986) Identification of ZO-1: A high molecular weight polypeptide associated with the tight junction (zonula occludens) in a variety of epithelia. J Cell Biol 103:755–766

Suttorp N, Flöer B, Schnittler HJ, Seeger W, Bhakdi S (1990) Effects of Escherichia coli hemolysin on endothelial cell function. Infec Immun 58:3796–3801

Suttorp N, Nolte A, Wilke A, Drenckhahn D (1993a) Human neutrophil elastase increases permeability of cultured pulmonary endothelial cell monolayer. Int J Microcirc Clin Exp 13:187–203

Suttorp N, Polley M, Seybold J, Schnittler HJ, Aktories K (1991) Adenosine diphosphate ribosilation of G-actin by botulinum C2 toxin increases endothelial permeability in vitro. J Clin Invest 87:1575–1584

Suttorp N, Seeger W, Dewein E, Bhakdi S, Roka L (1985) Staphylococcal alpha-toxin induced PGI$_2$ production in endothelial cells: role of calcium. Am J Physiol 248:C127–C134

Suttorp N, Weber U, Schudt C (1993b) Role of phosphodiesterases in the regulation of endothelial permeability in vitro. J Clin Invest 91:1421–1428

Unanue ER, Allen PM (1987) The basis for the immunoregulatory role of macrophages and other accessory cells. Science 236:551–557

Valiron O, Chevrier V, Usson Y, Breviario F, Job D, Dejana E (1996) Desmoplakin expression and organization at human umbilical vein endothelial cell-to-cell junctions. J Cell Science 109:2141–2149

Vestweber D (1992) Selectins: cell surface lectins which mediate the binding of leukocytes to endothelial cells. Seminars in Cell Biology 3:211–220

Vestweber D (1993) The selectins and their ligands. Curr Top Microbiol Immunol 184:65–75

Villinger F, Rollin PE, Brar SS, Chikkala NF, Winter J, Sundstrom JB, Zaki SR, Swanepoel R, Ansari AA, Peters CJ (1998) Markedly elevated levels of IFN-γ/α, IL-2, IL-10, TNF-α associated with fatal Ebola virus infection. J Infect Dis (in press)

Volchkov VE, Blinov VM, Netesov SV (1992) The envelope glycoprotein of Ebola virus contains an immunosuppressive-like domain similar to oncogenic retroviruses. FEBS Lett 305:181–184

Volchkov VE, Feldmann H, Volchkova VA, Klenk HD (1998a) Processing of the Ebola virus glycoprotein by the proprotein convertase furin. Proc Natl Acad Sci USA 95 (in press)

Volchkov VE, Feldmann H, Volchkova VA, Klenk HD (1998a) Processing of the Ebola virus glycoprotein by the proprotein converatse furin. Proc. Natl. Acad. Sci. USA 95:5762–5767

Volchkov VE, Volchkova VA, Slenczka W, Klenk HD, Feldmann H (1998b) Release of viral glycoproteins during Ebola virus infection. Virology 245:110–119

Volchkov VE, Volchkova VA, Slenczka W, Klenk HD, Feldmann H (1998b) Release of viral glycoproteins during Ebola virus infection. Virology (in press)

Weiss L, (1983) Cell and tissue biology. A textbook of histology. Urban Schwarzenberg, Baltimore. ISBN 0-8067-2176-6

Weiss SJ, Regiani S (1984) Neutrophils degrade subendothelial matrices in the presence of alpha-1-proteinase inhibitor: cooperative use of lysosomal proteinases and oxygen metabolites. J Clin Invest 73:1297–1303

West E, Schnittler HJ, Sprenger H, Klenk HD, Feldmann H (1996) Filoviral replication in human macrophages. 15th Annual Meeting of the American Society for Virology, London, Ontario, Canada, W14-3 (p 107)

White GE, Gimbrone Jr MA, Fujiwara K (1983) Factors influencing the expression of stress fibers in vascular endothelial cells in situ. J Cell Biol 97:416–424

Will C, Mühlberger E, Linder D, Slenczka W, Klenk HD, Feldmann H (1993) Marburg virus gene four encodes the virion membrane protein, a type I transmembrane glycoprotein. J Virol 67:1203–1210

Willot E, Balba MS, Heintzelman M, Jameson B, Anderson JM (1992) Localization and differential expression of two isoforms of tight junction protein ZO-1. Am J Physiol 262:C1119–C1124

Wong MKK, Gotlieb AI (1988) The reorganization of microfilaments, centrosomes, and microtubules during in vitro small wound reendothealialization J Cell Biol 107:1777–1783

Wysolmerski RB, Lagunoff D (1990) Involvement of myosin light-chain kinase in endothelial cell retraction. Proc Natl Acad Sci USA 87:16–20

Yang Z, Delgado R, Xu L, Todd RF, Nabel EG, Sanchez A, Nabel GJ (1998) Distinct cellular interactions of secreted and transmembrane Ebola virus glycoproteins. Science 279:1034–1037

Zahraoui A, Joberty G, Arpin M, Fontaine JJ, Hellio R, Tavitian A, Louvard D (1994) A small rab GTPase is distributed in cytoplasmic vesicles in nonpolarized cells but colocalizes with tight junction marker ZO-1 in polarized epithelial cells J Cell Biol 124:101–105

Zaki SR, Peters CJ (1997) Viral hemorrhagic fevers. In: Connor DH, Schwartz DA, Manz HJ, Lack EE (eds) Diagnostic pathology of infectious diseases. Appleton and Lange, Stamford, CT, pp 347–364

Zhong Y, Saitoh T, Minase T, Sawada N, Enomoto K, Mori M (1993) Monoclonal antibody 7H6 reacts with a novel tight junction associated protein distinct from ZO-1, cingulin and ZO-2. J Cell Biol 120:477–483

Zlotnik I (1969) Marburg agent disease: pathology. Trans Royal Soc Trop Med Hyg 63:310–327

Immune Response to Filovirus Infections

G.M. IGNATYEV

1 Introduction 205
2 Cell Immunity and Filoviridae 206
3 Involvement of Cytokines 207
4 Humoral Immunity in Filoviral Fevers 209
5 Immunogenic Characteristics of Filoviral Proteins 210
6 Immunogenic Properties of Inactivated Preparations of Filoviruses 212
7 Conclusions 214
References 214

1 Introduction

A syndrome referred to as viral hemorrhagic fever is caused by several RNA viruses including members of the families Arenaviridae, Bunyaviridae, Filoviridae and Flaviviridae (PETERS et al. 1991). Pronounced hemorrhagic manifestations are characteristic of these fevers as well as disseminated intravascular coagulation (DIC), generalized shock, and the highest mortality rate (30%–90%) (FISHER-HOCH et al. 1985, 1993; MURPHY et al. 1990). However, the data on immune response during filoviral hemorrhagic fevers are scarce. The general notion of immunity suggests the following scheme of immune response to the filoviral infection: viral activation of macrophages, T and B lymphocytes; production of mediators by mononuclear cells, e.g., interleukin (IL)-1 and IL-2, interferon (IFN), tumor necrosis factor (TNF); changes of the proliferative activity of the cells; alterations of lymphocyte subpopulations (CD4 and CD8); propagation of virus in immunocompetent cells. Since hemorrhagic manifestations, generalized shock, and disseminated intravascular coagulation (DIC) are characteristic of all hemorrhagic fevers, it seems reasonable to study and analyze the immunity indices which may be

Institute of Molecular Biology, State Research Centre of Virology and Biotechnology "Vector," Novosibirsk region, 633159, Russia

the cause of the above mentioned clinical manifestations and have already been studied in the other hemorrhagic fevers caused by special pathogens, for example, arenaviruses. Thus, decrease of lymphocyte proliferative activity in response to mitogen stimulation, decrease of the number of T and B lymphocytes, and inversion of CD4\CD8 lymphocytes ratio (FISHER-HOCH et al. 1987; VALLEGOS et al. 1985; ENRIA et al. 1986) were demonstrated in arenaviral hemorrhagic fevers. Participation of immunity mediators, and especially IFN (LEVIS et al. 1984, 1985) and TNF, in the development of clinical manifestation of arenaviral hemorrhagic fevers was also demonstrated. In this connection we focused our attention on the role and significance of these indices in filoviral hemorrhagic fevers.

2 Cell Immunity and Filoviridae

Investigations of various authors testify to the ability of filoviruses to propagate in the human endothelial cells and macrophagal cells of different origins (FELDMANN et al. 1996; SCHNITTLER et al. 1993; GEISBERT et al. 1992; SKRIPCHENKO et al. 1991; STRELTSOVA et al. 1991; PEREBOEVA et al. 1993). Thus, macrophages, along with endothelial cells, are the target cells where the reproduction of filoviruses occurs. The in vitro propagation of Marburg virus in cell culture of both human and guinea pig mononuclear cells is accompanied by considerable suppression of the proliferative reaction of these cells to mitogen stimulation, in particular concanavalin A (Con A) and LPS (IGNATYEV et al. 1996b). When guinea pigs and monkeys were infected with Marburg and Ebola viruses, a lethal infection developed. In these experiments, an increase of spontaneous lymphocyte proliferation and a decrease of lymphocyte proliferation indices in response to activation with mitogens were recorded (IGNATYEV et al. 1994a, b; 1995; 1996a). For example, the infection of monkeys *Papio papio* with Ebola virus resulted in a considerable (over fivefold) drop of mitogen-induced (Con A) lymphocyte proliferation indices as compared to before infection. At the same time considerable fluctuations of the activity of natural killer (NK) cells were recorded; this index increased significantly on days 5–7 post-infection (as compared to the pre-infection value). The activity of the NK cells is virtually absent at the terminal stage of the disease (G.M. Ignatyev and V.P. Luchko, unpublished data). Average survival time of guinea pigs and rhesus monkeys *Macaca mulatta* during lethal development of Marburg infection was 8–9 and 10–11 days, respectively. In addition, a pronounced decrease of mitogen-stimulated lymphocyte proliferative activity was recorded on days 5–7 post-infection. However, the activity of NK cells, which was determined in these experiments, increased during the course of the disease (IGNATYEV et al. 1994b; 1995; 1996a). Studies of Marburg fever in a *Macaca mulatta* model demonstrated that the ratio of CD4 to CD8 lymphocytes changed over the course of the disease. The ratio was 1.5 before the infection, decreasing to 1.25 and 1 on days 5 and 7 post-infection, respectively. At the terminal stage of infection on day 9, the index was 0.75 (IG-

NATYEV et al. 1996a). The observed immunosuppression during experimental filoviral fevers may be connected with the "immunosuppressive domain" localized in the GP protein of Marburg and Ebola viruses (VOLCHKOV et al. 1992; BUKREYEV et al. 1993). This possibility will be considered later. Propagation of filoviruses in immunocompetent cells and increase in lymphocyte proliferative activity may be accompanied by production of cytokines such as IFN, TNF, and IL-1, 2.

3 Involvement of Cytokines

Pronounced production of serum IFN was seen during experimental infection of guinea pigs and monkeys with Marburg and Ebola viruses with lethal outcome. Maximum IFN concentrations (1:1240 in guinea pigs and 1:2000 in monkeys) were recorded 1 day before the death of the animals (IGNATYEV et al. 1994a; 1995; 1996a).

The infection of human macrophages with Marburg virus led to increased release of TNF-α, which is one of several cytokines typically secreted by macrophages (FELDMANN et al. 1996). Increased TNF-α in blood serum was also recorded in experimental infection of guinea pigs and rhesus macaques with Marburg virus. The dynamics of this index coincided with the development of clinical manifestations and reached its maximum 1 day before the beginning of the terminal stage (IGNATYEV et al. 1994a; 1995; 1996a). Infection of monkeys with Ebola virus was also accompanied by increased serum TNF-α levels (G.M. Ignatyev and V.P. Luchko, unpublished data). The TNF-α concentration in serum of the animals infected with either Marburg, or Ebola viruses did not exceed 4 ng/ml. In the case of infection of macrophages with Marburg virus, the TNF-α concentration in cultural medium did not exceed 3 ng/ml (FELDMANN et al. 1996). Medium concentrations of TNF-α a in clinical cases not connected with filoviruses, including those of fulminating endotoxin and cardiac shock, are relatively low (0.33 ng/ml) and have never exceeded 4 ng/ml (ISHII et al. 1992). However, it was shown in vitro that endothelial permeability did not increase in response to TNF-α concentrations

Table 1. Effect of the treatment of Marburg virus-infected guinea pigs with anti-TNF-α serum and DFO

Preparation	Infection dose (Marburg virus, strain Popp)	No. of animals: surviving/total	Mean time to death (days)
Anti-TNF-α serum[a]	10 LD_{50}	3/6	8.9 ± 0.7
DFO[b]	10 LD_{50}	3/6	8.6 ± 0.7
Control	10 LD_{50}	0/6	7.7 ± 0.1

TNF, tumor necrosis factor; DFO, desferrioxamine.
[a] Anti-TNF-α serum administered intramuscularly on days 3, 4, 5, 6, 7 post-infection.
[b] DFO administered in 5 mg doses intramuscularly on days 4, 5, 6, 7, 8, 9 post-infection.

below 4 ng/ml (ISHII et al. 1992). This indicates that several secretory products (e.g. TNF-α, ILs, protease, oxygen radicals) are released from virus-infected macrophages and that the effect on the permeability is cumulative. Experiments involving administration of anti-TNF-α antibodies and TNF-α and IL-1 antagonists to Marburg virus-infected animals may confirm the cumulative effect of TNF-α and ILs. Half of the animals receiving anti-TNF-α antibodies, neutralizing 3 ng/ml of TNF, starting from day 3 post-infection with Marburg virus survived; the rest lived 2 days longer than animals not receiving anti- TNF-α (Table 1). Moreover, after the beginning of the treatment, no increase in TNF-α concentration was recorded in the serum of the animals that survived. An increased average survival time and the survival of three animals out of six was also recorded in experimental treatment of Marburg virus-infected animals with desferrioxamine (DFO) (the lethality in the control group was 100%) (IGNATYEV et al. 1996c). This effect may be explained by the specific properties of this preparation. DFO itself is known to suppress the expression of adhesion receptors; at the same time the cytokines TNF-α and IL-1 do not cause the expression of adhesion receptors in the presence of this preparation (CINATL et al. 1995). This suggests that DFO acts as an antagonist of TNF-α and IL-1. Thus, one may suggest that the cytokines TNF-α and IL-1 play a part in the development of filoviral hemorrhagic fevers. However, the TNF-α concentration in the "in vitro" and "in vivo" experiments did not exceed 4ng/ml. Even at low concentrations (4ng/ml), which had no direct permeability-increasing effect, TNF-α was capable of priming the effect in the presence of the factors or agents such as IL-2 and H_2O_2 (ISHII et al. 1992; FELDMANN et al. 1996).

Participation of IL-2 in development of filoviral hemorrhagic fevers was demonstrated while modeling Marburg fever in guinea pigs. Animals receiving carrier-conjugated IL-2 preparation introduced together with Marburg virus at a dose of 10 LD_{50} died earlier (on day 6, 8 ± 0.4) than animals injected only with Marburg virus (average survival time 8.3 ± 0.5 days). Reliable increases of TNF-α and IFN were recorded on day 5 in response to administration of IL-2 preparation and Marburg virus compared to injection of virus alone. The effect of IL-2 in the course of Marburg fever in the infected animals is confirmed as well by experiments on extracorporeal administration of the IL-2 inducer Diuciphone. In these experiments, the lymphocytes of Marburg-infected animals were in vitro treated with Diuciphone with subsequent reintroduction of the cells into the same animal donors. In this case, TNF-α and IFN levels in serum exceeded in those of the animals whose lymphocytes were not subjected to treatment with IL-2 inducer. Furthermore, death occurred earlier when lymphocytes were treated with Diuciphones (IGNATYEV et al. 1991; 1994b).

The data obtained in the "in vivo" and "in vitro" experiments suggest that immunity mediators play an important role in the development of filoviral hemorrhagic fevers. Sufficient production of TNF-α may cause damage of endothelial cells (hemorrhagic manifestations), development of shock, and DIC. TNF-α has been previously shown to interact with specific endothelial cell receptors (NAWROTH et al. 1986) and, among other factors, to enhance endothelial permeability (CYBULSKY et al. 1988; LAMPUGNANI et al. 1992; ISHII et al. 1992), loss of anticoagulant activity (SCHLEEF et al. 1988; 1991), and expression of adhesion molecules for leukocytes and platelets (BEVILAQUA et al. 1986; WERTHEIMER et al. 1992). One of the most important events during shock development is the increase of para-endothelial permeability, in which TNF-α is involved as a mediator (HORVATH et al. 1988). The combination of viral replication in endothelial cells (SCHNITTLER et al. 1993; GEISBERT et al. 1992) and virus-induced cytokine release from mononuclear cells may also promote a distinct proinflammatory endothelial phenotype that then triggers the coagulation cascade and, as a result, the development of DIC (SCHLEEF et al. 1988; BEVILAQUA et al. 1986; 1987).

4 Humoral Immunity in Filoviral Fevers

Data on the immune reaction during lethal outcome of Ebola and Marburg fevers are lacking in the available literature. Three major techniques – immunofluorescent assay (IFA), ELISA, and Western blot (WB) (ELLIOT et al. 1993; BECKER et al. 1992) – are used to determine antibody levels in hemorrhagic fever survivors. Virus-specific antibodies in convalescents are recorded as early as a month after the beginning of infection (SUREAU 1989; NIKIFOROV et al. 1994). IFA-determined antibody titer indices of those convalescing from Ebola fever vary considerably (1:28–1:1024) (SUREAU 1989). In case of an employee infected while performing laboratory work with Marburg virus, the ELISA-determined titer of antibodies a month after the infection were 1:64000 and were retained for 6 years (after 6 years the IFA titer was 1:1280) (NIKIFOROV et al. 1994; IGNATYEV et al. 1996b). Of considerable importance are IFA, ELISA, and WB examinations of sera of healthy monkeys imported to Germany and Switzerland from various regions of the world (Philippines, China, Uganda). The investigations performed have demonstrated that 43.3% (of 120 sera used) reacted positively with one of the three distinct antigens of filoviruses (Ebola, Marburg, Reston). Analogous examination of 1288 human sera (obtained from different donors including those who were in direct contact with Marburg fever patients) revealed antibodies to one of the three above mentioned viruses in 6.9% of the sera (BECKER et al. 1992). IFA, ELISA, and WB examination of 550 serum samples of US quarantine service staff with direct contact with monkeys revealed antibodies to either Ebola, or Marburg, or Reston in 7.6% (CENTERS FOR DISEASE CONTROL AND PREVENTION 1990a, b). A comparative study of sera of different donors by various techniques demonstrates that both IFA and ELISA can produce false-positive results, whereas WB is the more reliable method for confirmation of diagnosis. Filoviral proteins of major diagnostic importance are NP, VP40, VP30, VP35, and VP24 (BECKER et al. 1992; ELLIOT et al. 1993). The presence of antibodies to one of the filovirus antigens in sera of various donors (BECKER et al. 1992) and inhabitants of African countries (BLACKBURN et al. 1982; KNOBLOCH et al. 1982; MEUNIER et al. 1987) suggests that filoviral hemorrhagic fevers may sometimes have a subclinical course. Thus, both the subclinical course of filoviral hemorrhagic fevers and the course with clinical manifestations may result in formation of humoral immunity. However, a survey of the immunity status of employees who had worked with inactivated filoviruses for more than 3 years revealed no antibodies in either of the three reactions used, although sensibilization of lymphocytes to Marburg virus determined through blast-transformation reaction was demonstrated in the same employees (IGNATYEV et al. 1996b). Summing up, a long contact with the inactivated virus may cause specific cellular memory, manifesting itself in a delayed hypersensitivity reaction. The role of virus-specific antibodies in development of the acute form of filoviral fevers and their prognostic value are ambiguous. For instance, the retrospective analysis of the sera of monkeys which were exported to the USA from the Philippines in 1989 and which were carriers of the Ebola-like Reston virus demonstrated definite correlation between disease outcome and initial IFA-determined antibody titers. Thus, the mortality among the originally seropositive animals (IFA-determined antibody titer over 1:16) was reliably lower than among the seronegative animals (HAYES et al. 1992). Nonetheless, animals (guinea pigs and monkeys) immunized with inactivated Marburg virus also display ELISA-determined antibodies (IGNATYEV et al. 1991; 1994a; 1995; 1996a). However, these antibodies do not possess virus-neutralizing activity, and death or survival of the animals after Marburg virus infection did not depend on the initial antibody titers.

5 Immunogenic Characteristics of Filoviral Proteins

Bulk quantities of the proteins NP and VP40 of Marburg virus were obtained in order to study their immunogenic properties through reverse-phase chromatography on the carrier Polysyl-ODS 500. Immunization of guinea pigs with a preparation of VP40 at doses of 10µg and 20µg caused no cellular or humoral immune responses. Immunization of the same animals with the same doses of NP protein, in contrast, resulted in cellular and humoral immunity (the former was determined through blast-transformation). Antibody titers determined by ELISA on day 28 after double (days 0 and 14) intramuscular immunization with the protein preparation at a dose of 10µg were 1:128 (AGAFONOV et al. 1992, 1993). Data on the protective characteristics of proteins NP and VP40 are presented in Table 2. Summing up, immunization of guinea pigs with the preparation of NP protein causes formation of protective immunity and protects the animals against infection with a dose of 50 LD_{50}. In contrast, immunization with VP40 fails to cause either cellular or humoral immunity or to provide any protection.

Recombinants of vaccinia virus expressing the genes coding for the proteins NP, VP35, VP30, membrane-anchored GP, secreted form of GP (sGP), VP40, and VP24 of Ebola were obtained to study their protective characteristics. Subcutaneous immunization of guinea pigs with the recombinants expressing either VP40, or NP, or sGP at a dose of 10^7 PFU failed to provide protective immunity against infection

Table 2. Protective immunity after immunization of guinea pigs with various preparations of Marburg virus

Preparation	Immunization dose/scheme of immunization	Infection dose (Marburg virus strain Popp) and days of infection after 1 immunization	Number of animals: surviving/total
FIAMV[a]	1 µg (days 0 and 14)	10 LD, 28 days	1/10
	5 µg (days 0 and 14)	10 LD, 28 days	2/10
	10 µg (days 0 and 14)	10 LD, 28 days	6/10
	25 µg (days 0 and 14)	10 LD, 28 days	7/10
	50 µg (days 0 and 14)	10 LD, 28 days	8/10
Control[a]		10 LD	0/10
NP	10 µg (days 0 and 14)	50 LD	3/10
	2 × 10 µg (days 0 and 14)	50 LD	6/10
	20 µg (days 0 and 14)	50 LD	5/10
	2 × 20 µg (days 0 and 14)	50 LD	7/10
VP40	10 µg (days 0 and 14)	50 LD	0/10
	2 × 10 µg (days 0 and 14)	50 LD	1/10
	20 µg (days 0 and 14)	50 LD	0/10
	2 × 20 µg (days 0 and 14)	50 LD	1/10
Recombinant vSC-GP	5 × 10^6 PFU (days 0 and 30)	50 LD, 60 days	0/6
		5 LD, 60 days	1/6
	5 × 10^6 PFU (days 0 and 30)	50 LD, 30 days	0/6
		5 LD, 30 days	2/6
	5 × 10^6 PFU (days 0 and 30)	50 LD, 15 days	0/6
		5 LD, 15 days	0/6

FIAMV, formalin-inactivated antigen of Marburg virus; NP and VP40, Marburg viral protein; vSC-GP, recombinant strain of vaccinia virus expressing the surface protein (GP) of Marburg virus.
[a]Control animals which were placebo-immunized on days 0 and 14.

with 100 PFU of Ebola performed 30 days later. However, three of five animals immunized with the recombinant expressing membrane-anchored GP survived (GILLIGAN et al. 1996). To study the immunogenic properties of Ebola VP24 and VP30, recombinant vaccinia viruses were constructed which expressed the mentioned genes under the control of the 7.5K promoter. The study was carried out in a guinea pig model. Immunization of animals with the recombinant strains of vaccinia virus did not provide protective immunity against subsequent infection with Ebola virus (I.P. Dmitriev and A.A. Chepurnov, personal communication). A recombinant strain of vaccinia virus expressing the surface protein (GP) of Marburg virus (vSC-GP) was also constructed to study its protective properties. vSC-GP was used to immunize guinea pigs at different time points before challenge infections with Marburg virus at 5 or 50 LD$_{50}$. All immunized animals developed antibodies against GP. None of the vSC-GP-immunized animals survived a challenge infection with Marburg virus at a dose of 50 LD$_{50}$. Surprisingly, animals which were immunized twice with vSC-GP had the shortest average survival time. However, some of the immunized animals infected with 5 LD$_{50}$ managed to survive the challenge infection, but there was no correlation with immunization with vSC-GP, since control animals immunized with a nonspecific recombinant vaccinia virus also survived 5 LD$_{50}$ (BECKER et al. 1996).

An area of amino acid homology found in the GPs of Ebola and Marburg viruses that may play a crucial role in their pathogenesis lies in the COOH-terminal region. The COOH-terminal 160 residues of the Ebola and Marburg GPs have a strong homology with the COOH-terminal regions of the glycoproteins of oncogenic retroviruses, particularly with respect to an immunosupressive motif (VOLCHKOV et al. 1992; BUKREEV et al. 1993). This amino acid sequence (EBO 585–610) has the highest degree of identity seen in the alignment with the Marburg GP (80.9%). This motif is also present in the GPs of Reston and Reston-like filoviruses (SANCHEZ et al. 1993). Researchers have shown that a synthetic peptide corresponding to the immunosuppressive motif found in oncogenic retroviruses can: (1) inhibit blastogenesis of lymphocytes; (2) decrease monocyte chemotaxis and macrophage infiltration; (3) inhibit the activity of human NK cells; (4) block the activity (>95%) of protein kinase C, a cellular messenger involved in T cell activation (CIANCIOLO et al. 1985; HARRIS et al. 1987; KADOTA et al. 1991).

Peptides simulating the structure of the presumptive immunosuppressive domain of Ebola and Marburg viruses were synthesized for particular investigation of the properties of the GP protein. These peptides inhibited blastogenic lymphocyte proliferation and activity of human NK cells, and increased the activity of macrophages. At the same time they did not inhibit the production of IL-1, nor did they activate the production of TNF (AGAFONOV et al. 1994; IGNATYEV et al. 1994c). The potentialimmunosuppressive role of the immunosuppressive domain of GP protein is indirectly confirmed by the observation that guinea pigs and monkeys experimentally infected with Ebola and Marburg viruses are in an immunosuppressed state. In addition, the proliferation of filoviruses in macrophages and monocytes in vivo and in vitro (GEISBERT et al. 1992; FELDMANN et al. 1996; PEREBOEVA et al. 1993; IGNATYEV et al. 1996b) suggests that immunosuppression may be an important feature of the disease in humans and in nonhuman primates.

6 Immunogenic Properties of Inactivated Preparations of Filoviruses

It is not known if vaccines can be used as prophylactic means to provide protection against filoviral infections. Long-term investigations devoted to this question have generated considerably ambiguous results. Protective effects have been observed when immunized guinea pigs and monkeys were challenged with Ebola virus (LUPTON et al. 1980; MIKHAILOV et al. 1994). However, the protective effects in monkeys were recorded only after double immunization with large doses of the specific protein, namely 7.1mg per one immunization (MIKHAILOV et al. 1994). Low doses of inactivated Ebola virus antigen did not cause a protective effect (LANGE et al. 1985). Immunization of guinea pigs with various doses of inactivated antigen of Marburg virus also testified to the dependence of protective immunity indices on the dose (IGNATYEV et al. 1991; AGAFONOV et al. 1993). Data from the study on the protective effect of various doses of inactivated Marburg virus antigen are presented in Table 2. An increase of the immunizing dose over 10μg/animal did not result in a valid increase in the percentage of the surviving animals. Administration of the inactivated Marburg virus antigen in immunogenic doses (over 3μg/animal) caused activation of nonspecific immunity factors such as phagocytic activity of macrophages, NK cell activity, and production of IFN and TNF in both guinea pigs and monkeys. Maximum activity of these indices was recorded within the first 3–5 days after administration of the preparation. Virus-specific antibodies were revealed by ELISA 14 days after the first immunization (at a titer of 1:80) and a clone of the cells was formed which responded with pronounced in vitro proliferation to stimulation with inactivated viral antigen. A second administration of inactivated antigen on day 14 after the first immunization resulted in a more marked secondary immune reaction, as seen by an increase of virus-specific antibody titer, 1:320–1:640 on day 28. At the same time analysis of CD4/CD8 markers revealed an increase in CD4 lymphocytes during the first 14 days post-immunization and an increase in CD8 by days 25–32 post-immunization, while no decrease in in vitro lymphocyte proliferative activity in response to stimulation with mitogens was recorded. Summing up, immunization of the animals with inactivated preparations of filoviruses results in formation of humoral and cellular immune response. (IGNATYEV et al. 1991; 1994a; 1995; 1996a).

An important characteristics of immunity reactions in immunized animals after challenge with Marburg virus is the phenomenon of "early death", manifested as the earlier death, compared to nonimmune animals infected with the same dose of virus, of those animals which, after immunization with inactivated antigen, demonstrated both humoral and cellular immunity. In the guinea pig model, the immune reaction of the animals which then died was accompanied by the more pronounced production of TNF-α, IFN and spontaneous lymphocyte proliferation as compared to nonimmunized animals. By contrast, animals that survived the challenge demonstrated marked TNF-α and IFN activity only during the first 3 days; the indices then decreased and reached pre-infection levels. An increase of ELISA-determined antibody titers was recorded in the surviving immune animals. Virus in the blood of the these animals was detected, using biosampling techniques, only in the first 24 h post-infection.

Similar results were obtained in experiments using rhesus monkeys. In these experiments, six animals were immunized twice (on days 0 and 14) with inactivated Marburg virus at a dose of 7 μg/animal. On day 35 after the first immunization, the immunized and two placebo-immunized control monkeys were intramuscularly infected with Marburg virus at a dose of 250 LD_{50} for guinea pigs. Three immunized monkeys and the control monkeys died. In the surviving immunized animals, an increase of IFN, TNF-α, and NK activities was recorded on days 2–3 post-infection. In the control nonimmune animals these indices steadily increased over the course of the disease. A similar pattern was observed in immunized animals that died. However, on days 5–7, the TNF-α and IFN activity in the immune animals exceeded as in of the control animals. The death of the immune animals occurred earlier, on days 7, 8, 9. Changes in the CD4:CD8 ratio took place in both the control and immune animals, namely, reversion of the index from 1.75 and 1.5 before the challenge to 0.75 post-infection. It is likely that the increased number of lymphocytes carrying CD8 receptors is due to an increase in the T suppressor population, since all dead animals had decreased lymphocyte proliferative activity in response to mitogen stimulation (IGNATYEV et al. 1995; 1996a). Summing up, immunization of experimental animals with preparations of inactivated filoviruses results in formation of both humoral and cellular immunity. However, no correlation exists between the immunity formed and a protective effect while challenging the immune animals with the active virus.

7 Conclusions

The immune response in filoviral infections has so far not been adequately studied. It is known that filoviruses are capable of propagation in mononuclear and endothelial cells. Involvement of cells of the immune system in the infectious process is accompanied by decreased mononuclear proliferative activity in response to mitogen stimulation, and marked production of cytokines (especially TNF). However, the mechanisms of the immunosuppressive state that develops during filoviral infections are not understood. Immunization of animals with both inactivated preparations of filoviruses and preparations of subunit proteins was demonstrated to result in formation of specific (humoral and cellular) immunity. Nevertheless, the indices of resistance of the immunized cross-bred animals to challenge are independent on the intensity of specific immunity, as it was demonstrated that animals with the same indices of humoral and cellular immunity may either survive or die after infection with the same dose of the virus. Thus, further investigation into the immune response in filoviral fevers is necessary, and first and foremost, into the mechanism of virus interactions with immunocompetent cells.

References

Agafonov AP, Ignatyev GM, Akimenko ZA, Volchkov VE (1993) Study of immunogenic and protective properties of Marburg virus GP, NP and VP40 proteins. IXth International Congress of Virology, Glasgow, Scotland, PW 52-7:300

Agafonov AP, Ignatyev GM, Kuz'min VA, Akimenko ZA, Kosareva TV, Kashentseva EA (1992) The immunogenic properties of Marburg virus proteins (in Russian). Voprosy Virusologii 37:58–61

Agafonov AP, Streltsova MA, Kashentseva EA, Ignatyev GM (1994) Possible mechanism of immunosuppression for filoviruses infection. Inter J Immunorehab 1-Suppl:26

Becker S, Feldmann H, Will C, Slenczka W (1994) Evidence for occurrence of filovirus antibodies in humans and imported monkeys: do subclinical filovirus infections occur worldwide? Med Microbiol Immunol 181:43–55

Becker S, Volchkov VE, Mühlberger E, Klenk HD, Agafonov AP, Ignatyev GM (1996) A recombinant vaccinia virus expressing the surface protein of Marburg virus does not protect guinea pigs against Marburg-virus infection. Xth International Congress of Virology, Jerusalem, Israel, PW60-40:259

Bevilaqua MP, Pober JS, Majeau GR, Cotran RS, Gimbrone MA Jr (1986) Recombinant tumor necrosis factor induces procoagulant activity in cultured human vascular endothelium: characterization and comparison with the actions of interleukin 1. Proc Natl Acad Sci USA 83:4533-4537

Bevilaqua MP, Pober JS, Mendrick DL, Cotran RS, Gimbrone MA Jr (1987) Identification of an inducible endothelial-leukocyte adhesion molecule. Proc Natl Acad Sci USA 84:9238-9242

Blackburn NK, Searle L, Taylor P (1982) Antibody against haemorrhagic fevers in Central African human population. Trans Roy Soc Trop Med Hyg 76:803-805

Burkreyev AA, Volchkov VE, Blinov VM, Netesov SV (1993) The GP-protein of Marburg virus contains the region similar to the "immunosuppressive domain" of oncogenic retrovirus P15 E proteins. FEBS Lett 323:183-187

Centers for Disease Control and Prevention (1990a) Update: filovirus infection among persons with occupational exposure to nonhuman primates. MMWR 39:266-267, 273

Centers for Disease Control and Prevention (1990b) Update: filovirus infection associated with contact with nonhuman primates or their tissues. MMWR 39:404-405

Cianciolo GJ, Copeland TJ, Oroszlan S, Snyderman R (1985) Inhibition of lymphocyte proliferation by a synthetic peptide homologous to retroviral envelope protein. Science 230:453-455

Cinatl J, Schlotz M, Weber B, Cinatl J, Rabenau H, Markus BH, Encke A, Doerr HW (1995) Effects of desferrioxamine on human cytomegalovirus replication and expression of HLA antigens and adhesion molecules in human vascular endothelial cells. Transplant Immunology 3:313-320

Cybulsky MI, Chan MKW, Movat HZ (1988) Biology of disease. Acute inflammation and microthrombosis induced by endotoxin, interleukin-1, and tumor necrosis factor and their implication in Gram-negative infection. Lab Invest 58:365-378

Elliot LH, Bauer SP, Perez-Oronoz G, Lloyd ES (1993) Improved specificity of testing methods for filovirus antibodies. J Virol Meth 43:85-100

Enria D, Garsia Franco S, Ambrosio A, Vallejos D, Levis ɔ, Maiztegiu J (1986) Current status of the treatment of Argentine hemorrhagic fever. Med Microbiol Immunol 175:173-176

Feldmann H, Bugany H, Mahner F, Klenk HD, Drenckhahn D, Schnittler HJ (1996) Filovirus-induced endothelial leakage triggered by infected macrophages. J Virol 70:2208-2214

Fisher-Hoch SP (1993) Stringent precautions are not advisable when caring for patients with viral haemorrhagic fevers. Rev Med Virol 3:7-13

Fisher-Hoch SP, Mitchell SW, Sasso DR, Lange JV, Ramsey R, McCormick JB (1987) Physiologic and immunologic disturbances associated with shock in a primate model of Lassa fever. J Infec Dis 155:465-474

Fisher-Hoch SP, Platt GS, Neild GH, Southee T, Baskerville A, Raymond RT, Lloyd G, Simpson DIH (1985) Pathophysiology of shock and hemorrhage in a fulminating viral infection (Ebola). J Infect Dis 152:887-894

Geisbert TW, Jahrling PB, Hanes MA, Zack PM (1992) Association of Ebola-related Reston virus particles and antigen tissue lesions of monkeys imported to the United States. J Comp Pathol 106:137-152

Gilligan KJ, Geisbert J, Jahrling PB, Anderson K (1996) Protective efficacy of recombinant vaccinia viruses expressing individual Ebola virus structural proteins. Xth International Congress of Virology, Jerusalem, Israel, W60A-4

Harris DT, Cianciolo GJ, Snyderman R, Argov S, Koren HR (1987) Inhibition of human natural killer cell activity by a synthetic peptide homologous to a conserved region in the retroviral protein, p15 E. J Immunol 138:889-894

Hayes CG, Burans JP, Ksiazek TG, Del Rosario RA, Miranda MEG, Manaloto CR, Barrientos AB, Robles CG, Dayrit MM, Peters CJ (1992) Outbreak of fatal illness among captive macaques in the Philippines caused by an Ebola-related filovirus. Am J Trop Hyg 46:664-671

Heller MV, Saavedra MC, Falcoff R, Maiztegui JI, Molinas FC (1992) Increased tumor necrosis factor-a levels in Argentine hemorrhagic fever. J Infect Dis 166:1203-1204

Horvath CJ, Ferro TJ, Jesmok G, Malik AB (1988) Recombinant tumor necrosis factor increases pulmonary vascular permeability independent of neutrophils. Proc Natl Acad Sci USA 85:9219-9223

Ignatyev GM, Agafonov AP, Streltsova MA, Managarova GI, Spirin GV, Cherni NB (1991a) A comparative study of the immunological indices in guinea pigs administered an inactivated Marburg virus (in Russian). Voprosy Virusologii 36:421-423

Ignatyev GM, Chepurnov AA, Prosorovsky NS, Streltsova MA, Agafonov AP (1991b) The influence of interleukin-2 inducer "diucifon" in Marburg haemorrhagic fever. In: International Symposium "Interferon-92", Moscow, pp 160–164

Ignatyev GM, Streltsova MA, Agafonov AP, Zhukova NA, Kashentseva EA, Vorobieva MS (1994a) Study of some immunity parameters in animals immunized with inactivated Marburg virus after challenge with the homologous virus (in Russian). Voprosy Virusologii 39:13–17

Ignatyev GM, Streltsova MA, Agafonov AP, Prozorovsky NS, Zhukova NA, Kashentseva EA, Vorobieva MS (1994b) Immunologic characteristics of guinea pigs with Marburg hemorrhagic fever (in Russian). Voprosy Virusologii 39:169–171

Ignatyev GM, Volchkov VE, Blinov VM, Samukov VV, Agafonov AP, Streltsova MA, Kashentseva EA (1994c) Phenomenon of immunosuppression by Filoviruses. Inter J Immunorehabil 1-Suppl:138

Ignatyev GM, Streltsova MA, Agafonov AP, Kashentseva EA (1995) Mechanisms of protective immune response in monkeys with Marburg fever (in Russian). Voprosy Virusologii 40:109–113

Ignatyev GM, Agafonov AP, Streltsova MA, Kashentseva EA (1996a) Inactivated Marburg virus elicits a nonprotective immune response in Rhesus monkeys. J Biotechnol 44:111–118

Ignatyev GM, Streltsova MA, Kashentseva EA, Patruschev NA, Ginko SI, Agafonov AP (1996b) Immunity indices in the personnel involved in hemorrhagic virus investigation .In: Berg DA (ed) Proceedings of the 1996 ERDEC scientific conference on chemical and biological defense research. November 19–22, 1996, pp 323–330

Ignatyev GM, Streltsova MA, Agafonov AP, Kashentseva EA, Prosorovsky NS (1996c) Experimental study on the possibility of Marburg hemorrhagic fever treatment with Desferal, Ribavirin, and homologous interferon (in Russian). Voprosy Virusologii 41:206–209

Ishii Y, Partridge CA, Del Vecchio PJ, Malik AB (1992) Tumor necrosis factor-a-mediated decrease in glutathione increases the sensitivity of pulmonary vascular endothelial cells to H_2O_2. J Clin Invest 189:784–802

Kadota J, Cianciolo GJ, Snyderman R (1991) A synthetic peptide homologous to retroviral transmembrane envelope proteins depresses protein kinase C-mediated lymphocyte proliferation and directly inactivated protein kinase C: a potential mechanism .. r immunosuppression. Microbiol Immunol 35:443–459

Knobloch J, Albiez E, Schmitz H (1982) A serological survey on viral haemorrhagic fevers in Liberia. Ann Virol 133-E:125–128

Lampugnani MG, Resnati M, Raiteri M, Pigott R, Pisacane A, Houen G, Ruco LP, Dejana E (1992) A novel endothelial-specific membrane is a marker of cell-cell contacts. J Cell Biol 118:1511–1522

Lange JV, McCormick JB, Walker DH (1985) Vaccination of Rhesus monkeys with gamma-inactivated Ebola virus and results of live-virus. Annual Meeting of the American Society of Microbiology, Las Vegas, Nevada, 3–7 March, CDC, Atlanta, pp 295–303 [Abstr]

Levis SC, Saavedra MC, Ceccoli C, Falcoff E, Feuillade MR, Enria DAM, Maiztegui JI, Falcoff R (1984) Endogenous interferon in Argentine hemorrhagic fever. J Infect Dis 149:428–433

Levis SC, Saavedra MC, Ceccoli C, Feuillade MR, Enria DA, Maiztegui JI, Falcoff R (1985) Correlation between endogenous interferon and the clinical evolution of patients with Argentine hemorrhagic fever. J Interferon Res 5:383–389

Lupton HW, Lambert RD, Bumgardner DL, Moe JB, Eddy GA (1980) Inactivated vaccine for Ebola virus efficacious in guinea pig model. Lancet 2:1294–1295

Meunier DMJ, Johnson ED, Gonzalez JP (1987) Serological evidence for Marburg virus antibodies in human of Central Africa. Bull Soc Path Exot 80:51–61

Mikhailov VV, Borisevich IV, Chernikova NK, Potryvaeva NV, Krasnyansky VP (1994) Evaluation of the possibility of Ebola fever specific prophylaxis in Baboons (Papio hamadryas) (in Russian). Voprosy Virusologii 39:82–84

Murphy FA, Kiley MP, Fisher-Hoch SP (1990) Filoviridae – Marburg and Ebola viruses. In: Fields BN, Knipe DM (eds) Virology. Raven, New York, pp 936–942

Nawroth PP, Bank I, Handley D, Cassimeris J, Chess L, Stern DM (1986) Tumor necrosis factor/cachectin interacts with endothelial cell receptors to induce release of interleukin 1. J Exp Med 163:1363–1375

Nikiforov VV, Turovskiy YI, Kalinin PP, Akinfeeva LA, Katkova LR, Barmin VS, Ryabchikova EI, Popkova NI, Shestopalov AM, Nasarova VP, Vedishchev SV, Netesov SV (1994) Case of an employee infected while performing the laboratory work with Marburg virus (in Russian). J Microb Epidemiol Immunobio 3:104–106

Pereboeva LA, Tkachev VK, Kolesnikova LV, Krendeleva LYA, Ryabchikova EI, Smolina MP (1993) Ultrastructural changes of guinea pig organs in sequential passages of Ebola virus (in Russian). Voprosy Virusologii 38:179–182

Peters CJ, Jonson ED, McKee KT (1991) Filoviruses and management of viral hemorrhagic fever. In: Belshe RB (ed) Textbook of human virology. Mosby Year Book, pp 699–712

Sanchez A, Kiley MP, Holloway BP, Auperin DD (1993) Sequence analysis of the Ebola virus genome: organization, genetic elements, and comparison with the genome of Marburg virus. Virus Res 29:215–240

Schleef RR, Bevilaqua MP, Sawdey M, Gimbrone MA Jr, Loskutoff DJ (1988) Cytokine activation of vascular endothelium. Effects on tissue-type plasminogen activator and type 1 plasminogen activator inhibitor. J Biol Chem 263:5797–5803

Schleef RR, Loskutoff DJ, Podor TJ (1991) Immunoelectron microscopic localization of type I plasminogen activator inhibitor on the surface of activated endothelial cells. J Cell Biol 113:1413–1424

Schnittler HJ, Mahner F, Drenckhahn D, Klenk HD, Feldmann H. (1993) Replication of Marburg virus in human endothelial cells. A possible mechanism for the development of viral hemorrhagic disease. J Clin Invest 91:1301–1309

Skripchenko AA, Shestopalov AM, Yaroslavtseva OYa (1991) Comparative study of Marburg virus interaction with macrophages of different animal species in vitro (in Russian). Voprosy Virusologii 36:503–506

Streltsova MA, Agafonov AP, Ignatyev GM (1991) The sensitivity of different cell culture lines to Marburg virus (in Russian). Voprosy Virusologii 36:437–438

Sureau PH (1989) Firsthand clinical observations of hemorrhagic manifestations in Ebola hemorrhagic fever in Zaire. Rev Infec Dis 11:790–793

Vallegos DA, Ambrosio AM, Gamboa G, Briggiler AN, Maiztegui JI (1985) Alterations in lymphocyte subpopulations in Argentine haemorrhagic fever. Medicina (Buenos-Aires) 45:407

Volchkov VE, Blinov VM, Netesov SV (1992) The envelope glycoprotein of Ebola virus contains an immunosuppressive-like domain similar to oncogenic retroviruses. FEBS Lett 305:181–184

Wertheimer SJ, Myers CL, Wallace RW, Parks TP (1992) Intercellular adhesion molecule-1 gene expression in human endothelial cells. Differential regulation by tumor necrosis factor-a and phorbol myristate acetate. J Biol Chem 267:12030–12035

Subject Index

A
actin-filament system 192, 194
actin/myosin filament concentrations 192
actin/myosin-mediated contraction 194
actinomycin D 55
activation 180
acylation 10, 31
adherens junctions 192
adrenals 149, 158, 161
Aedes albopictus 57
aerogenic
– infection 60
– transmission 61, 70
age, influence on disease 65
ALAT 80
alveolar macrophages 177
Anopheles
– gambiae 57
– maculipennis 56
– stephensi 57
anterior eye chamber 67
antibody
– antiviral 54
– determine 209
– – ELISA 209, 210, 213
– – immunofluorescent assay (IFA) 209, 210
– – Western blot (WB) 209
– neutralizing 63, 69
antigen
– MHC class II 43
– presentation 43
antigenome 17
antiviral therapy 66
Arenaviridae 97, 98
artificial replication system 28
ASAT 80

B
baboons 150, 152, 153, 155, 159–162, 166–168
barriers 177
Belgrade 50, 61, 62, 65, 66, 69, 70
biohazard 4
bleeding 177

blood vessel permeability 61
blood-liquor barrier 179
bone marrow 150
Boniface 3
Bornaviridae 2
brain 159
bromalaine, treatment of virus 55
bromo-deoxyuridine 55
budding 17
Bunyaviridae 97, 98

C
cadherin family 194
cadherin/catenin-complex 185
carbohydrate
– N-glycan 10, 39
– O-glycan 10, 39
– structure 10
case fatality 87, 91
– rate 50
α-catenin 194
β-catenin 194
cause of death 60
CD 4 206, 213
CD 8 206, 213
cell
– culture (*see* propagation)
– entry
– – fusion 11, 14
– – receptor 14
– lysis 183, 184
cell-to-cell
– adhesion 192
– junctions 192
cell-to-substance adhesion 192
Cercopithecus aethiops 50, 56, 60, 69, 70, 120
CF test (*see* complement fixation test)
chemokines 188
chimpanzees 120
– behavior 77
– behavioral questions 82
– mortality 78, 82
– troop 78

chlamydia 52
chloroform, viral sensitivity to 55
circulating
– macrophages 180
– monocytes 180
classification 3
clotting factors 62
coagulopathy, disseminated intravascular (DIC) 62, 63, 192
– syndrome 147, 166, 167
collagenase, treatment of virus 55
colobus 83
complement
– fixation test 67, 68, 70
– proteins 179
conjunctivitis 64
convalescence 89
convalescent(s)
– human 61, 65, 66
– phase 67
– serum 66
convertase 39–41
creatine kinase 80
cross reactivity, serologic 4, 69
cyclophosphamide treatment 63
cynomolgus macaques 155–158, 160, 166–168
cytokines 176, 179, 180, 183, 184, 188, 191
– release 184
cytolysis 180, 190
cytolytic 188
cytopathic effects 58
cytopathogenic effect 14
cytotoxicity, antibody dependent cellular 63

D

dense peripheral band (DPB) 191, 192
deoxycholate, viral sensitivity to 55
detoxification 152, 169
diagnosis 90
– differential 50
dialysis, peritoneal 66
diapedesis 61
diarrhea 64
diathesis, hemorrhagic 49, 58, 59, 61, 62, 64
DIC (see disseminated intravascular coagulopathy)
dimer 42
disease, duration of 65
disseminated intravascular coagulation (DIC) 190
dissociation of tight and adherens junctions 192
domain
– fusion 40, 41
– immunosuppressive 43
doughnuts 52
DPB (dense peripheral band) 191, 192

E

Ebola virus (EBOV) 52, 54–56, 60, 85, 97, 98, 108, 112, 175
– antigens 109, 114
– cytopathic effect 79
– electron
– – micrograph 100
– – microscopy 79
– hemorrhagic fever 97, 98, 104, 109, 110
– – diagnosis 113
– – pathologic findings in liver 103
– – pathologic findings in spleen 109
– – skin 114
– infection
– – central nervous disorders 79
– – clinical syndroms 88
– – hair loss 80
– – myocardium 80
– – necrosis 82
– – rash 79
– natural cycle 78
– reservoir 83, 85
– species 2
– subspecies 3, 35, 41
– – Ivory Coast 3
– – Reston 3
– – – Philippines 3
– – – Reston 3
– – – Siena 3
– – Sudan 3
– – – Boniface 3
– – – Maleo 3
– – Zaire 3
– – – Eckron 3
– – – Gabon 3
– – – Kikwit 3
– – – Mayinga 3
– transmission
– – barrier nursing 81
– – direct contact 81
– – projection of droplets 81
– – simple contact 83
Eckron 3
editing 35–38
elastase, treatment of virus 55
electron microscopy 50, 51, 57, 67, 99, 113
ELISA test 67, 68, 71
enanthema 64
encephalitis 59, 60, 69
endocytosis 14
endoplasmic reticulum (ER) 39
endothelial cells 61, 176, 183, 185, 188
– destruction 190
endothelium 179, 183, 184
envelopes 54
– empty 54

epidemics 86
ER (*see* endoplasmic reticulum) 39
ether, viral sensitivity to 55
ethology 84

F
family infections 62, 65
fibrin 151, 152, 154-160, 166, 167
fibrinogen 64, 66
fibrinogenolysis 62
fibrinolysis 62
filoviral HF 176
Filoviridae 1, 2, 97, 98
filovirus(es) 2, 85, 99, 100, 102, 175
- diseases 85-93
- - nonhuman primates as a model 118
- hemorrhagic fever, diagnosis 106
- human, pathologic features 98, 99
- infections
- - diagnosis 99, 113
- - experimental 117-141
- - - biosafety 139
- - hepatic ultrastructural features 108
- - molecular pathogenesis 175-197
- - pathogenesis 122-125, 179-196
- - pathology 97-115
- particles, length distribution of 53
- persistence 132-136
- virion-proteins 55
Flaviviridae 97, 98
Frankfurt 50, 61, 62, 66, 69, 70
function
- hepatic 152
- kidney 155
- respiratory 156
furin 17, 39-41
fusion 11, 14, 40, 41

G
Gabon 3
gas chromatographic analysis 31
gender, influence on disease 65
genome
- gene order 6
- gene overlap 6, 16
- intergenic region 6
- 3'-leader 6, 15
- negative-strand RNA 6
- organization 2, 6
- 5'-trailer 6
- transcriptional signal 6, 15
glycoprotein (GP, $GP_{1,2}$) 25, 29, 31, 35-45
- cleavage subunit (GP, $GP_{1,2}$) 10, 11, 16, 17, 35, 39-41
- function 10, 11
- kinetics 30

- processing 10, 11, 35, 39-41
- recombinant 39
- small (sGP) 12, 36, 37, 42, 43
- - function 38, 42, 43
- - secretion 42
- - structure 42
- - synthesis 36, 37
- structure 10, 11, 39, 42
- synthesis 10, 11, 35, 39-41
- transport 10, 11, 30, 35, 39-41
- viral 68
glycosylation 10, 16, 29
GP (*see* glycoprotein)
granules, intracytoplasmic 51
green monkeys 151, 152, 154, 155, 158-160, 165, 167
guinea pigs 50, 51, 56, 57, 60-63, 65, 67, 68, 146-150, 161, 162, 165, 166

H
H_2O_2 191
hamster 146-148
- Syrian 56, 57, 60
heart 159
hemagglutination 57
hematological changes 150, 160, 161, 164, 167, 169
hemodialysis 66
hemorrhages 59, 61, 147, 149, 150, 155-160, 168, 190
- focal 60
hemorrhagic
- complications 66
- disease 176
- fevers 71, 90, 97, 99, 105, 113, 176
- manifestations 176
Henle-Koch's postulates 51
heparin treatment 63, 66
hepatitis 64, 65
HF (*see* hemorrhagic fever)
histamine 188
histopathologic
- changes 147, 154, 155, 167
- diagnosis 112
- features 102
human sera, convalescents 69
hyaluronidase, treatment of virus 55

I
ICAM-1 184
IHC (immunohistochemistry) 99, 106, 113
immune response 149, 163-165
immunity, protective 63, 65
immunofluorescence (IF) 51, 57, 58
immunofluorescent staining, indirect 68
immunohistochemistry (IHC) 99, 106, 113
immunosuppresive domain 11

immunosuppression 147, 165, 187
immunosuppressive domain 43, 207, 212
inclusion 17
- bodies 68, 69
- - intracytoplasmic 51, 57
- eosinophilic cytoplasmic 59
incubation time 63
infection(s)
- laboratory 64
- nonlethal 147, 166
- nosocomial 64
- source of 69
- terminal stages 148, 151, 154, 156, 160, 65-167
inflammation
- granuloma-like 147
- intravascular 169
- systemic 191
inflammatory reaction 59, 166
inhibition 38, 43
injection (inoculation, challenge, route, infection)
- aerosol 148, 149, 156
- intramuscular 162
- intraperitoneal 148-150, 162
- subcutaneous 150, 162
intercourse, sexual 65
interendothelial
- gap 191, 192
- - formation 195
interendothelial junctions 184, 191, 192
interferon 207, 213
interleukin-2 208
intestine 149
intracellular signal cascades 195
iododeoxyuridine 55
Ivory Coast 3, 41

J
Johannesburg 62, 63, 66, 67

K
Kenya 50
kidney 147, 149, 152, 161
- insufficiency 59
Kikwit 3
Kupffer cells 59, 82, 177, 179

L
L protein 12, 25
laboratory animals 126-129
LDH 80
lethality 64
leucopenia 64
leukocytes 185
- response 147, 148, 164

Liberia 81
liver 146-148, 152, 153, 161
- tropism 179
London 70
lung 156, 177
lymph nodes 148, 154, 164, 165, 177
lymphatic system 177
lymphocyte proliferation 206, 210, 212-214
lymphoid (lymphatic) tissue 148, 164, 165
lymphopenia 150

M
Macaca macaca 120
macaques (*see* rhesus) 56
macrophages 60, 61, 177, 179
- activated 180
- circulating 180
- peritoneal 177
- pleural 177
malaise 64
Maleo 3
Marburg virus (MBGV) 50, 61, 62, 69, 85, 97, 98, 104, 108, 175
- cleavage 41
- clinical syndroms 88
- hemorrhagic fever 105
- - pathologic findings of liver 105
- - renal pathologic findings 113
- particle 24
- protein NP 210
- protein VP40 210
- species 2
- strain
- - Musoke 3
- - Ozolin 3
- - Popp 3
- - Ratazyczak 3
- - Ravn 2, 3
matrix protein (*see* VP40)
- second (*see* VP24)
Mayinga 3
MBGV (*see* Marburg virus)
mediators 191
megakaryocytes 62
membrane
- anchor 37, 39, 42
- - minus 42
- fusion 40, 41
- particle 42
messenger RNA
- cap structure 15
- polyadenylation 15
- synthesis 14, 15
- structure 15
- transcriptional editing 16
mice 56

Subject Index

microcirculation (microcirculatory vasculator) 149, 151, 157, 160, 161, 168, 169
microglia 177
minigenome 29
minireplicon 18
minor nucleoprotein (see VP30)
monkey
- rhesus (see also macaques) 60
- shipment 69, 70
- vervet (see also Ceropithecus aethiops) 50
monocytes 61
- circulating 180
monocytes/macrophages 176, 179, 180, 187
Mononegavirales
- Bornaviridae 2
- Filoviridae 1, 2
- genome organization 2
- Paramyxoviridae 2
- Rhabdoviridae 1, 2
mononuclear phagocyte system (MPS) 176, 177
morphology 5, 6
mosquitoes, Aedes egypti 56
MPS (mononuclear phagocyte system) 176, 177, 179
Musoke 3
myocarditis 64
myosin 192
- light chain (MLC) activation 195

N

natural killer 206, 213
necrosis (necrotic changes, foci) 59, 146, 148-150, 152, 154-158, 164, 177
- follicular 59
negative-stranded RNA viruses 175
nervous system 177
neutralizaton test 70
neutrophils, degenerated 64
nonhuman primates
- Ebola virus infection 121
- immune response to filoviruses 125, 126
- infected with filoviruses, handling 119
- Marburg virus infection 120
- transmission of filoviruses among 130-132
nosocomial transmission 90
Novalgin 67
NP (see nucleoprotein)
nucleocapsid 52, 54
- proteins 26
- - function 27
nucleoprotein (NP) 8, 9, 25, 26

O

occludin 193
orchitis 64, 65
Ozolin 3

P

pancreas 157, 159
pancreatitis 64
panencephalitis 60
paraendothelial permeability 191, 195
Paramyxoviridae 2
particle 5, 6
- basophilic 59
passages 146, 147
pathogenesis (pathogenetic mechanisms) 152, 157, 161, 164, 169, 176
pathophysiological changes 176
peptide
- fusion 40, 41
- signal 38
pericytes 188
peritoneal macrophages 177
permeability 176, 191
- paraendothelial 191
peroxide 179, 188
person-to-person transmission 90
petechia 64
Philippines 3
phosphoprotein
- major 8
- minor 11
phosphorylation 8, 9, 17, 26
plakoglobin 194
plaque reduction assay 69
platelets (see also thrombocytes) 62, 64
pleural macrophages 177
poliomyelitis 50
 vaccines 50, 51, 71
polyacrylamide gel electrophoresis 55
polymerase
- cofactor (see VP35)
- DNA dependent RNA 37, 38
- RNA dependent RNA 37, 38
polypeptides, viral 55
Popp 3
postmortem
- knife 65
- specimens 67
postulates, Henle Koch's 51
precursor 35, 39, 42
prednisone treatment 67
preGP (see precursor)
prevention 92
procoagulant activity 176
proconvertin treatment 67
proinflammatory agents 188
propagation, cell lines 13, 14
β-propiolactone 55, 68
proprotein 39
proteases 179

protein(s)
- glycoprotein (GP, GP$_{1,2}$) 8, 10, 11, 35–45
- L protein (polymerase) 8, 12
- nonstructural 37, 38, 42
- nucleoprotein (NP) 8, 9
- processing 16, 17, 39, 40
- recombinant 23
- release 41, 42
- secretion 38, 42
- small glycoprotein (sGP) 12, 36, 37, 42, 43
- structural 24
- translation 16, 17
- transport 16, 17, 38, 39
- VP24 8, 11, 12
- VP30 8, 11
- VP35 8, 9
- VP40 8, 10
proteinuria 64
prothrombin treatment 67

Q
quarantine 51

R
rash 64
Ratayczak 3
receptor 14
recombinant proteins 23
relapse 64, 65, 67
release 41, 42
replicator 17
reproduction (replication) 180
- EBOV 147, 160, 167
- MBGV and EBOV 155, 156, 165
- virus (viral) 151, 157, 158, 163, 169
Reston 3, 41
Rhabdoviridae 1, 2
rhabdovirus
- family 56
- plant 56, 57
rhesus monkeys (see also macaques) 60, 150–152, 155–160, 165–167
ribonucleoprotein complex 6, 8, 9, 11, 17
rickettsia 52
RNA
- dependent RNA polymerase 37, 38
- editing 16, 35-38
- genome
- - nonsegmented 55
- - polarity of 55
- - single stranded 55
- viruses, negative-stranded 175

S
second small glycoprotein (ssGP) 36, 37
secondary cases 64–66
secreted glycoprotein (sGP) 187, 188

semen, virus in 65, 67
seroepidemiological studies 70, 71
serological surveys 70
serous cavities 177
SGOT 59, 64
SGOT/SGPT ratio 62
shock 176, 180, 183
Siena 3
signaling pathways 195
simian viruses 51
SIRS (systemic inflammatory response syndrome) 176, 180, 183
skin 151, 168
small glycoprotein (sGP) 12, 36, 37, 42, 43
- function 38, 42, 43
- secretion 42
- structure 42
- synthesis 36, 37
sodium lauryl sulfate 68
soluble 41, 42
- glycoprotein 187
South Africa 50
space of Disse 179
species (see Ebola virus, Marburg virus)
specimens, blood 67
spleen 146, 148, 154, 161, 165, 177
spreading 184
ssGP (see second small glycoprotein)
structural proteins 24
Stuart factor treatment 66
subgenomic RNA (see messenger RNA)
subspecies (see Ebola virus)
Sudan 3, 41, 97
surface
- glycoprotein 187
- projections 52
survivors 65
susceptible (target) cells 154, 161–163
symptoms, gastrointestinal 64
systemic inflammation 191
systemic inflammatory response syndrome (SIRS) 176

T
targets 188
taxonomy 2
terminal shock 168, 169
therapy (see also treatment) 91
- antiviral 66
thrombin, treatment 55
thrombocyte(s) 62
- agglutination of 57
- concentrates 66
- counts 64
thrombopenia 64
thrombosis 148, 155, 156, 165

thromboxytopenia 150
tight junctions 192
titration of virus 57
TNF-α (*see* tumor necrosis factor-α)
transcriptional
- editing 16, 35–38
- signal 6, 15
- stop signal 38
translation 16, 17
transmission 177
- nosocomial 90
transmission electron microscopy 180
treatment
- antibiotic 66
- antiviral 66
- antipyretic 67
- bromalaine 55
- collagenase 55
- convalescent serum 66
- cyclophosphamide 63
- elastase 55
- fibrinogen 66
- heparin 66
- hyaluronidase 55
- prednisone 67
- proconvertin 66
- prothrombin 66
- *Stuart* factor 66
- thrombin 55
- trypsin 55
tumor necrosis factor-α (TNF-α) 176, 180, 183, 188, 190, 191, 207, 208, 213
- concentration 207, 208

U
uveitis 64, 67

V
vaccinia virus 210, 211
vascular
- leakage 190
- permeability 168
VCAM-1 184
vehicles 42
Vero cells 52, 57, 67, 69
vesicular stomatitis virus (VSV) 52, 54
viral hemorrhagic fever (HF) 97, 90
virion
- envelope 39
- spikes 42
- structural protein (*see* proteins)
virosomes 42, 43
virus
- assembly 17, 18
- budding 17, 18
- concentration 147
- enveloped 55
- nonsegmented negative-strand RNA 1, 2
- particles 42
- – diameters of 52
- – standard length 52
- release 17, 18
- replication 179
- simian 51
- spread 190
- titer 162
vomiting 64
VP24 11, 12, 26
VP30 11, 25
VP35 9, 25
VP40 10, 25
VSV (vesicular stomatitis virus) 52, 54

W
Western blotting 70, 71
white blood cells 64

Z
Zaire 3, 35, 41, 97, 98, 100, 113
ZO-1 194

Current Topics in Microbiology and Immunology

Volumes published since 1989 (and still available)

Vol. 195: **Montecucco, Cesare (Ed.)**: Clostridial Neurotoxins. 1995. 28 figs. XI., 278 pp. ISBN 3-540-58452-8

Vol. 196: **Koprowski, Hilary; Maeda, Hiroshi (Eds.)**: The Role of Nitric Oxide in Physiology and Pathophysiology. 1995. 21 figs. IX, 90 pp. ISBN 3-540-58214-2

Vol. 197: **Meyer, Peter (Ed.)**: Gene Silencing in Higher Plants and Related Phenomena in Other Eukaryotes. 1995. 17 figs. IX, 232 pp. ISBN 3-540-58236-3

Vol. 198: **Griffiths, Gillian M.; Tschopp, Jürg (Eds.)**: Pathways for Cytolysis. 1995. 45 figs. IX, 224 pp. ISBN 3-540-58725-X

Vol. 199/I: **Doerfler, Walter; Böhm, Petra (Eds.)**: The Molecular Repertoire of Adenoviruses I. 1995. 51 figs. XIII, 280 pp. ISBN 3-540-58828-0

Vol. 199/II: **Doerfler, Walter; Böhm, Petra (Eds.)**: The Molecular Repertoire of Adenoviruses II. 1995. 36 figs. XIII, 278 pp. ISBN 3-540-58829-9

Vol. 199/III: **Doerfler, Walter; Böhm, Petra (Eds.)**: The Molecular Repertoire of Adenoviruses III. 1995. 51 figs. XIII, 310 pp. ISBN 3-540-58987-2

Vol. 200: **Kroemer, Guido; Martinez-A., Carlos (Eds.)**: Apoptosis in Immunology. 1995. 14 figs. XI, 242 pp. ISBN 3-540-58756-X

Vol. 201: **Kosco-Vilbois, Marie H. (Ed.)**: An Antigen Depository of the Immune System: Follicular Dendritic Cells. 1995. 39 figs. IX, 209 pp. ISBN 3-540-59013-7

Vol. 202: **Oldstone, Michael B. A.; Vitković, Ljubiša (Eds.)**: HIV and Dementia. 1995. 40 figs. XIII, 279 pp. ISBN 3-540-59117-6

Vol. 203: **Sarnow, Peter (Ed.)**: Cap-Independent Translation. 1995. 31 figs. XI, 183 pp. ISBN 3-540-59121-4

Vol. 204: **Saedler, Heinz; Gierl, Alfons (Eds.)**: Transposable Elements. 1995. 42 figs. IX, 234 pp. ISBN 3-540-59342-X

Vol. 205: **Littman, Dan R. (Ed.)**: The CD4 Molecule. 1995. 29 figs. XIII, 182 pp. ISBN 3-540-59344-6

Vol. 206: **Chisari, Francis V.; Oldstone, Michael B. A. (Eds.)**: Transgenic Models of Human Viral and Immunological Disease. 1995. 53 figs. XI, 345 pp. ISBN 3-540-59341-1

Vol. 207: **Prusiner, Stanley B. (Ed.)**: Prions Prions Prions. 1995. 42 figs. VII, 163 pp. ISBN 3-540-59343-8

Vol. 208: **Farnham, Peggy J. (Ed.)**: Transcriptional Control of Cell Growth. 1995. 17 figs. IX, 141 pp. ISBN 3-540-60113-9

Vol. 209: **Miller, Virginia L. (Ed.)**: Bacterial Invasiveness. 1996. 16 figs. IX, 115 pp. ISBN 3-540-60065-5

Vol. 210: **Potter, Michael; Rose, Noel R. (Eds.)**: Immunology of Silicones. 1996. 136 figs. XX, 430 pp. ISBN 3-540-60272-0

Vol. 211: **Wolff, Linda; Perkins, Archibald S. (Eds.)**: Molecular Aspects of Myeloid Stem Cell Development. 1996. 98 figs. XIV, 298 pp. ISBN 3-540-60414-6

Vol. 212: **Vainio, Olli; Imhof, Beat A. (Eds.)**: Immunology and Developmental Biology of the Chicken. 1996. 43 figs. IX, 281 pp. ISBN 3-540-60585-1

Vol. 213/I: **Günthert, Ursula; Birchmeier, Walter (Eds.)**: Attempts to Understand Metastasis Formation I. 1996. 35 figs. XV, 293 pp. ISBN 3-540-60680-7

Vol. 213/II: **Günthert, Ursula; Birchmeier, Walter (Eds.):** Attempts to Understand Metastasis Formation II. 1996. 33 figs. XV, 288 pp. ISBN 3-540-60681-5

Vol. 213/III: **Günthert, Ursula; Schlag, Peter M.; Birchmeier, Walter (Eds.):** Attempts to Understand Metastasis Formation III. 1996. 14 figs. XV, 262 pp. ISBN 3-540-60682-3

Vol. 214: **Kräusslich, Hans-Georg (Ed.):** Morphogenesis and Maturation of Retroviruses. 1996. 34 figs. XI, 344 pp. ISBN 3-540-60928-8

Vol. 215: **Shinnick, Thomas M. (Ed.):** Tuberculosis. 1996. 46 figs. XI, 307 pp. ISBN 3-540-60985-7

Vol. 216: **Rietschel, Ernst Th.; Wagner, Hermann (Eds.):** Pathology of Septic Shock. 1996. 34 figs. X, 321 pp. ISBN 3-540-61026-X

Vol. 217: **Jessberger, Rolf; Lieber, Michael R. (Eds.):** Molecular Analysis of DNA Rearrangements in the Immune System. 1996. 43 figs. IX, 224 pp. ISBN 3-540-61037-5

Vol. 218: **Berns, Kenneth I.; Giraud, Catherine (Eds.):** Adeno-Associated Virus (AAV) Vectors in Gene Therapy. 1996. 38 figs. IX, 173 pp. ISBN 3-540-61076-6

Vol. 219: **Gross, Uwe (Ed.):** Toxoplasma gondii. 1996. 31 figs. XI, 274 pp. ISBN 3-540-61300-5

Vol. 220: **Rauscher, Frank J. III; Vogt, Peter K. (Eds.):** Chromosomal Translocations and Oncogenic Transcription Factors. 1997. 28 figs. XI, 166 pp. ISBN 3-540-61402-8

Vol. 221: **Kastan, Michael B. (Ed.):** Genetic Instability and Tumorigenesis. 1997. 12 figs. VII, 180 pp. ISBN 3-540-61518-0

Vol. 222: **Olding, Lars B. (Ed.):** Reproductive Immunology. 1997. 17 figs. XII, 219 pp. ISBN 3-540-61888-0

Vol. 223: **Tracy, S.; Chapman, N. M.; Mahy, B. W. J. (Eds.):** The Coxsackie B Viruses. 1997. 37 figs. VIII, 336 pp. ISBN 3-540-62390-6

Vol. 224: **Potter, Michael; Melchers, Fritz (Eds.):** C-Myc in B-Cell Neoplasia. 1997. 94 figs. XII, 291 pp. ISBN 3-540-62892-4

Vol. 225: **Vogt, Peter K.; Mahan, Michael J. (Eds.):** Bacterial Infection: Close Encounters at the Host Pathogen Interface. 1998. 15 figs. IX, 169 pp. ISBN 3-540-63260-3

Vol. 226: **Koprowski, Hilary; Weiner, David B. (Eds.):** DNA Vaccination/Genetic Vaccination. 1998. 31 figs. XVIII, 198 pp. ISBN 3-540-63392-8

Vol. 227: **Vogt, Peter K.; Reed, Steven I. (Eds.):** Cyclin Dependent Kinase (CDK) Inhibitors. 1998. 15 figs. XII, 169 pp. ISBN 3-540-63429-0

Vol. 228: **Pawson, Anthony I. (Ed.):** Protein Modules in Signal Transduction. 1998. 42 figs. IX, 368 pp. ISBN 3-540-63396-0

Vol. 229: **Kelsoe, Garnett; Flajnik, Martin (Eds.):** Somatic Diversification of Immune Responses. 1998. 38 figs. IX, 221 pp. ISBN 3-540-63608-0

Vol. 230: **Kärre, Klas; Colonna, Marco (Eds.):** Specificity, Function, and Development of NK Cells. 1998. 22 figs. IX, 248 pp. ISBN 3-540-63941-1

Vol. 231: **Holzmann, Bernhard; Wagner, Hermann (Eds.):** Leukocyte Integrins in the Immune System and Malignant Disease. 1998. 40 figs. XIII, 189 pp. ISBN 3-540-63609-9

Vol. 232: **Whitton, J. Lindsay (Ed.):** Antigen Presentation. 1998. 11 figs. IX, 244 pp. ISBN 3-540-63813-X

Vol. 233/I: **Tyler, Kenneth L.; Oldstone, Michael B. A. (Eds.):** Reoviruses I. 1998. 29 figs. XVIII, 223 pp. ISBN 3-540-63946-2

Vol. 233/II: **Tyler, Kenneth L.; Oldstone, Michael B. A. (Eds.):** Reoviruses II. 1998. 45 figs. XVI, 187 pp. ISBN 3-540-63947-0

Vol. 234: **Frankel, Arthur E. (Ed.):** Clinical Applications of Immunotoxins. 1999. 16 figs. IX, 122 pp. ISBN 3-540-64097-5

Printed by Publishers' Graphics LLC
MO20120905-299